International Scientific-Practical Conference of Students, Graduate Students and Young Scientists "Transport and Storage of Hydrocarbons" 2016

IOP Conference Series: Materials Science and Engineering
Volume 154

Tyumen, Russia
20 - 25 May 2016

ISBN: 978-1-5108-3177-3
ISSN: 1757-8981

Printed from e-media with permission by:

Curran Associates, Inc.
57 Morehouse Lane
Red Hook, NY 12571

Some format issues inherent in the e-media version may also appear in this print version.

Printed by Curran Associates, Inc. (2016)

For permission requests, please contact the Institute of Physics
at the address below.

Institute of Physics
Dirac House, Temple Back
Bristol BS1 6BE UK

Phone: 44 1 17 929 7481
Fax: 44 1 17 920 0979

techtracking@iop.org

Additional copies of this publication are available from:

Curran Associates, Inc.
57 Morehouse Lane
Red Hook, NY 12571 USA
Phone: 845-758-0400
Fax: 845-758-2633
Email: curran@proceedings.com
Web: www.proceedings.com

Table of contents

Volume 154

International Scientific-Practical Conference of Students, Graduate Students and Young Scientists "Transport and Storage of Hydrocarbons"

20–25 May 2016, Tyumen, Russia

Accepted papers received: 19 October 2016
Published online: 31 October 2016

Preface

Papers

IOP Conf. Series: Materials Science and Engineering **154** (2016) 011001 doi:10.1088/1757-899X/154/1/011001

PREFACE

This conference proceedings is a collection of articles and reports presented at the international research and technical conference for undergraduates, post-graduates, and aspiring researchers "transport and storage of hydrocarbons" held in the Industrial University of Tyumen in 2016. The materials present the results of research, design and development in the most important areas in Tyumen region. These are: design and operation of transport and storage systems of oil and gas, construction and repair of transportation and storage systems of oil and gas, diagnostics of the oil and gas complex, design and operation of systems for collecting and preparation of the well production, improvement of the operational reliability of pipeline transport facilities, challenges of safe and environmentally friendly operation of pipeline transportation systems.

The publication is intended for scientific, engineering and technical workers, as well as for graduate students and students of technical universities. During the conference, issues of scientific relevance, novelty and direct participation in scientific research were discussed.

Peer review statement

All papers published in this volume of *IOP Conference Series: Materials Science and Engineering* have been peer reviewed through processes administered by the proceedings Editors. Reviews were conducted by expert referees to the professional and scientific standards expected of a proceedings journal published by IOP Publishing.

IOP Conf. Series: Materials Science and Engineering **154** (2016) 012001 doi:10.1088/1757-899X/154/1/012001

Fuel Temperature Fluctuations During Storage

R E Levitin and Yu D Zemenkov

Industrial University of Tyumen, 38 Volodarskogo st., Tyumen, 625000, Russia

E-mail: 89028130230@mail.ru

Abstract. When oil and petroleum products are stored, their temperature significantly impacts how their properties change. The paper covers the problem of determining temperature fluctuations of hydrocarbons during storage. It provides results of the authors' investigations of the stored product temperature variations relative to the ambient temperature. Closeness and correlation coefficients between these values are given. Temperature variations equations for oil and petroleum products stored in tanks are deduced.

1. Introduction

Significant losses of liquid hydrocarbons happen due to their temperature fluctuations during storage. According to some estimates, minimum 5% of oil produced in Russia is lost on its way to consumers, which is approximately 25 million ton per year. About a quarter of these losses are caused by their temperature fluctuations during storage. [1]

Fuel temperature, due to a relatively slight thermal fluid expansion, does not affect the volume increase and for one degree temperature gradients amounts to a relative value of about 1 volume %. This refers to the gas-vapor mixture in a tank where the coefficient of volume expansion is an order of magnitude greater and temperature fluctuations cause significant pressure fluctuations.

2. Research object

2.1 Seasonal fuel temperature fluctuations

Measurements of seasonal fluctuations of temperature and pressure were taken at the tank battery. Together with the temperature were measured wind velocity, wind direction, cloudiness and humidity According to experts the most important is the ambient temperature and wind velocity. Empirical data were processed and curves were developed (figures 1-4).

Pressure fluctuations in the vicinity of the tank battery are insignificant, which is noted in other works [1]. The greatest pressure fluctuations – about 40-50 mm of mercury over 2-3 days – are observed in November and April.

This leads to a change in the pressure vent valves operation conditions. As a consequence, the pressure drop leads to the gas-vapor mixture emission from the tank. Such major blast emissions occur no more than 5 times a year. Pressure drop does not exceed 5% under these conditions. Moreover, the pressure drop is leveled by daily fluctuations (small breathings). Due to low impact and simplified correlation dependences, pressure fluctuations, humidity and wind direction were not further taken into account. What was considered is temperature. Data on internal petroleum product state in the tank were recorded together with data on ambient environment changes over the autumn-winter and spring-summer periods.

As seen in figure 2, the curve repeats a change in ambient temperature with some delay compared to figure 1. This inertia is due to the high heat capacity of oil products. The curve has a flattened shape and smaller highs and lows. This is explained by a complex model of heat return in the tank. Changing the temperature in the vessel leads to a change in the density and innage level (see figures 3-4). The curves repeat the temperature variations precisely which indicates a good convergence and high precision of measuring equipment.

IOP Conf. Series: Materials Science and Engineering **154** (2016) 012001 doi:10.1088/1757-899X/154/1/012001

Based on literature analysis and expert evaluation it was discovered that the primary cause of oil product temperature changes in the tank is ambient temperature. In this regard we will first find the dependence in which changing one value changes another, a static one.. In particular, the static dependence is shown in the fact that the average value changes with one of the other values. In this case, the static dependence is called correlation.

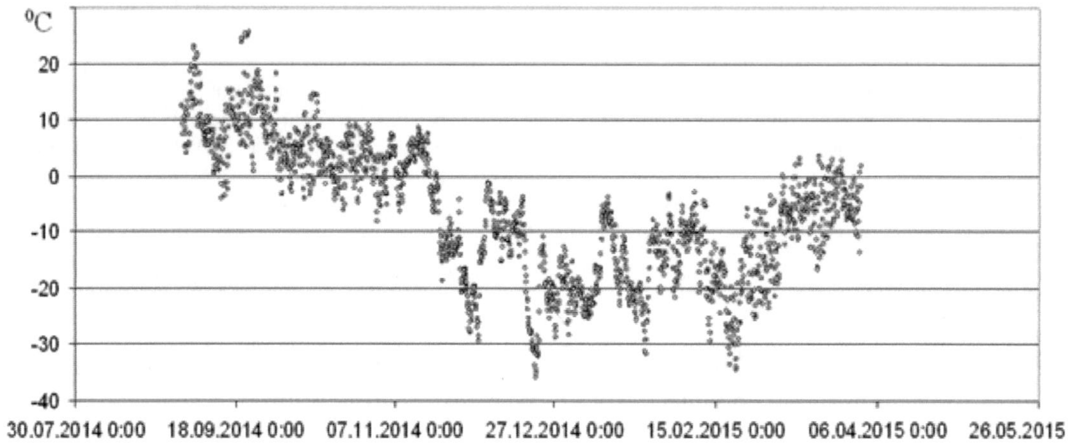

Figure 1. Ambient temperature variation in the vicinity of the tank battery.

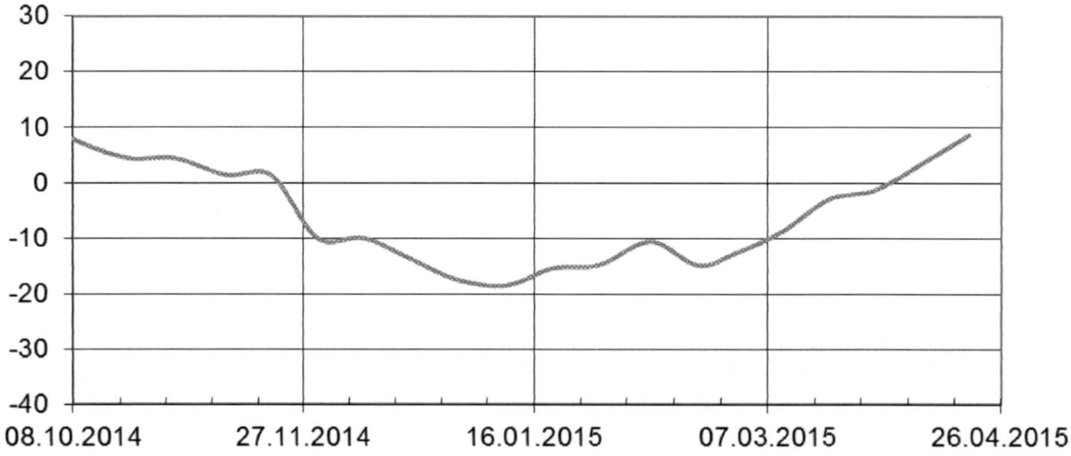

Figure 2. Average product temperature variations in the tank №0 of strategic storage.

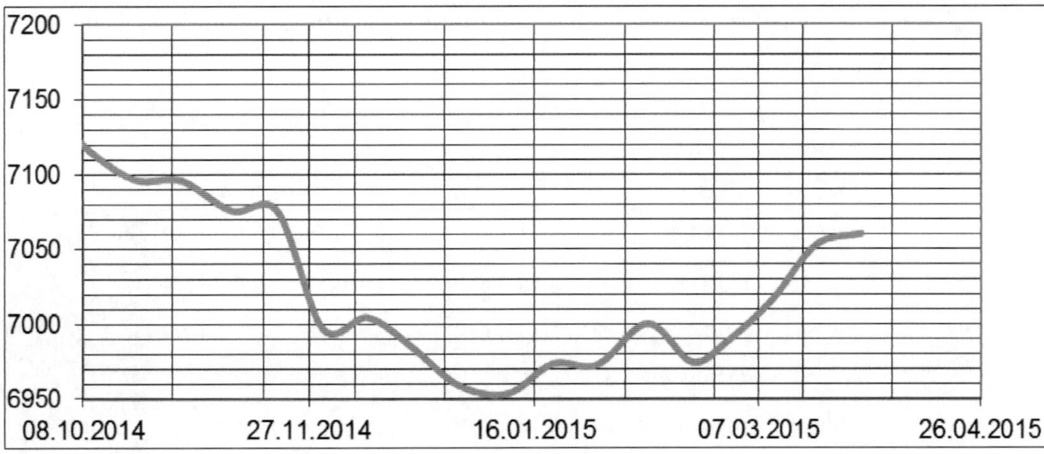

Figure 3. Product innage variations (mm) in the tank №0 of strategic storage.

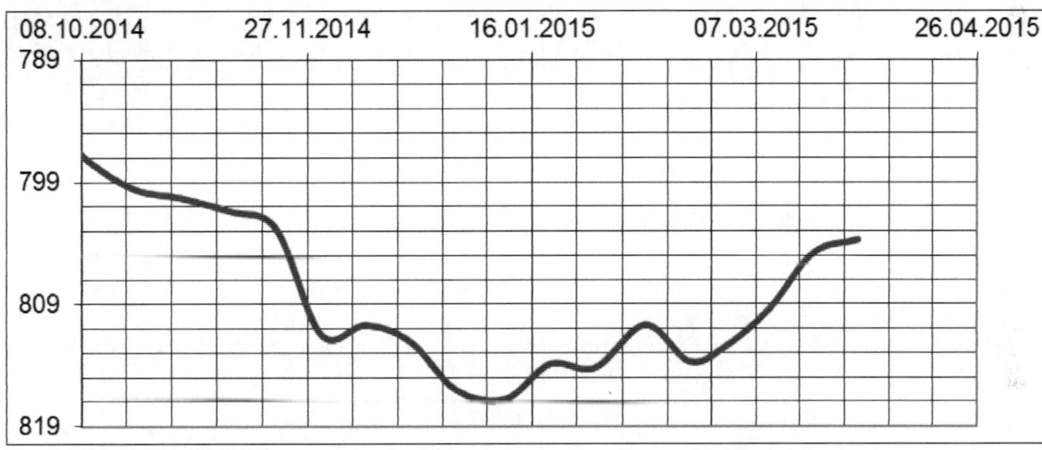

Figure 4. Average product density variations (kg/m^3) in the tank №0 of strategic storage.

2.2. Daily fuel temperature fluctuations

The rate of temperature change of the gas phase depends on the ability to absorb and emit radiant energy incident on the tank wall, on the outside temperature drop and velocity.

In this regard, there was a practical interest in determining the effect of daily temperature fluctuations on thermal oil product fluctuations [2]. To do this, at different levels of the tank were installed temperature sensors, with temperature measurement results shown in Table 1.

7 belts of the tank were filled with a petroleum product, belts 8 and 9 were left empty to measure the gas space temperature. Moreover, the radiant energy incident on the tank results in a non-uniform heating of the outer wall. For example, Table 2 shows the outer surface temperature depending on its orientation in space.

Table 1. Results of measuring temperature of gasoline A 92 at various tank belts.

Time	Innage height, mm	Capacity, liter	Weight, kg	Gasoline temperature at various tank innage levels								
				1st belt	2nd belt	3rd belt	4th belt	5th belt	6th belt	7th belt	8th belt	9th belt
08:00	9101	2605888	1938702	5.82	6.01	6.17	6.29	6.35	6.51	6.62	6.63	7.80
12:00	9103	2606452	1938756	6.01	6.16	6.29	6.44	6.57	6.64	6.74	7.91	14.65
16:00	9106	2607297	1937768	6.21	6.39	6.58	6.66	6.77	6.9	6.99	8.67	11.6
20:00	9106	2607297	1936022	6.16	6.33	6.49	6.64	6.68	6.78	6.94	7.57	3.49
23:00	9104	2606734	1936072	6.06	6.23	6.37	6.49	6.63	6.71	6.86	6.46	1.72
08:00	9097	2604760	1935679	6	6.12	6.16	6.2	6.25	6.23	6.35	5.89	0.86
12:00	9097	2604760	1935259	6.02	6.22	6.23	6.3	6.38	6.39	6.52	7.59	14.9
16:00	9101	2605888	1932292	6.23	6.4	6.48	6.56	6.62	6.73	6.83	8.89	17.15
20:00	9104	2606734	1932463	5.85	6.03	6.08	6.13	6.16	6.27	6.32	7.38	13.97
23:00	8740	2504122	1857783	7	7.15	7.29	7.42	7.55	7.66	7.86	11.91	10.79

Table 2. Temperature of the first tank wall belt depending on its location in space

Time	Temperature of the tank wall outside around the perimeter in different points													
	1	2	3	4	5	6	7	8	9	10	11	12	13	14
08:00	5.2	6.5	7	7.5	9	10,5	11.2	12	13	12	9	7	6	5
12:00	6	8	8.5	9	10	11	11	10.5	10	9.5	8	7	6.5	6
16:00	6.3	6.5	6.7	7.5	9	8.1	9.3	8.5	8.7	8.5	7	8.6	9.3	8.6
20:00	1.4	2.0	3.5	5.1	6.3	6.9	6.8	1.8	3.1	4.5	5.6	6,1	6	7.1
23:00	5.4	6.0	6.5	6.6	4.2	5.5	5.9	5.3	5.8	5.9	3.7	4.9	5.2	5.75
08:00	2.0	1.0	0.7	1.0	2.1	2.6	2.4	2.7	2.9	3.0	1.9	2.5	2.6	2.9
12:00	9.6	8.5	7.3	6.3	4.5	1	3.4	3.9	4.6	5.1	5.7	7	7.6	8.5
16:00	7.8	8.1	8.4	7.8	7.8	5.2	5.6	6	6.1	5.7	5.8	6.1	6	7.1
20:00	7.8	8.1	6.2	4.8	1,2	5.2	1.2	1.8	3.1	4.5	5.6	6.1	6	7.1
23:00	3.9	1.9	1.4	1.9	4.1	5.1	4.8	5.3	5.8	5.9	3.7	4.9	5.2	5.75

As a result of the correlation analysis of statistical data on the tank ambient temperature and innage level were prepared graphical and numerical dependences of daily temperature changes in the tank of strategic storage (figures 5, 6, 7).

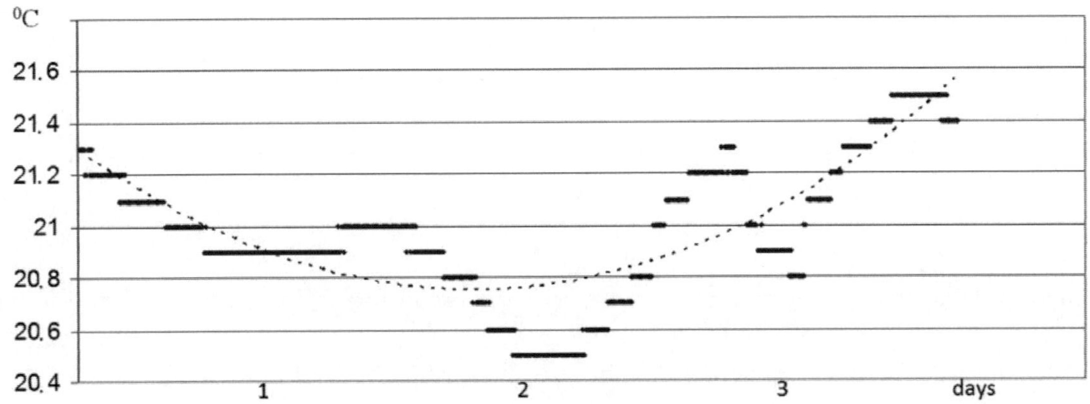

Figure 5. Average temperature variations in the tank.

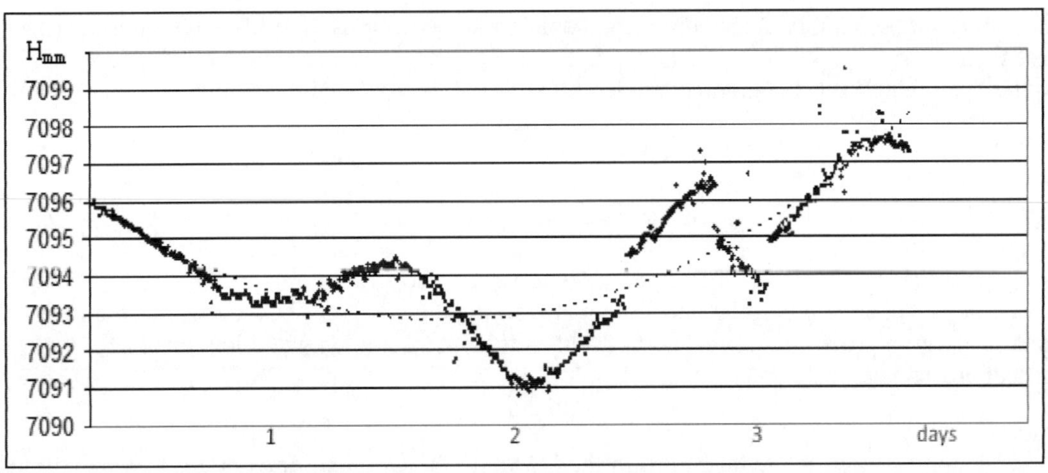

Figure 6. Product innage level fluctuations in the tank.

As seen from graphs 5 and 6 they have three clearly pronounced maximums and three minimums. These are maximum and minimum temperatures and tank innage levels corresponding to the three days of "still" storage. According to the formula and physical meaning, these curves repeat the ambient temperature changes (see figure 7). However, product temperature fluctuations in figure 5 have smaller amplitude than ambient temperature fluctuations in figure 6. This is explained by a high fill-up level and large capacity of a tank. We can also state that the curve will be even more smoothed if tank capacity is increased.

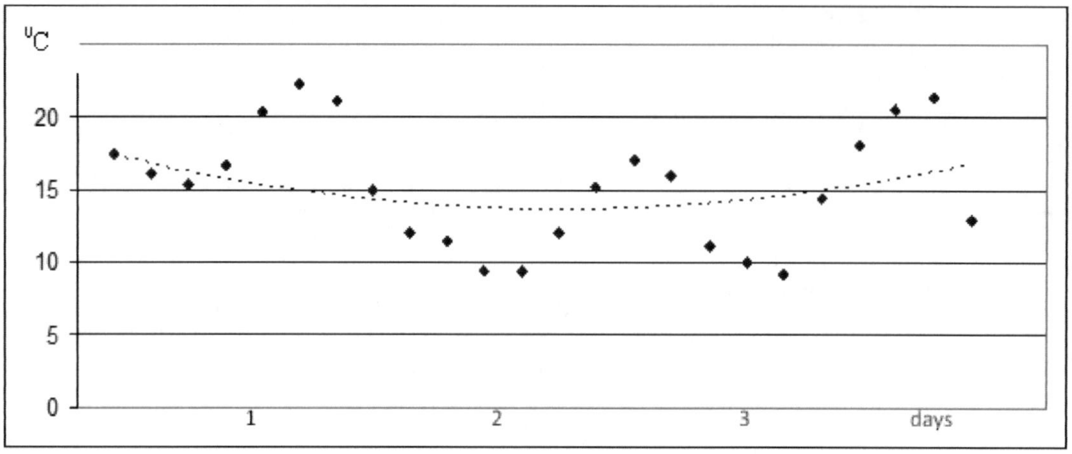

Figure 7. Ambient temperature variations

Polynomial (joint probability distribution of random variables) curves of average temperature (dash lines) were built.

$y = 4 \cdot 10^{-06}x^2 - 0.0028x + 21$ (average temperature in a tank) (1)

$y = 2 \cdot 10^{-05}x^2 - 0.0169x + 7096$ (product level in a tank) (2)

$y = 3.6 \cdot 10^{-02}x^2 - 0.9867x + 20$ (ambient temperature) (3)

As the figures show, for the last equation the approximation error is much greater, which is connected with the fact that it is difficult to forecast ambient temperature variations. It can also be noticed that for the test data the average ambient temperature axis is 1 degree Celsius below due to a higher average temperature prior to the period of investigation (operation range). On the other hand, the curves demonstrate a good correlation of input and output data. Moreover, a change in the level of the product in figure 6 proves high precision of temperature measurements and low error of raw data and statistical processing.

3. Summary

A close correlation between the temperature in the VST and the ambient temperature was identified, with the correlation coefficient and the sample correlation ratio being greater than 0.9 From the presented data of temperature variation we can make a conclusion on the degree of temperature change inertia at different tank belt levels. This allows efficient predicting hydrocarbon emissions in tanks.

References

[1] *Operation of Trunk and Process Oil and Gas Pipelines. Distribution and Accounting: Study Guide* ed Yu D Zemenkov 2014 (Tyumen: TSOGU) 370p

[2] Patent 2500486 RU, MPK B05C7/04, B65D88/74. *A device to improve the performance properties of vertical steel tanks* V O Nekrasov, R E Levitin, I V Tyrylgin and Yu D Zemenkov (RU) -№ 2012125478/05, filed 10.12.13

[3] API Manual of Petroleum Measurement Standards, Chapter 19 - Evaporative-loss Measurement, Section 1 - Evaporative Loss from Fixed-Roof Tanks Third Edition, March, 2002.

[4] API Manual of Petroleum Measurement Standards, Chapter 19 - Evaporative-loss Measurement, Section 2 - Evaporative Loss from Floating-roof Tanks, Second Edition, September, 2003.

[5] VDI 3479: *Emissionsminderung*. Raffinierieferne Mineralöltanklager, 2010.

IOP Conf. Series: Materials Science and Engineering **154** (2016) 012002 doi:10.1088/1757-899X/154/1/012002

Estimation of the Heat Balance of the Liquid Hydrocarbons Evaporation Process from the Open Surface During Geotechnical Monitoring

M Yu Zemenkova and Yu D Zemenkov

Industrial University of Tyumen, 38 Volodarskogo st., Tyumen, 625000, Russia

E-mail: muzemenkova@mail.ru

Abstract. Researchers in Tyumen State Oil and Gas University (TSOGU) have conducted a complex research of the heat and mass transfer processes and thermophysical properties of hydrocarbons, taking into account their impact on the reliability and safety of the hydrocarbon transport and storage processes. It has been shown that the thermodynamic conditions on the surface and the color of oil influence the degree of temperature rise in the upper layers of oil when exposed to direct solar radiation. In order to establish the nature of solar radiation impact on the surface temperature the experimental studies were conducted in TSOGU on the hydrocarbon evaporation and the temperature change of various petroleum and petroleum products on the free surface with varying degrees of thermal insulation of the side walls and bottom of the vessel.

1. Introduction

The consequences of accidents and incidents on oil pipelines continue to exacerbate environmental problems. When analyzing accidents or forecasting risks, it is advisable to use a systematic approach (figure 1) treating each of them as a complex system, where many of the factors included in it (the duration of repair and recovery period, natural and meteorological conditions, the nature of the damage, pumping mode, emissions and others) are interdependent due to the rapid dissipation of mass exchange transitions and anthropogenic processes.

The scope of expert calculations of leaks and hydrocarbon emissions during their pipeline transportation is very extensive. From the typical problems that can be solved by experts at various levels, it is necessary to note the following:

- evaluation and prediction of reliability and safety indicators;
- assessment of the degree of emission hazard, taking into account its volume, toxicity, physical and chemical properties;
- prediction of accident consequences;
- compilation of a list of possible events during the propagation of leakage and emission;
- determination of the most likely time intervals between failures;
- determination of the examination results reliability and creation of a databank;
- assessment of the possible terms of the accident elimination;
- calculation of financial, material and labor resources and their minimization, etc..

Forecasting the effects of accidents should be based on reliability indicators, maintainability in particular, since the period of spill elimination is both a probabilistic and deterministic function.

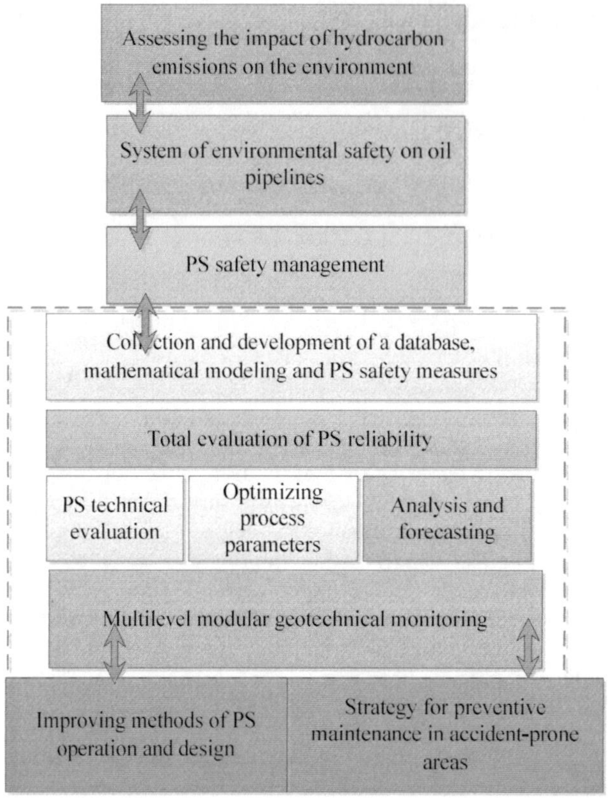

Figure 1. Functional structure of a multilevel modular geotechnical monitoring system for a PS (pipeline system).

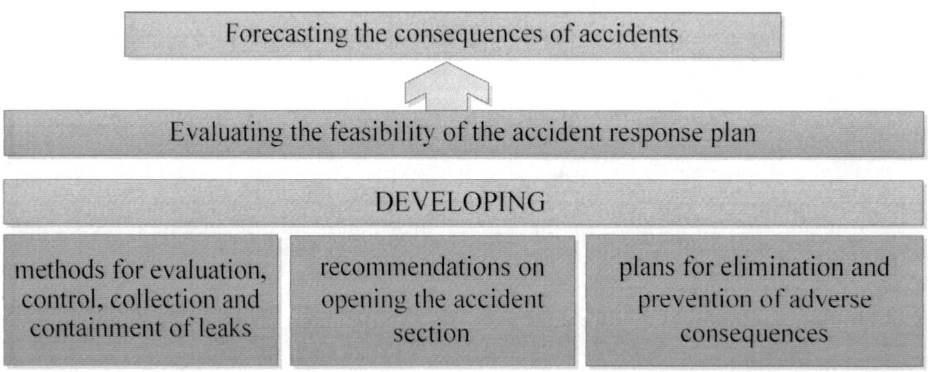

Figure 2. Stages of forecasting the consequences of accidents during an oil spill.

A method for planning works and forecasting consequences of an accident should take into consideration (figure 2) the evaluation of the feasibility of the accident response plan for a variety of reasons.

For instance, the maintainability indicator for the main pipeline facilities - the average time of unscheduled (emergency) recovery (repair) $\overline{T}_B$, according to statistics, according to current regulatory documents, is estimated by (1):

$$\overline{T}_{rep} = \frac{1}{r} \cdot \sum_{i=1}^{r} t_{repi}\,, \tag{1}$$

r - the number of failures that occurred during the observation period t;
t_{Bi} - the recovery duration time after the ith failure

Standard recovery time $T^{rep}_{i(rep)}$ for each ith accident should be estimated according to the formula in case of routine situations

$$T^{rep}_{reg} = \sum_{i=1}^{n} T^{rep}_{i(reg)} P_{(reg)i} \tag{2}$$

The authors propose to assess the forecasted recovery time $T^{rep}_{i(pred)}$ in total and for each ith accident $T^{rep}_{i(pred)}$ from the condition (figure 3):

$$T^{rep}_{pred} = \sum_{i=1}^{n} T^{rep}_{i(pred)} P_{(npred)i} + \sum_{i=1}^{n} T^{rep}_{i(pred)} P_{(pred)i} \tag{3}$$

where $P_{(nreg)i}$ - the likelihood of ith-type emergency situation occurrence during repair.

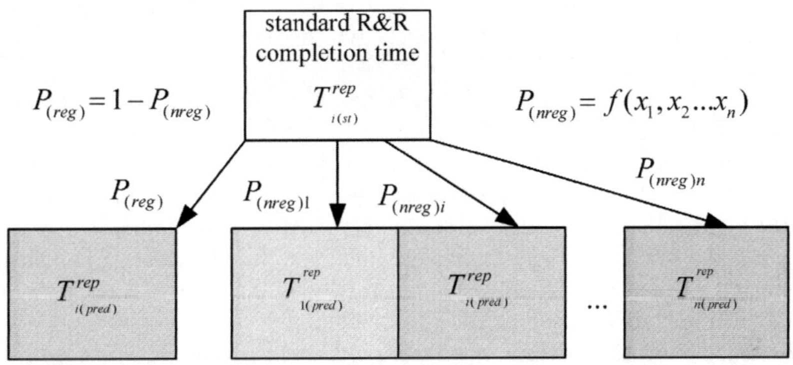

Figure 3. State chart diagram for system recovery considering the time spent.

In its turn, the likelihood $P_{(nreg)}$ will be determined by:

$$P_{(nreg)} = 1 - P_{(reg)} \,, \tag{4}$$

$P_{(reg)}$ - the likelihood of standard condition occurrence for repair and recovery works. $P_{(reg)}$ can be estimated statistically; however, conditions of emergency response may differ significantly from each other, which makes the error of calculations unreasonably high.

To estimate the probability of emergency situations it is advisable to establish a multi-factor model to estimate $P_{(nreg)}$ using deterministic components:

$$P_{(nreg)} = f(x_1, x_2 ... x_n), \tag{5}$$

where $x_1, x_2 ... x_n$ - factors that determine the possibility of carrying out the recovery work (technological, meteorological, human, environmental, economic, political, etc.).

In the case of an oil spill, the likelihood of an emergency situation is high enough and, therefore, a thorough analysis of the work conditions based on meteorological and technological factors is required. One of the most important technological problems during leakage and evaporation of petroleum is the occurrence of explosion hazardous areas. The main indicators for evaluating reliability and safety are the mass of evaporated oil and the intensity of evaporation, which are determined by the parameters of the heat balance of the evaporation process. Oil spilled during pipeline integrity loss or stored in emergency pits over the repair and recovery period has an open surface, evaporating into the atmosphere. Under the influence of solar radiation, nitrogen oxides and hydrocarbons can form photochemical oxidants - the components of photochemical smog that adversely affect not only the vegetation and animals, but also have a long-term effect on human health, depending on the intensity of the man-made stream. Nature and people are haphazardly experiencing the effects of all man-made physical and bio toxic factors in different combinations and as they accumulate

Oil spilled during the accident or stored in the earth storage for quite a long time is exposed to solar radiation.

Since 1980 in TSOGU the authors have been conducting research to assess the thermophysical parameters of evaporation from an open surface during spills and storage of petroleum and petroleum products. The research results are taken as a basis for the development of a standard document approved by the Ministry of Fuel and Energy of the Russian Federation «Method of determining the damage to the environment in case of accidents on main oil pipelines», are published as monographs and textbooks, and also elaborated and implemented by Ryabov N.N. when developing «Recommendations on fire safety during repairs on the line part of main oil pipelines» in the Ministry of Internal Affairs.

Analysis of the existing literature shows that research on this question is quite often qualitative in nature. Thermodynamic conditions on the surface and the color of oil influence the degree of temperature rise in the upper layers of oil when exposed to direct solar radiation. It is known that the phenomenon of thermal radiation, as a process of energy propagation in the form of electromagnetic waves, by nature differs from conduction and convection and is accompanied by the thermal energy conversion into radiant, and conversely, the radiant energy - into thermal. The energy radiation is the result of complicated intra-atomic processes occurring due to the thermal energy. When heated, part of the body energy is inevitably transformed into radiant, the value of which is determined by the temperature. To date, a lot of literature was published, such as monographs, both translated and by Russian authors, where at a high scientific level many provisions of the heat and mass transfer and the associated mass exchange are investigated. The process of liquid evaporation is usually accompanied by the heat and mass exchange in boundary layers. To summarize the experimental and calculated data of these conditions, different systems of similarity numbers are applied. Accordingly, the results obtained by researchers vary. Available data for evaporation within their accuracy so far do not allow finding a clear influence of the mixture heterogeneity on the heat and mass exchange coefficient.

2. General information

Carriers of the radiant energy are electromagnetic waves with a wavelength of from a fraction of a micron (sunbeams) to many kilometers (radio waves). In physics, they are known as ultraviolet, visible - light, infrared rays and electromagnetic waves. Here, of the greatest interest are light and infrared rays with a wavelength of 0.4 to 40 microns that are absorbed by the bodies as their energy passes into thermal. If all the radiant energy passes through the substance, it is called *absolutely transparent*, or *diathermic*. Solids and liquids are *virtually opaque* (*athermic*) for heat rays.

Just as it is the case with the concepts of absorption and reflection. A white surface reflects well only visible (solar) radiation. In practice, this feature is widely used - white-colored cars tanks, reservoirs, where sun exposure is not desirable. Notably, white paint absorbs invisible heat rays just as well as the dark one.

It is known that neutral resins give color to petroleum. They possess an intense color and a strong coloring power. In fact, 0.005% of heavy neutral resin is enough to dye colorless gasoline straw yellow. According to V. M. Rybak, there is no strict relation between the color and the amount of resinous substances. For example, for oils of the same petroleum color can to some extent be a criterion for resin extraction, but when it comes to oils derived from other petroleum, we must consider the fact that the resinous substances of different petroleum have different color. We should mention here that there are many different types of resin (oil, neutral, potential and actual), fundamentally different both in the physical and chemical properties. We can also single out carbamide, phenol-formaldehyde resins. From the optical standpoint, carbamide resins exhibit high lightfastness and are transparent, phenol-formaldehyde resins are colored brown or yellow which under the influence of sunlight can transform into yellow-brown or even black.

It is known that the opacity of petroleum layers is due to the presence of asphaltenes in it – the most macromolecular substances having a molecular weight in the range of 1600-6000. Asphaltenes do not melt, at high temperatures can form gases and hard-burning coke and are a hardly soluble component similar in the sense to various kinds of mechanical impurities.

It is expected that the color of oil will depend also on its content of fat, paraffin, ceresin, etc. Known methods for determining the color do not allow determination of the concentration coloring substances by comparing the color intensity of oil and a sample. The calorimeter practice suggests that petroleum and petroleum products color is not an additive property, and is close to the color of a darker component.

Thus, a color should be understood as a conditional oil quantity characterizing the color of oil and its degree of intensity compared to the color of standard solutions. It is also assumed that the petroleum emissivity factor ε depends on the color intensity. When analyzing the published literature, we haven't discovered any data on the emissivity factor of petroleum and petroleum products, although, for example, for sixteen different paints of all colors the emissivity factor varies from 0.92 to 0.96.

Under the direct solar radiation is meant radiation coming to the surface directly from the Sun at a radius of 5^0. In the SI system, the radiation intensity is measured in (W/m²). This value is usually called the irradiance. In actinometry, an equivalent term “radiation intensity” is used.

It is known that the free surface irradiance by the Sun determines the activity and position of the Sun above the horizon, as well as the state of the atmosphere. Solar activity according to the actinometrical observations from space remains constant for many years. Direct solar radiation J is measured by means of actinometers such as a Savinov-Janiszewski, AT-50 and others at actinometrical stations (AS). Distribution of AS in Russia is extremely non-uniform. For example, in the West Siberia region there is only one Omsk AS. Since data from actinometrical stations can be extended to the territory of up to 100 km, in some cases, the intensity of the direct solar radiation can be calculated using the average radiation intensity values, depending on the latitude, with an input probability of 0.95.

From the actinometrical standpoint, the state of the atmosphere is characterized by transparency, fluctuating from 0 to 100%: 0% - athermic atmosphere, 100%- diathermic. When the transparency of the atmosphere is about 0.8 or 80%, it means that 80% of the energy passes through the atmosphere, while the remaining 20% are dispersed. The transparency of the atmosphere depends on the cloudiness, which is the product of condensation and sublimation of water vapor. For engineering calculations, cloudiness may be quantified by various characteristics and features of cloudiness.

3. Mathematical modeling

If data on cloudiness is available, the average daily intensity of the direct solar radiation $J_{CP,}$ coming on a horizontal surface, can be found by:

$$J_{CP} = \left[\frac{K_1 J_1 \tau_{1-B}}{2} + \left(\frac{K_1 J_1}{2} + \sum_{i=2}^{4} K_i J_i + \frac{K_5 J_5}{2} \right) \tau_0 + \frac{K_5 J_5 \tau_{3-5}}{2} \right] \tau^{-1}{}_{DH}, \quad (6)$$

where τ_{1-B} - the time interval between the sunrise and the first observation, hour; τ_{3-5} - the time interval between the sunset and the last observation, hour; Ki – the cloudiness observed during corresponding observation periods; Ji - the intensity of the direct solar radiation coming on a horizontal surface, W/m²; τ_{DH} - the duration of daytime, hour;

$$\tau_{DH} = \tau_3 - \tau_B, \tag{7}$$

where τ_B, τ_3 - the sunrise and sunset times for a given location; τ_0 - the time interval between observations.

Technical handbooks usually list common absorptive or emissive properties of surfaces, the use of which can lead to significant errors. In fact, the total absorption capacity depends not only on the nature of the substance, temperature and pressure, but also on the distribution of the incident radiation along the wavelength. Most of the known bodies have a monochromatic absorption capacity. It is to be recalled that *a gray surface* is a surface the absorptive (reflecting) capacity of which is the same for a large wavelength range. Common water absorbs 50-60% of radiation with wavelengths of about 6500-9000 nm, wavelengths of 5800-6300 - up to 80-90%, 3500-5500 - only 25-30%, 2800-3200 - more than 90% and 800-2300 - less than 15%. Thus, a wide spectrum of wavelengths (3000 nm) of the solar radiation, on one hand, and the limitations of the measuring equipment and the complexity of the petroleum composition, on the other, held back the research into the absorptivity of hydrocarbon fluids, such as condensate, petroleum, LPG and etc.

In order to establish the nature of the influence of solar radiation on the surface temperature the scientists in TSOGU conducted experimental studies. Samples of petroleum from various West-Siberian fields, kerosene and diesel fuel were placed in chemical vessels from 0.01 to 0.30 cm high and put under the sun and in the shadow. The amount of evaporated hydrocarbons σ was determined by weighing. We also observed changes in the temperature of petroleum and petroleum products with the free surface $S = (0.003 \div 0.03)$ m^2 and varying degrees of thermal insulation of the side walls and bottom. Checking the repeatability of the experimental data by the value σ using Cochran's criterion showed the dispersions homogeneity.

Figure 4 provides data on evaporation of Shaimskaya (curves 1, 3) and Surgutskaya (curves 2, 4) oils at the ambient temperature t_B = 29°C and wind velocity ϑ_B = 0.6 m/s. Curves 3-4 characterize evaporation "in the shadow".

From these dependences we can see that the influence the solar radiation on the petroleum evaporation temperature is sufficiently large. So, over 9 hours of evaporation under the influence of solar radiation (curves 1-2) the value of losses σ was 2 times higher than during evaporation "in the shadow". This is due to the fact that the surface temperature $t_П$ "under the sun" of the Shaimskaya and Surgutskaya oils increased to 45 and 42°C, and the mean temperature of petroleum "in the shadow" was only 29°C. It should be noted that, when comparing the evaporation curves obtained when petroleum was exposed to the solar radiation with the evaporation curves of the same petroleum without the influence of the solar radiation (under condition that the temperatures are equal), they are fully identical. The change in temperature decreased with depth and almost stopped at a depth of 0.15÷0.20 m.

Analyzing the experimental data, we can note that the highly resinous Shaimskaya oil is slightly more exposed to solar radiation, although the difference and the surface temperature of the investigated oils are not significant. Thus, when determining the evaporative loss of oil from the open surface, the influence of direct solar radiation can be considered through the change in temperature in the daytime.

As a result of the regression analysis of numerous experimental data it was possible to obtain the formula for calculating the temperature of the upper layers of liquid

$$t_{SD} = 0.5(t_{HD} + t_{BD})(16.94 \times 10^{-9} J_{CP}^{2.55} \tau_{DH} + 1), \tag{8}$$

where t_{SD} and t_{BD} – the surface temperature and the average temperature in the daytime, °C; t_{HD} – the product temperature at the beginning of storage, J_{CP} – the average daily intensity of the direct solar radiation coming on a horizontal surface, W/m^2; τ_{DH} – the duration of daytime, hour.

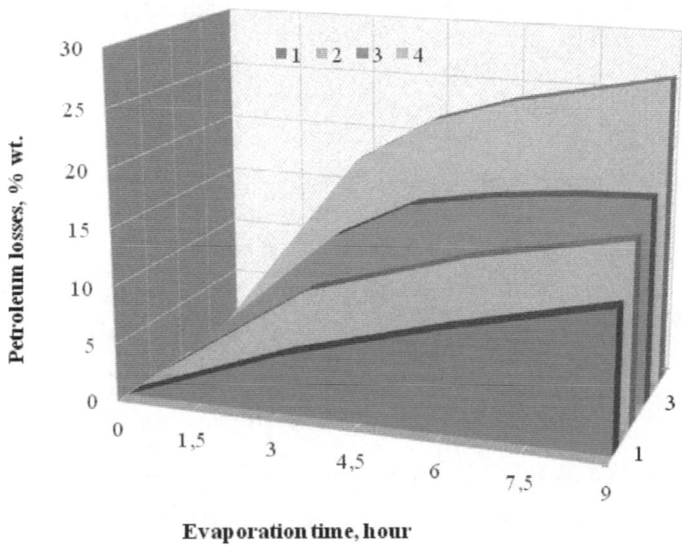

Figure 4. The impact of solar radiation on the dynamics of petroleum evaporation (oil: 1,3 - Surgutskaya; 2,4 – Shaimskaya).

We can recommend using the obtained dependence in engineering design, for example, when determining evaporative losses and emissions of liquid hydrocarbons from the open surface. The relative error in the determination of the surface temperature of the oil in the formula (4) is about 20%, at the radiation intensity $J_{CP} \geq 190$ W/m^2. The coefficient of multiple correlation was 0.94 and is with a confidence level of 0.95 by Student's test.

The following relationships are recommended to determine the surface temperature of petroleum and petroleum products:

$$t_S = t_H + 0.4 \div 0.6(t_B - t_H), \quad (9)$$

and

$$t_S = 0.5\,(t_B + t_H)$$

where t_B and t_H – the temperature of the air and the product, °C.

$$t_S = \frac{t_{SD}\tau_{DH} + t_{SH}(24 - \tau_{DH})}{24}, \quad (10)$$

However, formulae (9) are obtained with respect to tanks and storage facilities, i.e. when the direct solar radiation does not reach the surface of the oil. Therefore, the average temperature of the open liquid surface per day should be defined as follows:

$$t_{SH} = 0.5(t_{BH} + t_{HH}), \quad (11)$$

where t_{SH} – the average temperature of the liquid surface at night, °C.

$$t_B = \frac{t_{BD}\tau_{DH} + t_{BH}(24 - \tau_{DH})}{24},$$

where t_{BH}, t_{HH} – the temperature of the air and liquid at night, °C. The average daily ambient temperature can be determined by

$$t_{BH} = \frac{24t_B - t_{BD} \times \tau_{DH}}{24 - \tau_{DH}}. \quad (12)$$

$$t_{HH} = \frac{24t_H - t_{HD}\tau_{DH}}{24 - \tau_{DH}}. \quad (13)$$

Similarly to t_B we can write a formula for determining the average daily product temperature, therefore

$$t_{SH} = \frac{24(t_B + t_H) - (t_{BD} + t_{HD})\tau_{DH}}{2(24 - \tau_{DH})}. \quad (14)$$

Considering (12) and (13) after transformations, (11) can be presented as (14).
Using (14), the average daily temperature of the free surface can be determined as

$$t_S = \frac{t_{SD}\tau_D + 12(t_B + t_H) - 0.5(t_{BD} + t_{HD})\tau_{DH}}{24}$$

or

$$t_S = \frac{12(t_B + t_H) + \tau_{DH}\left[t_{SD} - 0.5(t_{BD} + t_{HD})\right]}{24} \quad (15)$$

After transformations using (8) we have

$$t_S = 0.5(t_B + t_H) + 0.354 \times 10^{-9} J_{CP}^{2.55} \tau^2_{DH}(t_{HD} + t_{BD}). \quad (16)$$

or

$$t_S = 0.5(t_B + t_H) + \frac{\tau_{DH}}{24}\begin{bmatrix} 0.5(t_{BD} + t_{HD})(16.97 \times 10^{-9} J_{CP}^{2.55} \times \\ \times \tau_{DH} + 1) - 0.5(t_{HD} + t_{BD}) \end{bmatrix}.$$

When calculating the average daily temperature of the free surface of liquid hydrocarbons stored in the earth storage, we assume $t_H = t_{HD}$ or determine it at a depth of: 2/3 from the fill-up height h at $t_H > t_B$, and 1/3 from h at $t_H < t_B$.

When petroleum and petroleum products spill on the water surface we can assume that t_H is the water temperature measured at a depth of 0.5 m where the daily fluctuations of its value practically attenuate. In case the oil is on the soil surface, instead of t_H we can use the soil temperature t_3 measured by means of Savinov or Ivanov thermometers at a depth of 15 cm.

The air temperature at the time of sunset τ_3 can be taken as the average daily air temperature t_B, allowing a sinusoidal temperature variation during the day. This calculation, of course, is characterized by significant errors (20÷30%), although in some cases it may be justified. For more accurate calculations, we recommend using the following formula:

$$t_B = \frac{1}{N}\sum_{i=1}^{N} t_{Bi}, \quad (17)$$

where t_{Bi} – the temperature of a single measurement, °C; $N = \frac{24}{\tau_N}$ - the number of measurements; τ_N – the time between measurements, hour.

The average daily temperature of the petroleum surface when it is stored for a number of days can be determined by the formula similar to

IOP Conf. Series: Materials Science and Engineering **154** (2016) 012002 doi:10.1088/1757-899X/154/1/012002

$$t_{SAV} = \frac{1}{n}\sum_{i=1}^{n} t_{Si}, \qquad (18)$$

where n – the number of storage days; t_{Si} – the average daily temperature of the petroleum surface determined by (16).

4. Summary

Experimental verification of the developed analytical relationships under production conditions in the study of the mass transfer process showed satisfactory statistical estimates. Models (8), (15) and (16) are adequate by Fisher's test with a calculation error of 10÷15% and the multiple correlation coefficient R=0.92÷0.95. The proposed methodological approach to determining the temperature of oil and its surface allows taking into account components of the heat balance and can be recommended for expert calculation of evaporative losses and emissions of hydrocarbons.

References

[1] Antipjev V N, Nevolin A P and Zemenkov Yu D 1981 Operation of intermediate pumping stations when pumping gas saturated oil *Oil Industry* **10** 46-8

[2] Zemenkov Yu D 1985 Evaporation of oil from the exposed surface in case of failure on the main oil pipelines *Thesis for the degree of candidate of technical sciences (Ufa, UNI)* 202

[3] Zemenkov Yu D 1999 Methodological support of expert calculations of leakage and emissions from pipelines when transporting liquid hydrocarbons *Thesis for the degree of Doctor of Technical Sciences (Tyumen, TSOGU)* 360

[4] Zemenkov, Yu.D., V.V. Shalay, M.Yu. Zemenkova. Immediate Analyses and Calculation of Saturated Steam Pressure of Gas Condensates for Transportation Conditions (2015) Procedia Engineering, Volume 113 (2015), pp. 254-258 doi:10.1016/j.proeng.2015.07.330

[5] RD 153-39.4P-124-02 Regulations on the procedure for technical inspection and extending the service life of process equipment of OPS MP

[6] Vakulin A A and Shabarov A B 1998 *Diagnostics of thermophysical parameters in oil and gas technologies* (Novosibirsk, Nauka) 249

[7] Vlasenko V S, Slesarenko V V, Gulkov A N and Lapshin V D 2015 Preparation of arctic oil and gas condensate deposit formation fluid for storage and transport in the form of hydrocarbon hydrate-containing dispersed system *Proceedings of the International Offshore and Polar Engineering Conference* 1863

[8] Stepanov S, Cheremisin N, Sokolov S, Altunin A and Shabarov A 2012 Computational method for relative permeability curves based on the bernoulli generalized porous media equations *ECMOR 2012 13th European Conference on the Mathematics of Oil Recovery*

[9] Chekardovskiy, M.,Chekardovskiy, S.,Ilyukhin, K.,Gladenko, A. Upgraded Algorithm for Calculating the Turbo-Expander of Gas Distribution Stations 2016 MATEC Web of Conferences 73

[10] Walas Stanley M 1998 *Phase equilibria in chemical engineering* Translated from English (Moscow, Mir) **2** 360

[11] Stepanov S, Cheremisin N, Sokolov S, Altunin A and Shabarov A 2012 Computational method for relative permeability curves based on the bernoulli generalized porous media equations *ECMOR 2012 13th European Conference on the Mathematics of Oil Recovery*

[12] Shantarin V D and Voitenko V S 1990 *Physical chemistry of disperse systems* (Moscow; Nedra) 315

IOP Conf. Series: Materials Science and Engineering **154** (2016) 012003 doi:10.1088/1757-899X/154/1/012003

Modeling Bottom Sediment Erosion Process by Swirling the Flow by Tangential Supply of Oil in the Tank

V O Nekrasov

Industrial University of Tyumen, 38 Volodarskogo st, Tyumen, 625000, Russia

[1]E-mail: nekrasov-sever@mail.ru

Abstract The article carries out a statistical data processing of quantitative and territorial division of oil tanks operating in Tyumen region, intended for reception, storage and distribution of commercial oil through trunk pipelines. It describes the working principle of the new device of erosion and prevention of oil bottom sediment formation with tangential supply of oil pumped into reservoir. The most significant similarity criteria can be emphasized in modeling rotational flows exerting significant influence on the structure of the circulating flow of oil in tank when operation of the device described. The dependence of the distribution of the linear velocity of a point on the surface along the radius at the circular motion of the oil in the tank is characterized, and on the basis of this dependence, a formula of general kinetic energy of rotational motion of oil and asphalt-resin-paraffin deposits total volume in the oil reservoir is given.

Introduction.

Trunk pipeline transport is one of the most important components of the Russian Federation transportation system. It performs more than half of cargo turnover by all means of transport.

A significant part of trunk oil pipelines, operating now, passes through the territories belonging to the regions of the Far North and localities equated to them. Thus, 172 tanks of VST mark (vertical steel tank) intended for the reception, storage and distribution of commercial oil are in operation in Tyumen region. Only 15% of the general fund of the tanks are installed in the south of region and 85% of tanks with total commodity capacity of 1.48 million m3 are located in the territories of the Far North and localities equated to them (Figure 1).

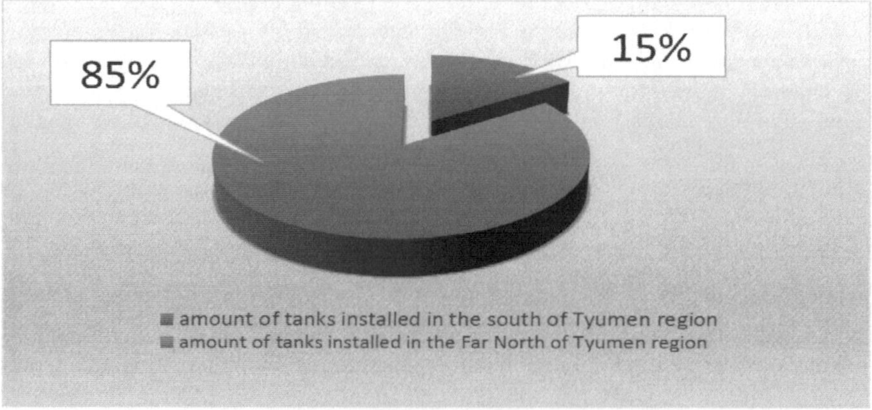

Figure 1. Territorial distribution of trunk oil pipelines tanks of Tyumen Region

IOP Conf. Series: Materials Science and Engineering **154** (2016) 012003 doi:10.1088/1757-899X/154/1/012003

According to the "Energy Strategy of Russia for the period until 2030", one of the most important directions of development of national economy is implementation of comprehensive programs on energy development of the Eastern Siberia and the Far East, the North-West region of Russia, the Yamal Peninsula and continental shelf. And in accordance with the RF Government Decree of 04.15.2014, No 321 "On approval of the state program of the Russian Federation "Energy efficiency and energy development", expansion of energy and raw material infrastructure of the Far North should be implemented in the nearest future. Moreover, OJSC "AK "Transneft", implementing strategic development program developed for the period until 2020, plans to build about 800,000 m3 of tank reservoirs for crude oil and about 720,000 m3 for oil products, spending about 40 billion rub on tanks construction. Therefore, now there stands out a particularly urgent problem of energy-efficient operation of oil tanks providing a high level of technical and economic parameters of oil pipeline system on the territory of the Far North.

Methods

An important indicator of the climatic characteristics of the Far North territory is reduced negative average annual ambient temperature. This factor acts as a catalyst of sedimentation rate of asphalt-resin-paraffin fractions (ARPF), that leads to the formation and accumulation of bottom sediment on the bottom and walls of oil tanks. ARPF deposits can reach ¼ of useful capacity of container per year.

The "Transport of Hydrocarbon Resources" department of Tyumen State Oil and Gas University has developed and patented a new device of erosion and prevention of bottom sediments formation in the oil reservoirs VST [1]. However, for conducting field experiments at operating reservoirs it is necessary to carry out a number of laboratory experiments aimed at studying and modeling of process of hydrodynamic flow of the multiphase and multi-component mixture in the tanks VST using the developed device. While conducting field experiments, one should take into account the different heights and the uneven formation of bottom sediments, quantitative and qualitative composition of stored oil products, as well as constructive conditions and modes of reservoirs operation.

Erosion and preventing deposit formation is carried out due to large peripheral speeds of oil flow in a tank, due to which the separation of ARPF particles from the bottom and turning them into suspension is carried out. Rotational motion of the total volume of oil in a tank is carried out due to swirling a flow when the tank is being filled by the tangential supply of liquid through several peripheral openings in the common collector, routed around the perimeter of the tank bottom [2].

For the study of rotational motion of fluid in a tank by experimentation in laboratory conditions, it is necessary to comply with conditions fulfillment of geometric, kinematic and dynamic similarity.

Geometric similarity of rotational (circulating) flows is ensured by fulfillment of values equality condition of relative rotation radii of particles in a tank $r_0 / R = idem$, and the relative roughness of tank walls $k_э / R = idem$ [3].

As a result of statistical data processing, a tank VST-20000 has become a modeling object, as this type of tanks is 61% of the total number of tank capacities operating now in Tyumen region on the territory of the Far North (Figure 2).

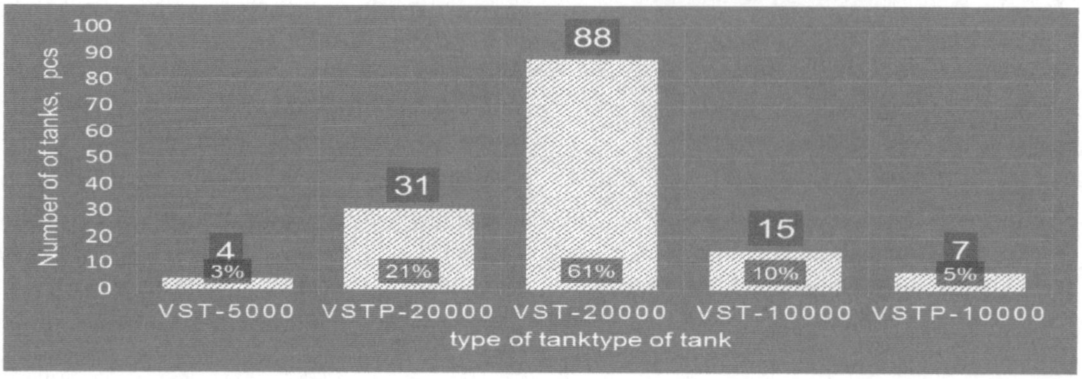

Figure 2. Specific gravity of tank types operating in Tyumen region on the territory of the Far North

In modeling the rotational motion of fluid, it is almost impossible to achieve strict simultaneous compliance of all the conditions of dynamic similarity because of their large-scale incompatibility for streams having different linear scale, but the same physical properties (density, viscosity) and moving in the same gravitational field. Therefore, the following criteria in modeling the circulating motion of oil in a tank were selected as dominant:

- Reynolds Criterion:

$$Re = \vartheta \cdot D/\nu = idem \ , \quad (1)$$

where ϑ – velocity of oil flow, m/s; D – tank diameter, m; ν – kinematic viscosity of oil, m^2/s;

- Euler criterion:

$$Eu = p_0/\rho \upsilon_0^2 = idem \ , \quad (2)$$

where p_0 – pressure on the rotation axis of fluid, Pa; ρ – oil density, kg/m^3; υ_o – peripheral flow velocity of oil in the tank, m/s;

The given criterion of considered totality defines the conditions (pressure) on the axis of flow rotation. The conditions significantly affect the structure of the circulation flow.

- Froude criterion:

$$Fr = \upsilon_0/gR = idem \text{ or } Fr_\Gamma = \Pi = Fr\left(\frac{u_0}{\upsilon_0}\right)^2 = idem \ , \quad (3)$$

where R – tank radius, m; u_0 – axial motion velocity of oil in tank, m/s.

For the considered circulating flow of an incompressible fluid, Euler and Froude criteria are compatible in scale, and their product can act as an additional or self-combined criterion, if necessary:

$$Eu \ Fr = p_0/\rho g R = idem \ . \quad (4)$$

IOP Conf. Series: Materials Science and Engineering **154** (2016) 012003 doi:10.1088/1757-899X/154/1/012003

In the second form of a fixation of the Froude number on(4), attention should be paid to the ratio of characteristic peripheral and axial velocity, which determines the condition of kinematic similarity, and is one of the most significant in the analysis of circulating flows:

$$Ro_* = \frac{u_0}{\upsilon_0} = idem \quad . \tag{5}$$

In modeling circulation flows of viscous incompressible fluid, three basic parameters are used as the kinematic similarity criteria: integral parameter of swirling of the full flow of quantity of motion (F), integral parameter of flow swirling in an arbitrary section (F_*) and limit swirl angle of flow (φω) [4].

Integrated swirling parameter of the full swirl of motion quantity of flow shows the ratio of rotational quantity of motion to the axial projection of the total flow motion quantity on the scale of R and is determined by the formula:

$$\Phi = \frac{M}{K \cdot R}, \tag{6}$$

where M – axial component of flow of angular momentum:

$$M = 2\pi \int_0^R \rho \omega u r^2 dr; \tag{7}$$

K – axial component of the total flow motion quantity:

$$K = 2\pi \int_0^R \left(p + \rho \omega^2\right) r dr, \tag{8}$$

where R – channel radius; r – current radius; p – pressure in the considered section; ρ – fluid density; u and ω – rotational and axial components of flow velocity.

The integrated flow swirling parameter in an arbitrary section is a ratio of rotational quantity of flow motion to an axial in a scale R, which is defined by the formula:

$$\Phi_* = \frac{M}{K_x \cdot R}, \tag{9}$$

where K_x – quantity motion flow axial component:

$$K_x = 2\pi \int_0^R \rho \omega^2 r dr \tag{10}$$

The limiting angle of flow swirling characterizes the deviation degree of the limiting stream line from the axial direction and is expressed by a dependence:

$$tg\varphi_\omega = \frac{\tau_{\varphi\omega}}{\tau_{x\omega}}, \tag{11}$$

where $\tau_{\varphi\omega}$ and $\tau_{x\omega}$ – projection of surface tension of friction on the coordinate axes.

As a result of the experiment and data processing, the dependence of the linear velocity distribution of a point on a surface along the radius at a circulating motion of oil in tank was determined (Figure 3).

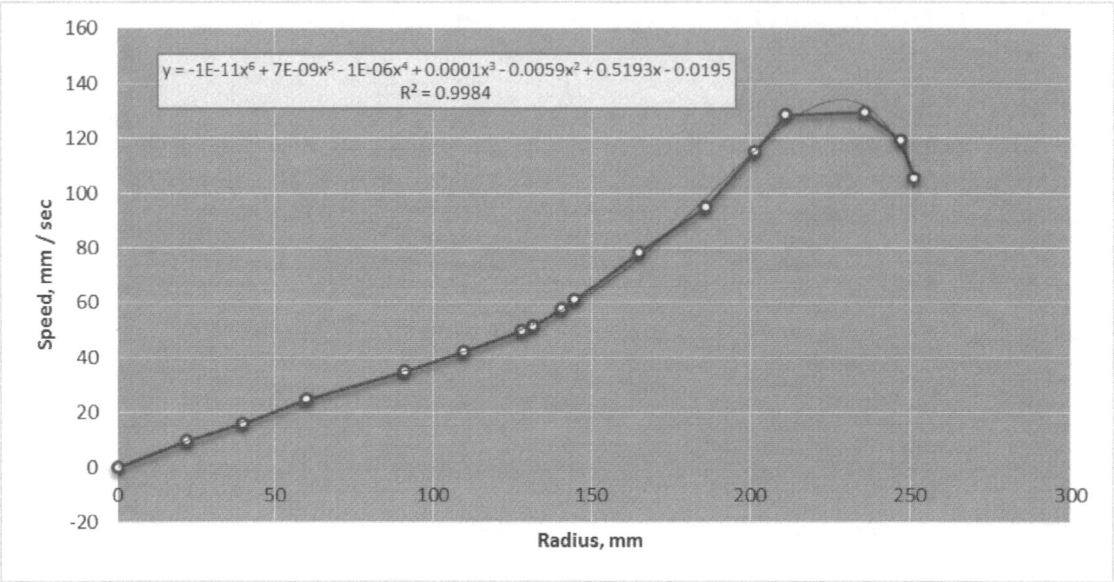

Figure 3. Distribution of a point speed on the surface of oil along the radius at circulating motion of oil in a tank

It can be seen from the graph that this dependence is described by a polynomial of sixth degree with a standard deviation R = 0.9984:

$$\upsilon(r) = -1E\text{-}11r^6 + 7E\text{-}09r^5 - 1E\text{-}06r^4 + 0.0001r^3 - 0.0059r^2 + 0.5193r - 0.0195. \quad (12)$$

During research, a formula was obtained of the kinetic energy of rotational fluid motion in case of indirectly proportional change of peripheral velocity within the vortex core:

$$W = \pi \rho h_{cp} \int_0^R \upsilon^2 r dr \, , \quad (13)$$

where $\upsilon = f(r)$, ρ – oil density, h_{cp} – the average level of oil innage in a tank.

Conclusion

As a result of analysis of the country's energy strategy, the strategic development program of the largest oil transport company in Russia and a number of legislative initiatives, the preservation and further development of energy efficiency and conservation vector in fuel and energy sector of the Russian Federation can be concluded.

The modeling experience of rotational motion of the fluid indicates the impossibility of achieving an absolute similarity of circulation flows due to their large-scale incompatibility. Thus, during the modeling of axisymmetric rotational motion of continuous media, one must perform preservation of

IOP Conf. Series: Materials Science and Engineering **154** (2016) 012003 doi:10.1088/1757-899X/154/1/012003

similarity of the most essential criteria, providing dominant influence on the structure of circulating flow.

The dependence of velocity distribution on the surface of oil from the radius at a flow swirl by the tangential supply of fluid caused by the design of a new device was established. This dependence is indirectly proportional manifested in view of non-equidistant location of pressure tubes of flow swirling device from the tank center. On the basis of this dependence, a formula was obtained to determine the total kinetic energy of rotational motion of oil and ARPF in the oil tank in operation of erosion and prevention of oil sediments formation.

References

[1] Pat. 2500486 RU, IPC B05C7/04, B65D88/74 A device for increasing operating properties of vertical steel tanks / Nekrasov V O, Levitin R E, Tyrylgin I V, Zemenkov Yu D (RU) -№ 2012125478/05, stated 12/10/13.

[2] Nekrasov V O, Zemenkov Yu D Promising methods of improving operating properties of oil tanks./ Nekrasov V O, Zemenkov Yu D // Pipeline transport [Theory and Practice].– Moscow: VNIIST, 2012. - №6 – 24-26 pp.

[3] Zuikov A L The dynamics of viscous circulation flow in pipes and surface funnels. Dis. ... Dr. tehn. sciences. M., 2010.

[4] Shchukin V K, Khalatov A A Heat exchange, mass exchange and hydrodynamics of swirled flows in axisymmetrical channels. M., Mechanical Engineering, 1982.

Control of Mechanical Stresses of High Pressure Container Walls by Magnetoelastic Method

S M Kulak, V F Novikov and A V Baranov

Industrial University of Tyumen, 38 Volodarskogost., Tyumen, 625000, Russia

E-mail: ksm-rabochi@rambler.ru

Abstract. Deformations of the walls of pressure vessels arising in the process of testing and operation, as well as reduce their thickness due to corrosion, to create the prerequisites for the growth of mechanical stresses which accelerating the processes of strain aging, embrittlement of the material and reducing its fatigue properties. This article is devoted to researches of the magnetoelastic demagnetization in the wall of steel vessel of loading by internal pressure. It is established that the increasing pressure on the vessel wall is accompanied by a monotonic decrease in the intensity of the magnetic stray field of local magnetization of steel. It is shown that a magnetic stray field of local magnetization of the wall of steel vessel is non-uniform due to differences in structure and stresses.

It is proposed to use the obtained results to control the stress state of vessels, experiencing multi-axial loads generated by internal pressure (pipelines, oil tanks, etc.)

The method of magnetoelastic of the demagnetization of the steel has a high sensitivity to mechanical stress, the simplicity of implementation and expressiveness compared to the strain gauge and method of coercive force.

1. Introduction

Deformation of vessel walls under pressure, arising in the process of their testing and operation, as well as corrosion decrease in their thickness, create preconditions to mechanical stresses growth accelerating processes of strain aging and embrittlement of the material and reducing its fatigue properties.

At the forefront of solving the problem of high pressure vessels reliability, calculation tasks for the strength, stability and durability are put forward. To solve them, information about the mechanical loads and impacts on the walls is required, analysis of stress-strain state (SSS)that eventually is used in the calculation of their reliability and resource [1,2].

The main factor determining the stress state of the vessel body is process pressure [3]. Under the influence of the internal pressure, vessel walls are in volume stressed state when along the generator the meridional (axial) stress σ_s operates, tangentially to a circle - hoop stress σ_τ, and at standards to the wall - the radial stress σ_r. Radial stress σ_r, acting in the direction of the wall thickness is assumed as equal to zero. Thus, the vessel walls are subjected to biaxial loading.

2. Models and methods

Magnetoelastic method (the method of magnetoelastic demagnetization or magnetoelastic memory MEM) [4-6] of mechanical stresses control has a high strain sensitivity, possibility of contact measurement without breaking insulation coatings, high efficiency, and simplicity of implementation. Magnetoelastic demagnetization is very important and with the application of the equal and unequivocal orthogonal stresses of one sign, and for this reason MEM is a more sensitive indicator of equivalent stresses than, for example, the coercive force.

MEM method studies were carried out on a steel tank of carbon dioxide extinguisher OU-3 in which biaxial stress state was produced by internal pressure of water.

Experimental setup scheme for studying the stress state of extinguisher cylinder by the magnetoelastic, coercimetric and strain-gauge methods is shown in Figure 1.

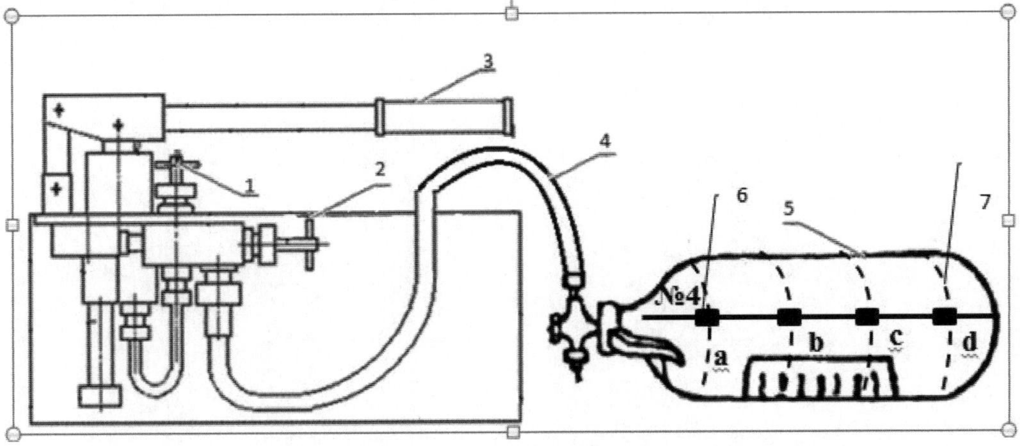

Figure 1. Scheme of laboratory installation of extinguisher cylinder for studying the stress state 1-manometer; 2-valve of pressure relief; 3-pressurization lever; 4-rubber hose; 5- extinguisher cylinder; 6 places of sticking four pairs of tensoresistive sensors on one of the cylinder №4 meridional lines; 7- circular lines (a, b, c, d).

Installation consists of the extinguisher cylinder and is connected to it with a manual testing pump RP 30 (hydraulic pump). Extinguisher cylinder is a hollow cylinder with convex bases, made of carbon steel. The height of the cylinder is 450mm, outer diameter is 107mm, wall thickness is determined by means of ultrasonic thickness gauge UDT-40 in five different points and is equal to 5.5 mm. The pressure in the container was produced with water under a maximum pressure $P = 3$ MPa.

On the lateral surface of cylinder along the generators four straight meridional lines (№1,2,3,4) were applied with a marker, connecting the top of its bases and four equidistant circular lines (a, b, c, d) perpendicular to them (Figure 1). In intersection points of the meridian line №4 with circular lines, tensoresistive sensors recording meridional ε_s and circular ε_τ deformation of the cylinder wall under internal pressure were pasted. With magnitude of these strains, respectively meridional σ_s, circular σ_τ and equivalent σ_{ekv} mechanical vessel wall stresses [7] were determined. The other three meridional lines (№1,2,3) were intended for magnetoelastic and coercimetric research.

Extinguisher cylinder was fixed in a non-magnetic body in a horizontal position away from possible sources of magnetic fields and was filled with water. Local magnetization of cylinder wall perpendicular to σ_s action was carried out, the initial value of normal Hn1 and tangential Hτ1 components of the magnetic stray field of the local magnetization (LM) in the areas of intersection of meridian and circular lines with a fluxgate magnetometer IKNM-2FP was measured. Gradually increasing the internal pressure in the cylinder by an equal amount, new indications of flux-gate magnetometer Hn2 and Hτ2 at the same points were recorded.

The magnetization of extinguisher cylinder wall was carried out by scanning its surface along the meridian lines of U-shaped magnetizing device (Figure 2). Thus, on the surface of the cylinder a band was made with a local residual magnetization $\overrightarrow{M_r}$ perpendicular to its generatrix (meridian line) (Fig.2b). With this method, demagnetization factor remains constant, which allows tracking changes in the structure and the stresses in the steel.

IOP Conf. Series: Materials Science and Engineering **154** (2016) 012004 doi:10.1088/1757-899X/154/1/012004

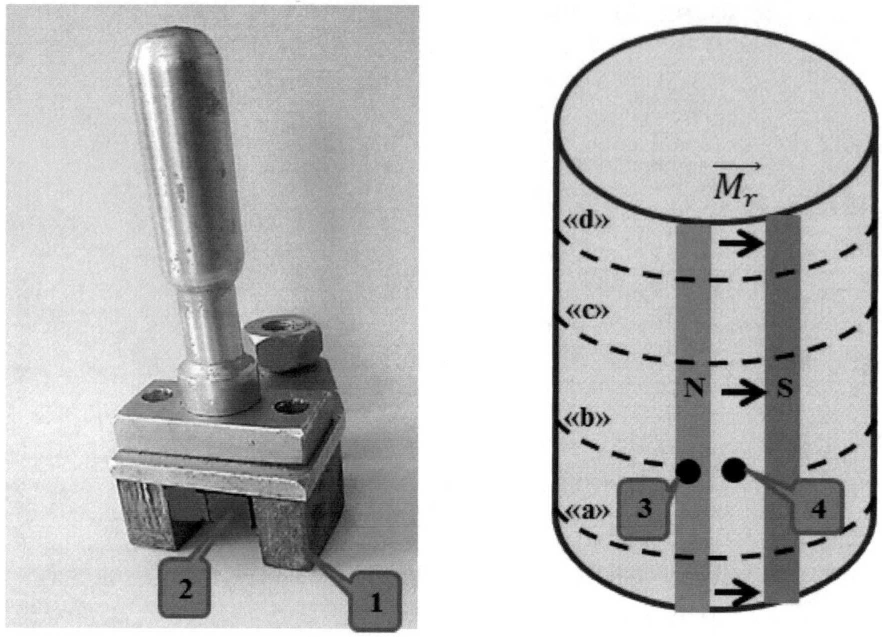

a b

Figure 2. "U-shaped" magnetizing device (a) and band of local residual magnetization on cylinder side wall (b): 1-pole pieces «N» and «S»; 2- magnets; 3,4-points of measurement of normal Hn and tangential Hτ components of the scattering field LN

Dependence of change normal ΔHn = Hn1 - Hn2 and tangential ΔHτ = Hτ1 - Hτ2 components of the magnetic stray field LN along the meridian line №3 in the areas of its intersection with the circular lines on the value of internal pressure in the cylinder are shown in Fig. 3. The full line approximates the results of measurements along the meridian line №3.

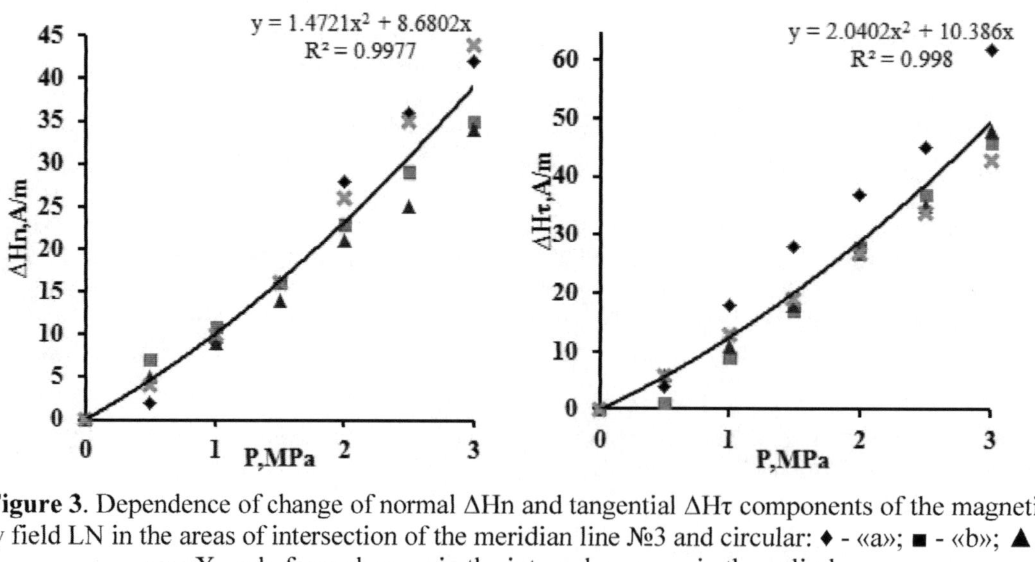

Figure 3. Dependence of change of normal ΔHn and tangential ΔHτ components of the magnetic stray field LN in the areas of intersection of the meridian line №3 and circular: ♦ - «a»; ■ - «b»; ▲ - «c»; X - «d» from changes in the internal pressure in the cylinder

The largest change in the tangential component of the magnetization field is observed in a zone of the ring "a" located at the bottom of the cylinder with outlet pipe and the minimum at the area of the ring «d» from another bottom. Obviously, this is due to the fact that in the central part of cylinder side wall, circular voltage is greater than the meridional [8]. It should be noted that in the zone of lines «a» and «d», located close to cylinder bottom ΔHn of magnetic field changes equally. This can be explained by the fact that the bottom of cylinder has a spherical shape, and therefore there should be implemented the same circular and meridional stresses, leading to equivalent steel magnetoelastic demagnetization.

To understand magnetic processes that occur in biaxial deformation of a cylinder steel shell, cylinder metal can be represented (excluding the shape factor) as the material in which two unequal light magnetic axes are formed: one - along a circular, and the other - along the meridian tensile stresses. With the growth of stresses the domain structure is changing, leading to the steel demagnetization.

Average values of coefficients Kn and Kτ of a magnetoelastic demagnetization of steel cylinder shell along the normal and tangential directions, calculated for the maximum value σ_s and σ_τ, made:

$$K_n = \frac{\Delta H_n}{H_{n1} \cdot \sigma_s} \approx 0.7 \cdot 10^{-8} Pa^{-1};\ K_\tau = \frac{\Delta H_\tau}{H_{\tau 1} \cdot \sigma_\tau} \approx 0.41 \cdot 10^{-8} Pa^{-1}$$

The magnitude of circular σ_τ and meridional σ_s stresses in controlled points on the surface was evaluated based on the results of tensometric measurements corresponding to deformation of container sidewall caused by internal pressure. The results of this evaluation are presented in Figure 4.

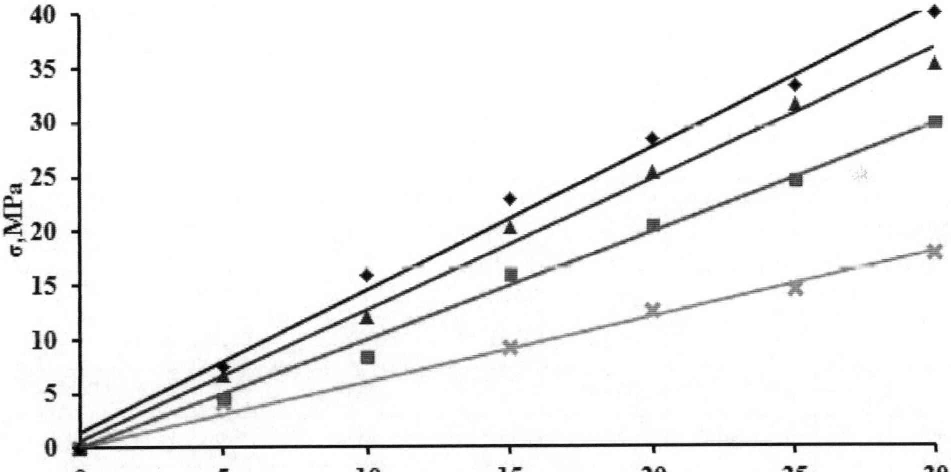

Figure 4. Dependence of mechanical stresses in the vessel wall, defined tensometrically on the value of internal pressure in it: ♦ -circular stresses σ_τ («c» line); ■ - meridional stresses σ_s («c» line); ▲ - circular stresses σ_τ (line «a»); × - meridional stresses σ_s («d» line).

In the central part of the vessel (line "c") the largest circular and meridional stresses and circular stresses near the upper base of the cylinder are generated. The latter can be explained by proximity of location of circular line «a» to the outlet pipe of cylinder. Minimum stresses are defined at another bottom (line «d»). In research [7] it is noted that based on the aggregation of a large number of experimental data obtained in our country and abroad the most loaded node in a container with pressure is pipes junction. The relative change in the cylinder wall deformation along the meridional and circular line is equal to $\varepsilon_s \sim 18 \cdot 10^{-5}$ and $\varepsilon_\tau \sim 32 \cdot 10^{-5}$, respectively, which is much smaller than the relative change of the magnetic stray field LN $\varepsilon_n \sim 13 \cdot 10^{-2}$ and $\varepsilon_\tau \sim 16 \cdot 10^{-2}$.

At biaxial loading as a criterion for strength the value of the equivalent stress is taken σ_{ekv} that in the offered work is determined by the energy of the theory [9]:

$$\sigma_{ekv} = \sqrt{\sigma_\tau^2 + \sigma_s^2 - \sigma_\tau \sigma_s}, \quad (1)$$

From the value σ_{ekv}, the reliability of the vessel wall is defined, determining its stress-strain state. For the central part of side wall of the vessel, the maximum σ_{ekv} ~ 36 MPa, and near its bottom - σ_{ekv} ~ 30 MPa.

Evaluation of cylindrical shell strength will be carried out according to the method of permissible stresses. According to this method, the maximum equivalent stresses in the dangerous section shall not exceed the permissible [σ]:

$$[\sigma] = \eta \cdot min\left\{\frac{\sigma_T}{n_T}; \frac{\sigma_B}{\sigma_B}\right\}, \quad (2)$$

where η - coefficient, which for welded vessels and devices is equal to 1; $\sigma_В$, $\sigma_Т$ - tensile strength and yield strength; $n_В$, $n_Т$ - margin of strength by tensile strength and yield strength at the estimated temperature, respectively. According to regulatory documents $n_Т$ = 1.5 and $n_В$ = 2.6. In this paper, a cylindrical shell material is steel St.3. The estimated temperature is 20 ° C. Limiting stresses for this steel at 20 ° C and shell thickness <20 mm are: $\sigma_В$ = 460 MPa, $\sigma_Т$ = 350 MPa. Condition (3) is satisfied.

In the areas of intersection of the circular (a, b, c, d) and meridian lines №1,2,3 of extinguisher cylinder sidewall, measurements were made of the coercive force of steel along the meridian and circular (perpendicular) directions by a coercimeter - structurescope KRM-C-K2M. The measurement results of coercive force Hc‖ along the meridian line №3 and Hc⊥ along the the circular lines are shown in Table 1.

Table 1. Results of measurement of the coercive force

Internal pressure in the container P, MPa	circular line «a»		circular line «b»		circular line «c»		circular line «d»	
	$H_{cl,}\cdot10^2$A/m	$H_{c}\perp\cdot10^2$ A/m	$H_{cl,}\cdot10^2$ A/m	$H_{c}\perp\cdot10^2$ A/m	$H_{cl,}\cdot10^2$ A/m	$H_{c}\perp\cdot10^2$ A/m	$H_{cl,}\cdot10^2$ A/m	$H_{c}\perp\cdot10^2$ A/m
0	12.3	10.3	11.5	9.6	11.6	9.9	11.9	10.3
0.5	12.2	10.3	11.5	9.5	11.5	9.8	11.6	10.2
1.5	12.2	10.3	11.5	9.5	11.5	9.8	11.7	10.2
2	12.1	10.3	11.5	9.4	11.5	9.7	11.7	10.1
2.5	12.2	10.3	11.5	9.4	11.5	9.7	11.6	10.0
3	12.2	10.2	11.5	9.4	11.6	9.6	11.6	10.0

It was found that in the initial state and with growth of the pressure in container, the coercive force measured along mutually perpendicular directions in all areas of intersection of the circular lines (a, b, c, d) and meridional lines (№1,2,3) on the surface of container, remained practically unchanged.

Conclusion

It is shown that the loss of the normal and tangential components of the magnetic stray field of the local residual magnetization of the container wall with an increase in internal pressure can be the metrological parameter of the equivalent mechanical stresses;

IOP Conf. Series: Materials Science and Engineering **154** (2016) 012004 doi:10.1088/1757-899X/154/1/012004

the coercive force of the cylinder steel shell at its biaxial deformation remains unchanged at low voltages, and can not be an indicator of their level;

It is shown that the magnetoelastic method of steel demagnetization is highly sensitive to mechanical stresses, easiness of implementation and expressiveness in comparison to tensometric method.

References

[1] RD 03-421-01 *Guidance for the diagnosis of technical condition and residual life of vessels and devices. [Text]* 2001 (Moscow, Gosgortechnadzor Russia) 86 p

[2] Gazprom corporate standard 2-2.3-491-2010 *Technical diagnostics of vessels working under pressure at the OJSC "Gazprom" facilities*

[3] Danikina T S, Turebaeva R D, Aktaukenova G S 2013 *Effect of changes in the pressure drop on the stress state of the inhomogeneous hollow cylinder / Building mechanics and calculation of constructions* **6** 46-49

[4] Gorkunov E S 2014 Different states of residual magnetization and their resistance to external influences. On the question of "method of magnetic memory" *non-destructive testing* **11** 3-21

[5] Novikov V F, Vazhenin Yu I, Baharev M S, Kulak S M, Muratov K R 2009 *Diagnosis of places of increased pipeline degradability (*M, "Nedra-business centers") 200 p

[6] Kostin V N, Tsarkova T P, Nichipuruk A P Loskutov V E, Lopatin V V, Kostin K V 2009 Irreversible changes in magnetization as indicators of stress - strain state of ferromagnetic objects *non-destructive testing* **11** 5 p

[7] Rudachenko A V Saruev A L 2011 A study of stress-strain state of pipelines: a tutorial *National Research Tomsk Polytechnic University (TPU)* (Tomsk, Publishing house TPU) 136 p

[8] Pershin V F, Selivanov Yu T 2004 Calculation of the strength of thin-walled shells of rotation and thick cylinders (Tambov, Publisher Tambov State Technical University) 20p

[9] Kurochkin V V, Malyushin N A, Stepanov O A, Moroz A A 2001 *Operational durability of pipelines* (M, Nedra) 231 p

IOP Conf. Series: Materials Science and Engineering **154** (2016) 012005 doi:10.1088/1757-899X/154/1/012005

Research on Thermodynamic Parameters of a Micro-Turbine for Standalone Cogeneration

M N Chekardovskiy, S M Chekardovskiy, I A Chekardovskaya, A I Mihajlenko

Industrial University of Tyumen, 38 Volodarskogo st., Tyumen, 625000, Russia

E-mail: zemenkov@tsogu.ru

Abstract. In order to advance the heating and hot water systems in the Russian Federation we need to address the problem of introducing new sources of heat and electric energy. The paper overviews the relationship between the reliability, efficiency and diagnostic maintenance of the GMTU, the studies of rated and operating conditions of the GMTU, the development and improvement of methods for calculating rated and operating thermodynamic and diagnostic parameter of the unit. This resulted in developing a passport of thermodynamic parameters that can be compared to the parameters of the same units when they operate for the purpose of diagnosing their state.

1. Introduction

The American firm Capstone indirectly introduces gas micro-turbines of the centrifugal type for combined production of heat and electricity. In the near future, as a result of import substitution, they can play the role of a new driving force in the evolution of engineering systems for decentralized production of electricity and heat (cogeneration).

Micro-turbines can create a reliable, year-round cogeneration for power of both sports, shopping, social facilities, and oil and gas facilities, such as compressors, pumping stations and oil and gas processing plants [1].

2. Research

The object of this research is equipment in the winter sports center "Pearl of Siberia" located 50 km from the city of Tyumen - gas micro-turbine units (GMTU).

The subject of this research is cogeneration processes in the GMTU CapstoneC65. The objective is development of algorithms for determining the parameters of control and diagnostic of operation modes and the technical state of the GMTU.

Figures 1 and 2 show a micro-turbine in the section and its exterior.

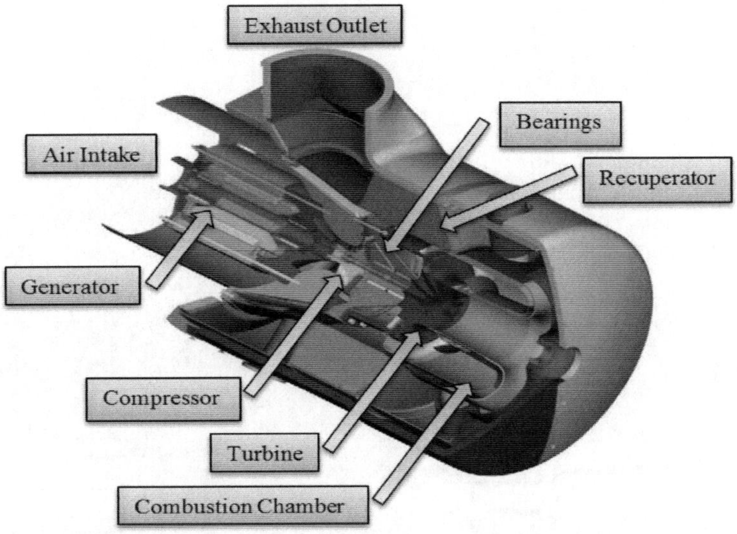

Figure 1. The GMTU in the section.

Figure 2. The exterior of a micro-turbine.

Operating modes are clustered and standalone. The article presents the results of the research on a standalone operating mode of micro-turbines.

A thermotechnical scheme of a unit with rated parameters is shown in Figure 3. Here are given rated (passport) temperature values for air and combustion products at the inlet and outlet of the generator, the centrifugal compressor and turbine, the combustion chamber and the integrated recuperator for heating the air entering downstream of the compressor into the combustion chamber.

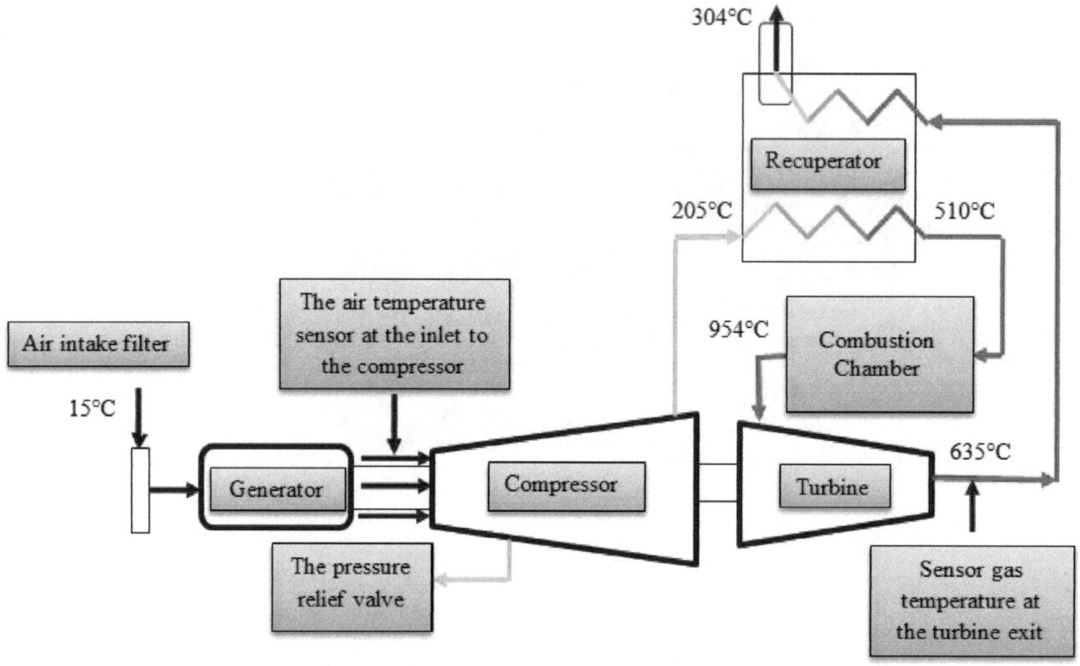

Figure 3. A thermotechnical scheme of the GMTU.

Figure 4 shows a device of the heat exchanger designed for heating water of the heat supply or hot water supply systems. Heating is provided by heat transfer from the combustion products through the walls of the ribbed tube heat exchanger to the water [2, 3].

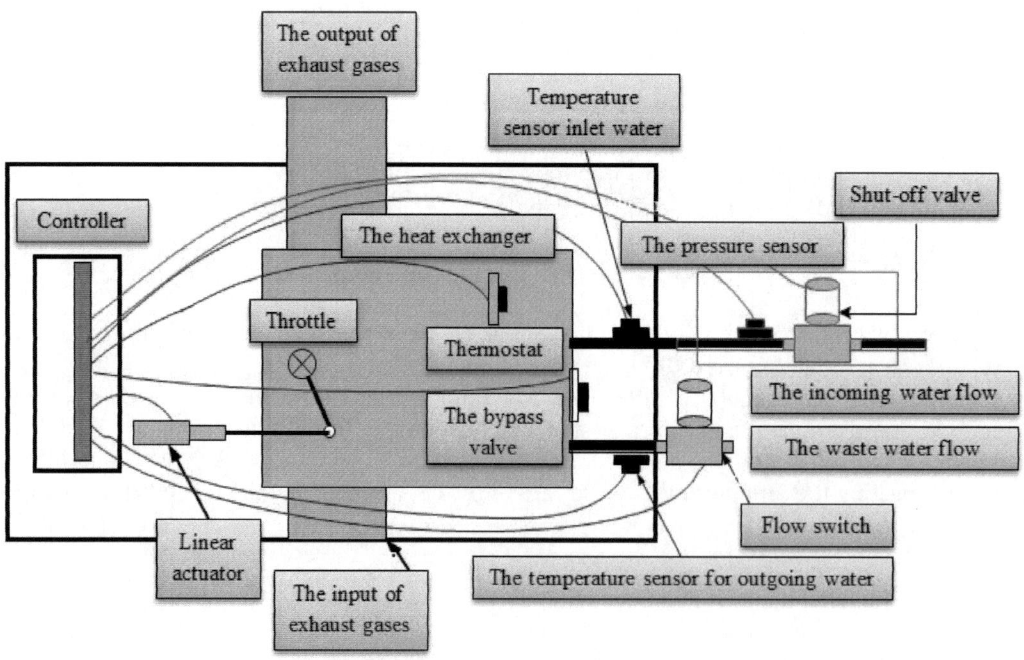

Figure 4. A scheme of the main components of the heat exchanger.

More information on the rated specifications of micro-turbines is presented in Table 1.

Table 1. Rated specifications of the GMTU.

Parameter	Notation	Units	Value
Ambient air temperature	T_a	K	288
Ambient air pressure	P_a	kPa	101.33
Effective capacity	N_e	kW	65
Mass air flow: into the combustion chamber (CC)	M_{CC}	kg/s	0.415
for cooling the CC	M_{cool}	kg/s	0.218
total, through the centrifugal compressor	M_{air}	kg/s	0.633
Air temperature upstream the centrifugal compressor	T_o	K	783
Air temperature downstream the centrifugal compressor	T_c	K	478
Air temperature downstream the recuperator	T_v	K	783
Temperature of combustion products upstream the PT	T_z	K	1227
Temperature of combustion products downstream the PT	T_s	K	908
Temperature of combustion products downstream the recuperator	$T_s^`$	K	577
Mass flow of combustion products	M_{CP}	kg/s	0.49
Mass flow of the fuel gas	B	kg/s	0.0047
Net calorific value of the fuel gas	$Q_н^p$	kJ/kg	47590
Effective efficiency of the GMTU	η_e	-	0.29
Combustion chamber efficiency	η_{CC}	-	0.98
Shaft rotation frequency	n	s^{-1}	1600

To calculate the additional rated characteristics of the unit we developed an algorithm that includes the sequence of known thermodynamic equations for calculating gas turbine units [4]. The algorithm is shown below.

The algorithm for calculating the additional rated characteristics of the GMTU

$$L_0 = \frac{Q_H^P}{2900}. \quad (1)$$

$$\alpha = \frac{\eta_{CC} \cdot Q_н^p}{L_o \cdot C_p \cdot (T_z - T_v)}. \quad (2)$$

$$N_{PT}=M_{CP}\cdot C_{P_{CP}}\left(T_z-T_s\right), \tag{3}$$

$$N_C=M_{air}\cdot C_{P_{air}}\left(T_c-T_a\right),\ \text{kW}. \tag{4}$$

$$N_{AG}=N_{PT}-N_C,\ \text{kW}. \tag{5}$$

$$\delta_{N_e}=\frac{N_{AG}-N_e}{N_{AG}}\cdot 100,\ \%. \tag{6}$$

$$\eta_{CC}=0.98. \tag{7}$$

$$\eta_{AG}=\frac{N_{AG}}{Q_н^p\cdot B} \tag{8}$$

$$\delta_{\eta_e}=\frac{\eta_{AG}-\eta_e}{\eta_{AG}}\cdot 100,\ \%. \tag{9}$$

$$P_c=P_o\left(\frac{T_c}{T_o}\right)^{\frac{\kappa}{\kappa-1}},\ \text{kPa}. \tag{10}$$

$$P_z=P_c\cdot\sigma_{CC},\ \text{kPa}. \tag{11}$$

$$\varepsilon_C=\frac{P_c}{P_o}. \tag{12}$$

$$\lambda_{PT}=\frac{P_z}{P_s}. \tag{13}$$

$$\varphi=\frac{T_v-T_c}{T_s-T_c}. \tag{14}$$

$$C_P=A+t\cdot 9.355\cdot 10^{-5}+t^2\cdot 3.694\cdot 10^{-7}-t^3\cdot 2.769\cdot 10^{-10},\ \frac{\text{kJ}}{\text{kg}\cdot\text{K}}. \tag{15}$$

Table 2 shows the results of calculating the additional rated characteristics of the unit such as: stoichiometric ratio (amount of air per 1 kg of fuel); excess air ratio; power generated by the PT; power consumed by the centrifugal compressor; air pressure behind the centrifugal compressor; pressure of combustion products upstream the PT; degree of regeneration. Completion of the full calculation of the rated operating mode allowed us to create a passport on thermodynamic parameters that is compared to the operating parameters of the same unit to diagnose its modes of operation and technical state.

Table 2. The results of calculating the additional rated characteristics of the GMTU.

№	Parameter	Notation	Units	Value
1	Stoichiometric ratio	L_0	kg/kg	16.41
2	Excess air ratio	α	-	5.38
3	Heat capacity of combustion products for the CT	$C_{P_{CP}}$	kJ/kg·K	1.189
4	Power generated by the PT	N_{PT}	kW	190.54
5	Heat capacity of air for the centrifugal compressor	C_{Pair}	kJ/kg·K	1.043
6	Power consumed by the centrifugal compressor	N_C	kW	125.54
7	Calculated effective capacity of the GMTU		kW	65
8	Calculation error N_e	δ_{N_e}	%	0
9	Combustion chamber efficiency	η_{CC}	%	0.98
10	Effective efficiency of the GMTU	δ_{η_e}	%	29
11	Calculation error η_e	δ_{η_e}	%	0
12	Air pressure behind the centrifugal compressor	P_c	kPa	573.1
13	Pressure of combustion products upstream the PT	P_z	kPa	555.9
14	Indicator of adiabatic expansion process of combustion products in the PT	K_T	-	1.2855
15	Degree of increase in air pressure in the centrifugal compressor	ε_C	-	5.89
16	Expansion ratio of combustion products in the PT	λ_{PT}	-	3.88
17	Degree of regeneration	φ	-	0.71

Below shows presents the developed algorithm for calculating the operating modes for different loads from 25 to 100 percent of the nominal value. The calculation results are presented in Table 3. During operation of micro-turbines the number of controlled operating mode parameters is limited and presented in Table 3 as power coefficients of the centrifugal compressor and the PT; temperature coefficients of the PT and the centrifugal compressor; temperatures of air and combustion products downstream the PT. Completion of the full calculation of variable operating modes of the unit allowed us to create a passport on thermodynamic parameters that is compared to the operating parameters of the same unit under the load of 25 to 100 percent of the nominal value to diagnose its modes of operation and technical state by comparison.

The algorithm for calculating the operating characteristics of the GMTU

$$\varphi_C = \frac{N_e^{II}}{N_C^H}. \quad (16)$$

$$\varphi_{CT} = \frac{N_e^H}{N_{CT}^H}. \quad (17)$$

$$\varphi_{\mathrm{TCT}} = \frac{T_Z}{\mathrm{T_S}}. \tag{18}$$

$$\varphi_{\mathrm{TC}} = \frac{T_C}{\mathrm{T_Z}}. \tag{19}$$

$$N_C^{\Phi} = \frac{N_e^P}{\varphi_C}, \mathrm{kW} \tag{20}$$

$$M_{AIR}^F = \frac{N_C^F}{C_{P_{AIR}}(T_C^P - T_O^P)}, \mathrm{kg/s}. \tag{21}$$

$$N_{CT}^F = \frac{N_e^P}{\varphi_{CT}}, \mathrm{kW}. \tag{22}$$

$$M_{CP}^F = \frac{N_C^F}{C_{P_{CP}}(T_Z^P - T_S^P)}, \mathrm{kg/s}. \tag{23}$$

$$T_v^P = \varphi \cdot (T_S^P - T_C^P) + T_C^P, \mathrm{K}. \tag{24}$$

$$\alpha^F = \frac{\eta_{CC} \cdot Q_H^P}{L_O \cdot C_P(T_Z^P - T_v^P)}. \tag{25}$$

$$B = \frac{M_{CC}}{\alpha^F L_O}. \tag{26}$$

$$\varepsilon_C = \left(\frac{T_C^P}{T_0^P}\right)^{\frac{k}{k-1}}. \tag{27}$$

$$\varepsilon_{CT} = \left(\frac{T_Z^P}{T_S^P}\right)^{\frac{k_T}{k_T-1}}. \tag{28}$$

Table 3. Initial data for various operating modes of the GMTU.

№	Parameters	Notation	Units	Power, kW			
				$0.25N_e^H$	$0.50N_e^H$	$0.865N_e^H$	N_e^H
1	Capacity of the GMTU	N_e^P	kW	16.25	32.5	56.23	65
2	Rated power factor of the centrifugal compressor CT	φ_C φ_{CT}	-	0.5178 0.3411	0.5178 0.3411	0.5178 0.3411	0.5178 0.3411

3	Rated temperature factor of the CT centrifugal compressor	φ_{TCT} φ_{TCK}	-	1.3513 0.3896	1.3513 0.3896	1.3513 0.3896	1.3513 0.3896
4	Air temperature	T_0^P	K	288	288	288	288
5	Temperature of combustion products downstream the CT	T_S^P	K	595	693	801	908

Table 4 shows the main results of calculating the operating modes of CapstoneC65

Table 4. The results of calculating the operating modes of CapstoneC65.

№	Parameters	Notation	Units	Power, kW			
				$0.25 \cdot N_E^P$	$0.50 \cdot N_E^P$	$0.865 N_E^P$	N_E^P
1	2	3	4	5	6	7	8
1	Actual capacity of the centrifugal compressor	N_C^F	kW	29.45	62.77	108.63	125.53
2	Air temperature downstream the centrifugal compressor	$T_{C_{AIR}}^P$	K	413.3	431.7	459.0	478.0
3	Average air temperature in the centrifugal compressor	$t_{CP_{AIR}}$	°C	79.15	86.85	100.50	110.00
4	Average heat capacity of air	$C_{P_{AIR}}$	kJ/kg·K	1.040	1.042	1.044	1.045
5	Air flow of the centrifugal compressor	M_{AIR}^F	kg/s	0.2412	0.4193	0.6090	0.6322
8	Actual capacity of the CT	N_{CT}^F	kW	47.64	95.28	164.91	190.56
9	Temperature of combustion products	T_Z^P	K	1061.0	1108.0	1152.6	1226.9
10	Temperature of combustion products	T_S^P	K	785	820	853	908
9	Average temperature in the CT	$t_{CP_{ПС}}$	°C	650.0	691.0	729.8	794.5
10	Average heat capacity	C_{CP}^P	kJ/kg·K	1.17	1.18	1.19	1.21
11	Flow of combustion products in the CT	M_{CP}^F	kg/s	0.1470	0.2804	0.4625	0.4930
12	Air temperature downstream the recuperator	T_v^p	K	677	707	738	783
13	Average temperature at T_Z and Tv	t_{aver}	°C	596.0	634.7	672.3	732.0
10	Average heat capacity	C_{CP}^P	kJ/kg·K	1.16	1.17	1.18	1.19

11	Excess air ratio	α^{F}	-	6.38	6.06	5.81	5.38
15	Flow of the fuel gas	B^{F}	kg/s	0.00151	0.00277	0.00420	0.00470
16	Degree of pressure increase	ε_{C}	-	3.50	4.12	5.11	5.89
17	Expansion ratio of combustion products in the CT	λ_{CT}	-	3.88	3.88	3.88	3.88

3. Conclusion

The studies and the results of calculations on the nominal operating parameters and operating modes of the GMTU allow us to create a passport of thermodynamic characteristics that, when compared to the parameters of the same units in operation, become diagnostic features. Furthermore, based on the results obtained design solutions can be taken on the development of systems with alternative sources of heat and power (cogeneration), for example when APG is used as a fuel.

References:

[1] Zemenkov Yu D 2015 *Design and operation of gas turbine units* Reference manual ed by Yu D Zemenkov (Tyumen; TyumGNGU) 434

[2] Zemenkov Yu D 2015 *Heat and mass transfer equipment and thermal processes in the transport and storage of oil and gas production systems* Reference manual ed by Yu D Zemenkov (Tyumen; TyumGNGU) 175

[3] Moiseev B V, Zemenkov Yu D and Toropov S Yu 2014 *Industrial power system* Reference manual (Tyumen; TyumGNGU)

[4] Ilyuhin K N, Shapoval A F, Chekardovskiy S M Organization of control and diagnostics of equipment in the system of heat and gas supply *Collection of materials of scientific-practical conference dedicated to the 30th anniversary of the TyumGASA* (Moscow)182-5

[5] Zemenkov, Yu.D., Shalay, V.V., Zemenkova, M.Yu. (2015) Expert Systems of Multivariable Predictive Control of Oil and Gas Facilities Reliability. Procedia Engineering, Volume 113, pp.312-315 DOI: 10.1016/j.proeng.2015.07.271

[6] Zemenkov, Yu.D., V.V. Shalay, M.Yu. Zemenkova. Immediate Analyses and Calculation of Saturated Steam Pressure of Gas Condensates for Transportation Conditions (2015) Procedia Engineering, Volume 113 (2015), pp. 254-258 doi:10.1016/j.proeng.2015.07.330

IOP Conf. Series: Materials Science and Engineering **154** (2016) 012006 doi:10.1088/1757-899X/154/1/012006

Research of Reservoir Rock Properties in Violation of Darcy's Linear Law

O S Tamer, E S Toropov, T E Shevnina and T I Vorobieva

Industrial University of Tyumen, 38 Volodarskogo st., Tyumen, 625000, Russia

E-mail: nashdoc@yandex.ru

Abstract: The paper presents the results of research to identify the dependence of reservoir rock properties in violation of the basic linear filtration law. The classification of macroroughness coefficient depending on the permeability is presented. Critical values of the coefficient of permeability when inertial force begins to exert are defined.

1. Introduction

Defining the filtration and capacity properties of geological materials (PES) obtained by laboratory researchof cores is an important factorin developing oil and gas fields. Determination of porosity and absolute permeability plays a special role in such research. Quantitative determination of the filtration characteristics is crucial because the permeability of geological materials plays an important role in determining oil recovery factor.

While exploiting oil fields the conditions of occurrence of oil, water and gas vary. This process is accompanied by considerable changes in properties of geological material, stratum fluids, gases, gas-condensatemixes. Therefore, these properties are considered in dynamics depending on the change of formation pressure, temperature and other conditions in the fields.

In this situation, the relevance of the paper is connected with research of permeability and porosity properties of geological material to account for violations of Darcy'slinear law in the development of gas and gas condensate fields and more informative laboratory tests.

2. Research

Absolute permeability testing of rock samples using helium takes a special role in the research. Permeability is a key parameter characterizing conductivity of manifold. According to Darcy's law, which establishes a linear relationship between mass filtration rate and pressure differential, absolute permeability is a factor of proportionality [1-3]:

$$v = \frac{Q}{S} = \frac{k}{\mu}\left(\frac{\Delta p}{\Delta L}\right), \qquad (1)$$

where v – linear filtration speed, m/s; k – coefficient of rock permeability; Q – volumetric flow rate, m^3/s; S – cross-sectional area of the core, m^2; μ – dynamic viscosity of gas, Pa·s; $\left(\frac{\Delta p}{\Delta L}\right)$- pressure differential per unit length of the core, Pa/m.

In case of violation of linear filtration law Forchheimer equation is used for fluid in interstitial space:

$$-\frac{dP}{dx} = \frac{\mu v}{k} + \beta \rho v^2, \qquad (2)$$

for describing the gas flow:

$$\frac{P_2^2 - P_1^2}{L} = \frac{2P_a Q\,\mu}{kS} + \frac{2\beta P_a Q^2}{S^2}, \qquad (3)$$

where P_a– pressure in atmospheric conditions, Pa; throughflow in atmospheric conditions, m^3/s; β – Forchheimer coefficient (macroroughness coefficient), 1/m; P– pressure, Pa; L – sample length, m.

Zhengwen Zeng and Reid Grigg [4], S.A. Holditch, R.A.Morse, R.D. Evans, C.S. Hudson and J.E.Greenlee [5], [6] have studied the behavior of the gas flow in porous and fractured media in violation of linear filtration law. One of the conclusions is the fact that macroroughness coefficient is not only the quantity characterizing porous medium, but it is also flow coefficient under influence of inertial forces upon condition of high rates of filtration and increasing of pressure differential.

The most thoroughly studied deviations from Darcy's law caused inertia effectsin the time of increasing rates of filtration. In this case the criterion of applicability of the linear law is Reynolds number which characterizes the ratio of inertial forces and resistance forces (viscosity);

$$Re = v\rho d/\mu, \tag{4}$$

where d – some specific linear size of porous medium; ρ – fluid density, kg/m^3.

The upper limit of applicability of Darcy's law is usually associated with the critical (maximum) value of the Reynolds number. In this case, if Reynolds number in the task is less than the critical value $Re<Re_{cr}$, then Darcy's law is done and if $Re \geq Re_{cr}$, then Darcy's law is violated. Here are the expressions to calculate Reynolds number:

a) Shchelkachev's formula:

$$Re = \frac{10 v \rho \sqrt{k}}{m^{2.3} \mu}, \quad 1<Re<12; \tag{5}$$

b) Millionshtchikov's formula

$$Re = \frac{\sqrt{k} v \rho}{\mu \sqrt{mm}}, \quad 0.022<Re<0.29. \tag{6}$$

The results in papers [4], [5], [6] allowed us to assume that there is a relationship between Forchheimer coefficient, porosity and permeability.

Experimental researches were carried out with permeameter-porosimeter AP-608, which allows to measure intercommunicating porosity and absolute air permeability of rock samples. Samples for research were prepared in accordance with GOST 26450.0-85 "Rocks. General requirements for the selection and preparation of samples for determination of reservoir properties". The measurements were conducted with AP-608 using helium as its properties are close to ideal gas. Measurements of samples' porosity with AP-608 were implemented using volumetrical method applying Boyle-Mariotte's law. 6000 core samples were measured; helium was passed through them under pressure. Output data obtained with AP-608 containing the information about fractional porosity, permeability and macroroughness coefficients are given below.

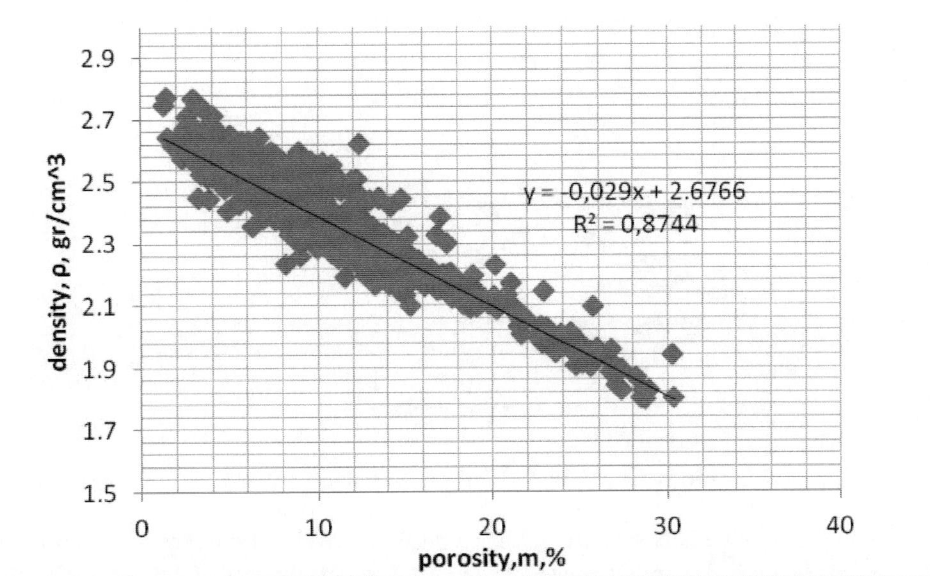

Figure 1. Density-porosity dependence.

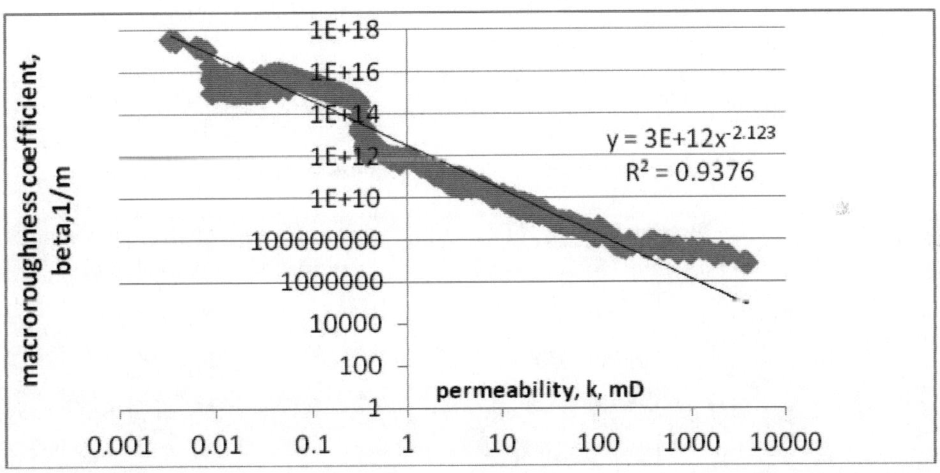

Figure 2. Macroroughness coefficient-permeability coefficient dependence.

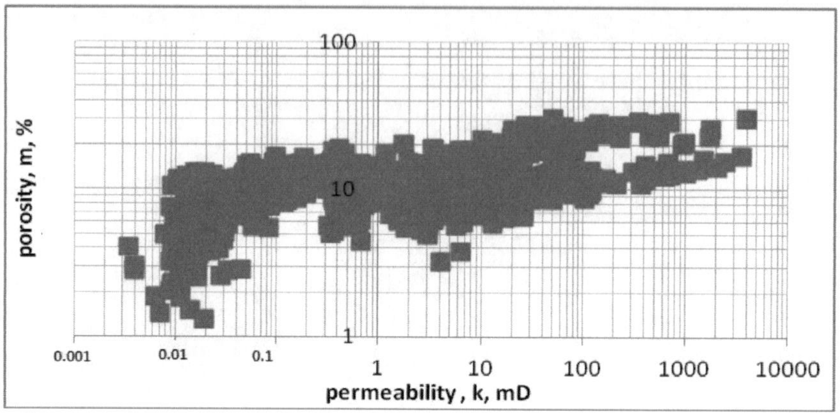

Figure 3. Porosity-permeability dependence.

The results of measurement allow drawing the following conclusions:

- dependence of density on porosity has a pronounced linear character, as expected;
- dependence of macroroughness coefficient on permeability is a power function, which is observed in permeability ranging from 0.5 mD. It becomes difficult to determine the form of dependence to that value.
- dependence of porosity on permeability is difficult to determine, because it is necessary to include lithological properties.

Due to large number of samples corresponding to different medium parameters, we have divided them into groups according to Khanin's classification of rock permeability [7] (Table 1):

Table 1. Dividing macro-roughness coefficient into groups

Group	Permeability k(mD)	Macro-roughness coefficient β (micrometers)
1	k> 1000	$\beta\sim \leq 10^7$
2	500<k< 1000	$\beta\sim \leq 10^7$
3	100 < k < 500	$10^7 < \beta < 10^8$
4	10 < k < 100	$10^8 < \beta < 10^{10}$
5	1 < k < 10	$10^{10} < \beta < 10^{11}$
6	0.1 < k <1	$10^{11} < \beta < 10^{14}$
7	k < 0.1	$\beta\sim \geq 10^{15}$

As an example, we present the research results for the fourth group of the classification, which is clearly seen as exponential dependence shown in figure 4, and, also the results for the seventh group, where the dependence β(k) cannot be determined accurately.

Forchheimer in his equation used dimensionless value β*, the formula is $\beta^* = \beta\sqrt{\mathbf{k}}$. In design equation using AP-608 we took into account the usual macro-roughness coefficient β. But comparing the graphs for β and β*, we can conclude that they are very similar and the further calculations can be carried out using Forchheimer coefficient.

Group 4 (10<k<100, mD)

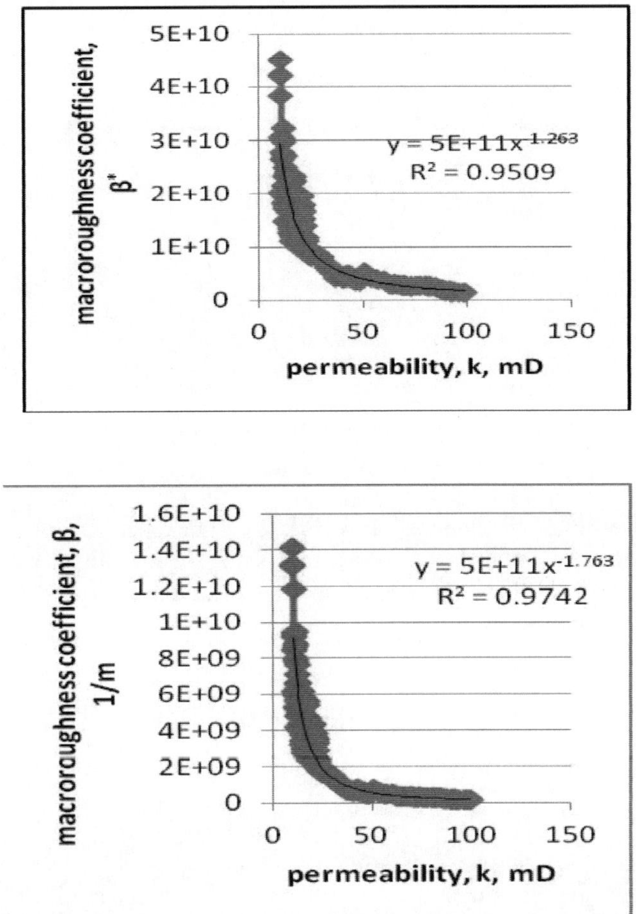

Figure 4. Changing Forchheimer coefficient* and Forchheimer coefficient for different permeability values of samples belonging to group 4 of the classification.

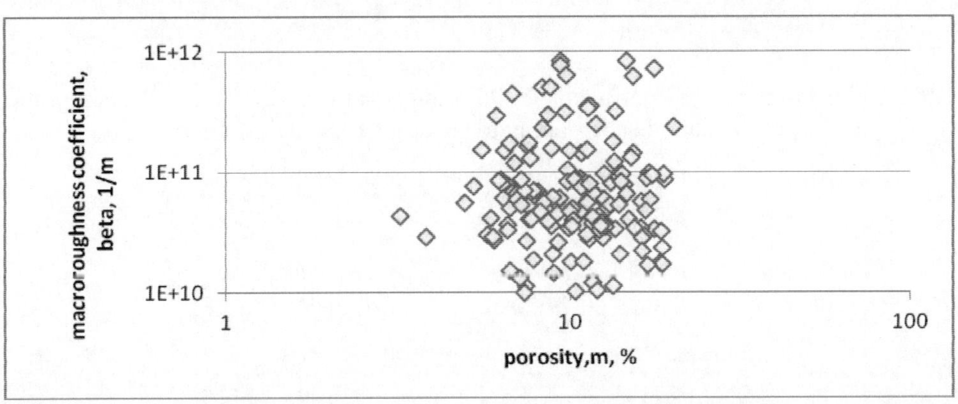

Figure 5. Changing Forchheimer coefficient for different values of porosity for samples belonging to group 4 of the classification.

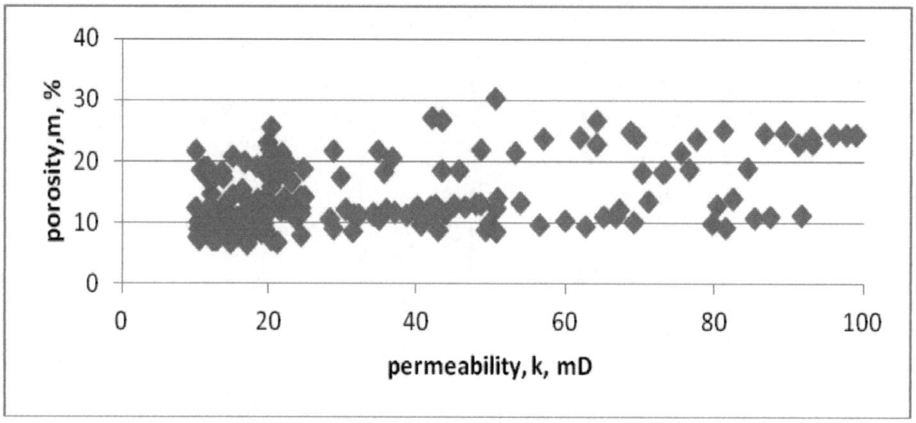

Figure 6. Dependence of porosity on permeability for the samples belonging to group 4 of the classification.

As can be seen from the dependence, shown in the figure 4 macroroughness coefficient is a function of permeability and has power-law dependence. Using data obtained with AR-608 (Picture 5, 6), a certain kind of dependence between Forchheimer coefficient and porosity and also between porosity and permeability cannot be identified.

Group 7 (k<0.5mD)

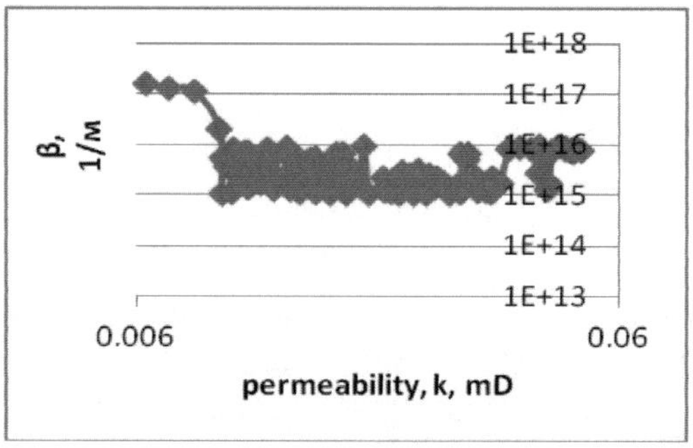

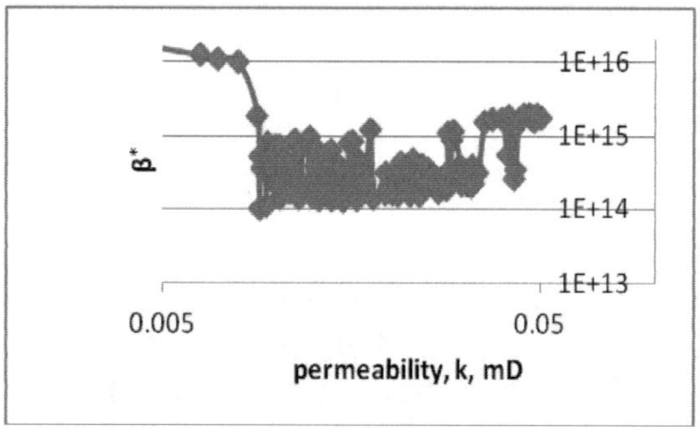

Figure 7. Changing Forchheimer coefficient* and Forchheimer coefficient for different permeability values of samples belonging to group 7 of the classification.

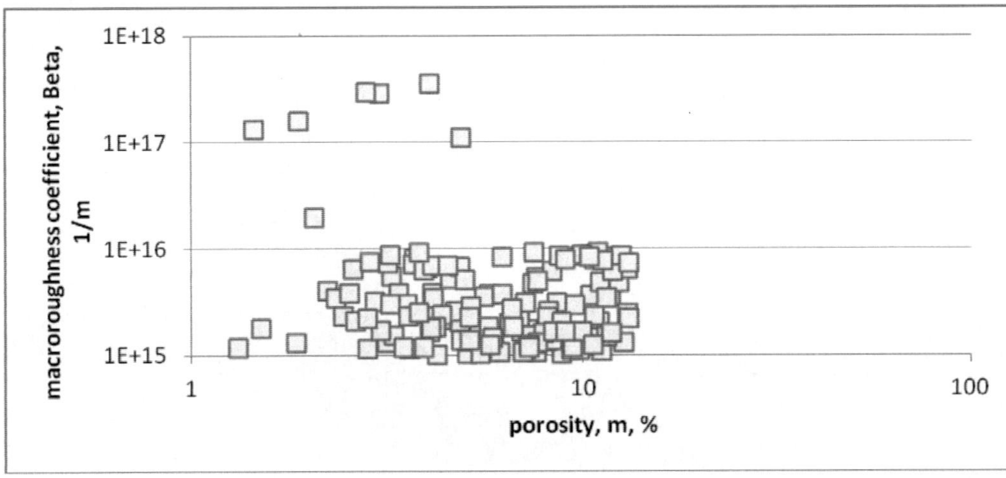

Figure 8. Changing Forchheimer coefficient for different values of porosity for samples belonging to group 7 of the classification.

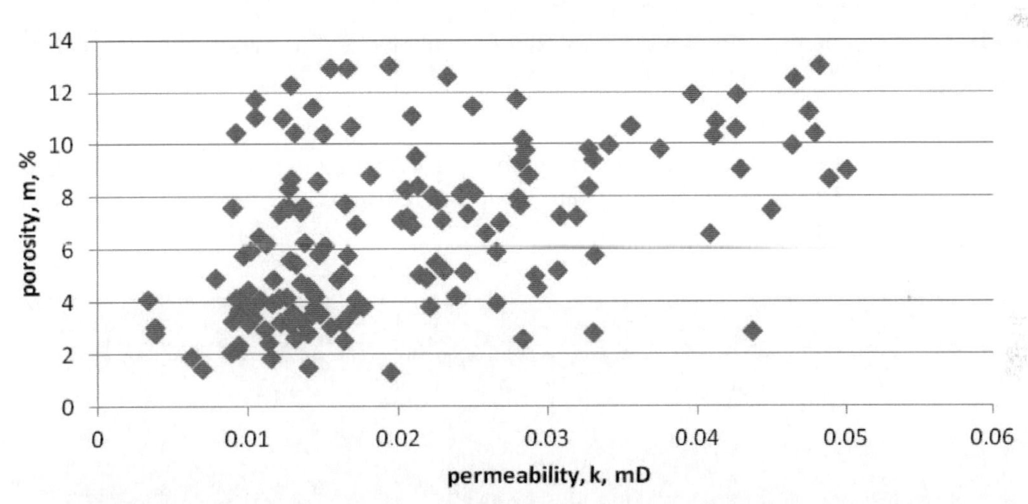

Figure 9. Dependence of porosity on permeability for the samples belonging to group 7of the classification

Calculating of Reynolds numbers using formulas (5) and (6) was carried out according to the results of laboratory tests of 6000 samples of rocks from various oil and gas fields in Western Siberia. The dependence of Reynolds number, calculated with Shchelkachev'sformula (5) on the absolute permeability of the samples is shown in figure 10. Similar dependence of Reynolds number, calculated with Millionshtchikov'sformula (6) is shown in figure 11: the minimum critical Reynolds number is indicated with the red line.The intersection of this line with the graph of dependence of Reynolds number permeability is the critical value of the absolute permeability, the excess of which leads to a violation Darcy's law.

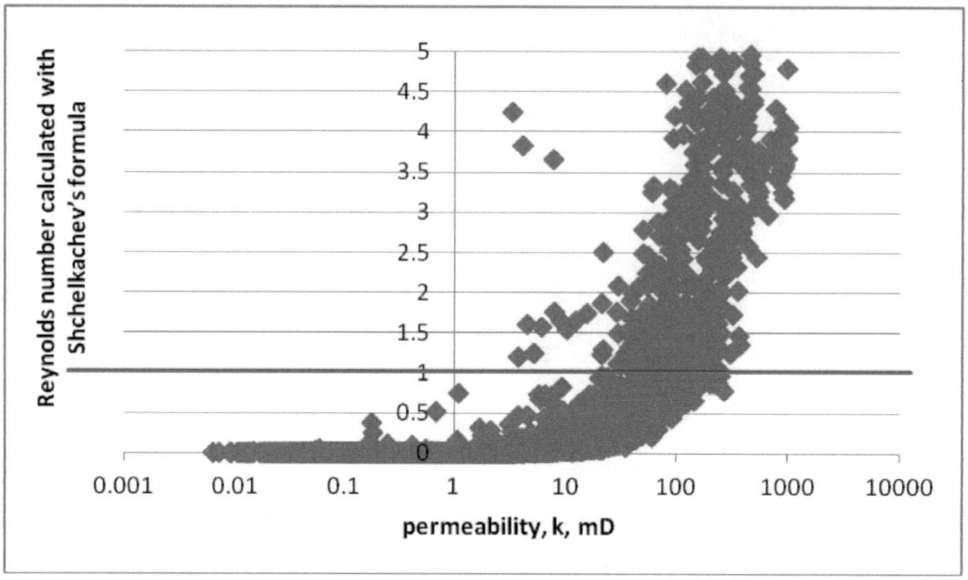

Figure 10. Changing permeability for different Reynolds numbers, calculated with Shchelkachev's formula.

It should be noted that the critical Reynolds number, as shown by the authors of the formulas (5) and (6) is not unequivocal, but varies in a fairly wide range of values, and therefore the critical permeability may also vary.

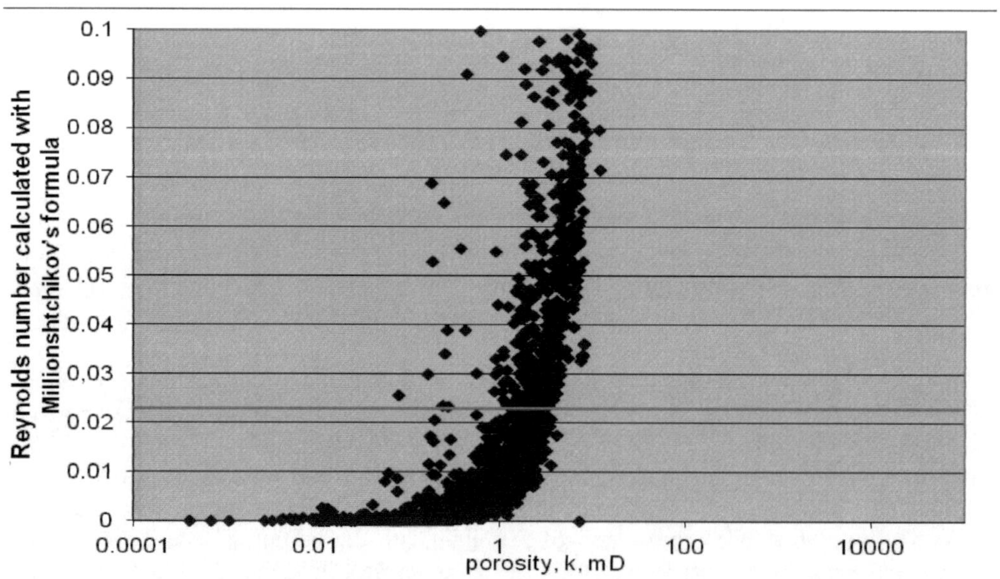

Figure 11. Changing permeability for different Reynolds numbers, calculated with Millionshtchikov's formula.

IOP Conf. Series: Materials Science and Engineering **154** (2016) 012006 doi:10.1088/1757-899X/154/1/012006

3. Conclusion

Thus, it was found that the determination of absolute permeability of rock samples in laboratory conditions using helium is significantly affected by inertial forces, characterized with Forchheimer coefficient. Also, it was determined that macroroughness coefficient depends on the absolute permeability: if permeability increases, macroroughness coefficient decreases. This dependence is expressed by a power formula with a high degree of correlation. The classification of macroroughness coefficient depending on permeability, according to Khanin's classification was developed.

It was found that the violation of Darcy's law occurs with the samples which permeability is more than 80 mD according to Shchelkachev's formula and more than 0.7 mD according to Millionshtchikov's formula. These values show that if rock samples' absolute permeability exceeds the specified critical values, inertial forces start to exert. This causes using of the power filtration laws, for example, Forchheimer equation.

References

[1] Basniev K S, Kochin I N and Maksimov V M 1993 Subsurface hydromechanics (Moscow; Nedra) 415

[2] Bzhitskih T G, Sandu S F, Pulkina N E 2008 *Determination of physical and filtration-capacitive properties of reservoir rocks* (Tomsk Polytechnic Institute) 92

[3] Gimatudinov S K and Shirkovsky A I 1982 *Physics of oil and gas reservoir* (Moscow; Nedra) 311

[4] Zhengwen Zeng and Reid Grigg 2005 A criterion for Non-Darcy flow in porous media *Socorro* 64-5

[5] Holditch S A and Morse R A 2005 *The effects of Non-Darcy flow on the behavior of* hydraulically *fractured gas wells* (Texas)

[6] Evans R D, Hudson C S and Greenlee J E 1987 *The effect of an Immobile Liquid Saturation on the Non-Darcy Flow Coefficient in Porous media* (SPE Production Engineering)

[7] Khanin A A 1969 Oil and gas reservoir rocks (Moscow; Nedra) 356

IOP Conf. Series: Materials Science and Engineering **154** (2016) 012007 doi:10.1088/1757-899X/154/1/012007

Development of a Method for Operational Vibration Diagnostics of a Combustion Chamber GPA-C-16C

A A Razboinikov and V V Vashchilin

Industrial University of Tyumen, 38 Volodarskogo st., Tyumen, 625000, Russia

E-mail: tmn.ers@mail.ru

Abstract. The paper identifies the main causes of the GCU combustion chamber vibration, as well as emerging for this reason failures and defects. It describes the main elements of the developed method for operative diagnosing the GCU combustion chamber on vibration signals in real time. The expected efficiency of the implementation of this method is described.

1. Introduction

Currently, manufacturers and customers tend to conduct operation of gas turbine engines (GTE) on the gas-compressor stations, depending on their technical condition. Among the most important performance indicators are the costs of maintenance and repairs. And their minimization is practically impossible without effective control, monitoring and diagnostics in cases where the technical system is maintainable.

For many years, the issue of increasing the reliability and efficiency of the equipment of different areas of the industry by means of control and vibration diagnostics has been the subject of study of various scientists and authors. And today at CS a large number measuring and control devices and diagnostic systems is tested and operated, the introduction of which has already produced a positive result. Typically, these systems are built on a parametric control and diagnostics of thermal and gas-dynamic parameters of GCU. However, to date the most effective recognized systems are those based on the analysis of vibration characteristics responsible for the operation of the GCU and its serviceable or fault condition.

Unlike aircraft engines, turbines at gas compressor stations are facilities where vibration diagnostics can be implemented in its entirety. Operational statistics confirm the high efficiency of vibration diagnostics methods in the control of gas turbine unit state. Return of funds spent on the purchase and operation of diagnostic systems, usually occurs during the first fault prevention [2]. Operational vibration diagnostics and the development of new methods for detecting the causes of vibrations in the combustion chamber of the GCU will provide new options for assessing the effectiveness of its work, which determines the relevance of this work.

2. Research

One of the main units of the engine is its combustion chamber. Its work has an impact on the longevity and reliability of the engine, as well as its cost-effectiveness and impact on the environment. The combustion chamber of a gas turbine engine is a system in which complex and diverse processes take place.

Developing vibration diagnostics methods of the engine combustion chamber is an urgent task for today due to the fact that the combustion chamber of the engine is a resource-limiting element.

The intensity of vibration is expressed in vibration velocity rms value (mm/s) and by a method for monitoring the technical condition of the engine. The measurement is performed in a frequency range of 10 to 5000Hz, since this is the most appropriate range for the determination of the general condition of mechanisms and is a good indicator of destructive forces that affect these mechanisms.

To diagnose the technical condition of machinery and troubleshoot problems there is a method of spectral vibration diagnostics, since the vibration intensity monitoring allows you to detect malfunction of the unit, but does not indicate the cause of the excessive vibration [7].

Spectral diagnostics of the combustion chamber helps to identify potential threats to the functioning of the engine at an early stage and is a preventive procedure in large enterprises.

Visual diagnostics brings visible results, since during the inspection it is very difficult to detect flaws and damage to parts of equipment, which in turn deteriorates the quality of work and takes too much time. Therefore, the spectral vibration diagnostics is the perfect way to ensure effective diagnosis of equipment.

Sensors read continuously emerging vibration waves, as well as their size and amplitude resulting from the entire CS operation. Thus, monitoring can be carried out directly during the operation and does not interfere with the production cycle.

Diagnostics is carried out using the analyzer built on the basis of computer complexes.

It includes:

- a multichannel analog-to-digital signal conversion device (ADC);
- - a personal electronic computer.

At present, equipping gas compressor units (GPU) with modern complexes is of great importance, allowing diagnosis of their technical condition during operation. Vibration diagnostics takes a special place among them, using modern means of digital and analog computers. Today, vibration methods are more efficient and the costs of their implementation have been declining. However, besides that their advantage lies in the fact that monitoring and diagnosis can begin at any time, including a few years since the equipment was put into operation, when the cost of routine maintenance and repairs will exceed the economically viable value [6].

As a result of years of research into the causes of failures and defects in the GCU equipment it has been revealed that a significant number of them is accompanied by a high level of vibration [1], in particular due to the inefficient operation of the combustion chamber and the resonance phenomenon. The combustion chamber is among the busiest elements of the GCU. Its parts are working under conditions of high static, dynamic and thermal loads, which directly affects the reliability of the mechanical part of the GCU. Its work has an impact on the longevity and reliability of the engine, as well as its cost-effectiveness and impact on the environment. Development of vibration diagnostics methods of the engine combustion chamber is an urgent task for today due to the fact that the combustion chamber of the engine is a resource limiting element.

The developed method is based on the identification of high-frequency vibration of the ship's engine DG-90 as part of GPA-C-16C using a piezoelectric transducer. According to the method, installation of three sensors is proposed in the combustion chamber housing in a circle every 120 ° (Figure 1). A vibration sensor MV-04 converts mechanical vibrations into an electrical signal proportional to the vibration acceleration.

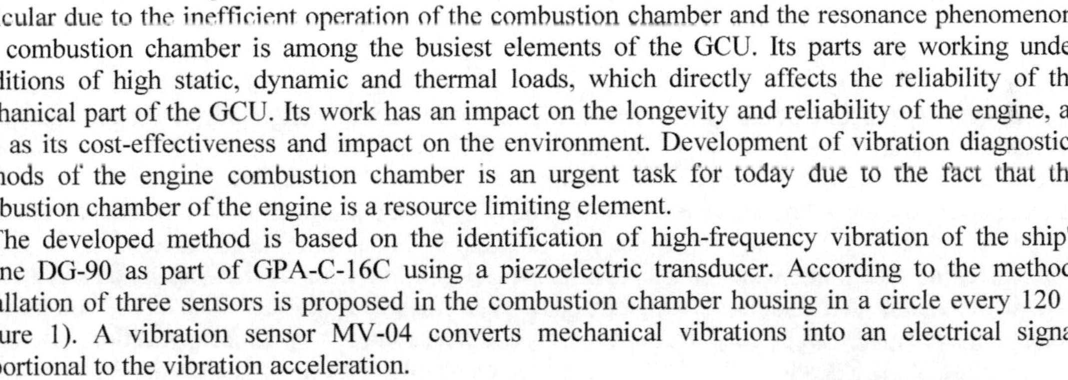

Figure 1. Scheme of the sensor installation (top down: combustion chamber housing, rotor shaft, mounting plane, locking washers (2 pcs), bolts (2 pcs), radially mounted sensor to improve sensitivity)

A vibration sensor is mounted on the combustion chamber housing and under oscillations produces an electrical signal that is proportional to the vibration velocity. An amplifying and conversion device as a module integrates and amplifies the received signal. As a result, the output signal is produced proportional to the vibration displacement. This signal can be observed on the screen of the module and the computer. At this stage, we applied the method of automatic amplitude-frequency decomposition of the time recording of the vibration signal by the method of fast Fourier transforms and the subsequent presentation of the amplitude-frequency spectrum of the signal as a result of the analysis. According to the obtained spectrum, we judge on the vibrating state of the GTU.

As a result of the measurement, each sensor produced a spectrum of vibration velocity of the CS housing at predetermined points.

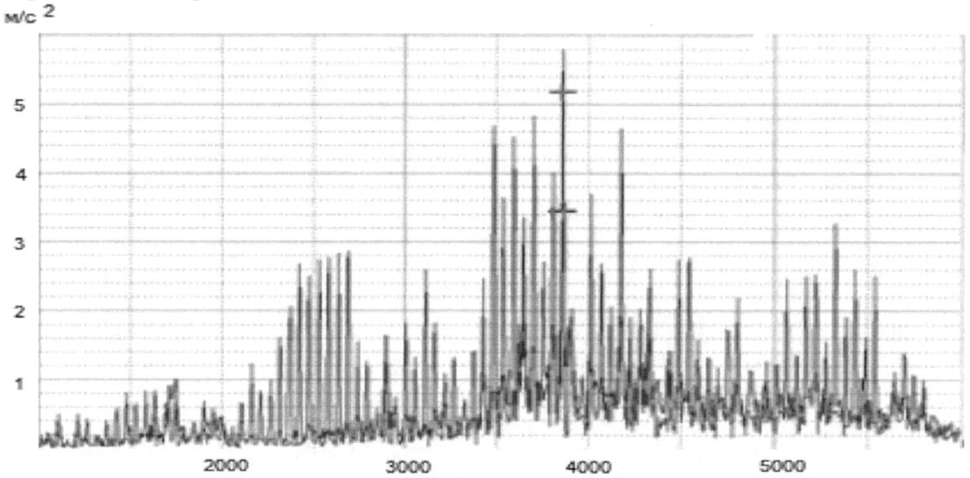

Figure 2. A spectrum of the GTU housing vibration (m/s^2).

The obtained spectrograms give the analysis of a frequency range of 2500-5000 Hz which corresponds to the operating frequencies of a combustion chamber [3]. The resulting spectrogram of the CS housing vibration at the same mode shows that the sign of the pressure pulsations are identified easily enough by the vibration sensor at a frequency of 3850 Hz with an amplitude of vibration velocity of 5.9 m/c^2. This is likely to indicate a malfunction of one of the flame tubes.

In order to determine which of the 16 flame tubes runs with deviations, made vectors are construction based on available spectrograms. The vector length is set according to the amplitude of vibration velocity. The angle between the vectors is 120° according to the sensor location. Vector “a” is 1.3, vector “b” is 6, vector “c” is 1.9. We find the resultant vector using the cosine theorem, which points to a faulty flame tube. As a result of calculations we obtain the resultant vector “e” 4.37 in length and at an angle of 76.89° of the first vector. This vector points to flame tube 1 that requires inspection for faults (Figure 3).

IOP Conf. Series: Materials Science and Engineering **154** (2016) 012007 doi:10.1088/1757-899X/154/1/012007

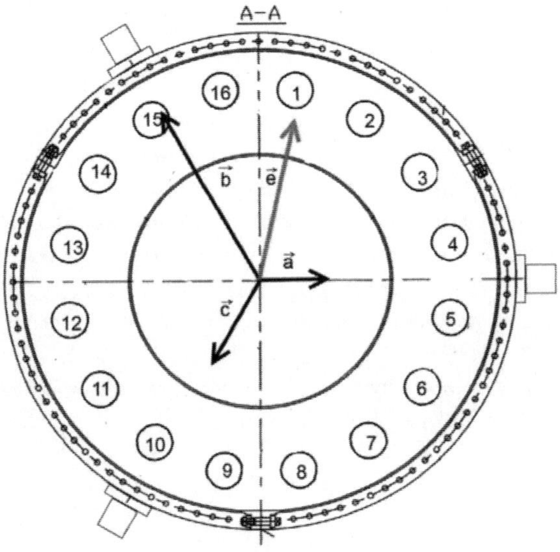

Figure 3. Identifying the fault by vectors.

3. Conclusion

It is vibration monitoring and diagnosis due to the specifics of vibrational signals that are primarily responsible for the overall control of the engineering state of the equipment and the prevention of accidents associated with the development of a variety of mechanical damages. However, if vibration diagnostics systems are paid much attention to down to the construction of automatic and adaptive systems of vibration diagnostics and systems of expert analysis of the vibration information, vibration monitoring systems are often referred to as just a necessary part of the overall system ensuring the operational reliability of the equipment. Organizations operating gas turbine engines lack confidence that the result they obtain by the vibration diagnostics systems are correct. This and the possibility of false removal and the lack of functionality and reliability of hardware-software complexes prevented the promotion of vibration diagnostics systems operation to the full.

The emergence of microprocessor technology, personal computers, compact computers for industrial and military use, powerful operating systems, advanced multi-channel, multi-bit analog-to-digital converters (ADCs), new tools, software solutions has intensified the process and led to the creation and implementation of numerous diagnostic complexes both for a special purpose and universal for a wide range of application. The use of vibration monitoring systems is one of the main solutions to the problem of sustainable functioning of the GCU, as it allows assessing the actual situation and on this basis taking informed decisions on the continuation of its operation.

The effectiveness of the implementation of the method of vibration diagnostics is made up of several indicators. First of all, a real turnaround time increases due to the fact that the unscheduled repairs are eliminated (according to statistics, unreasonable repairs reduce the required turnaround time by 15-30%). In addition, the rate of repair cost is reduced due to accurate detecting of defects. Eliminating these defects does not require dismantling the equipment and reduces the amount of backup equipment in separate processes and the amount of work as a whole is significantly minimized. Savings are also achieved by reducing costs unrelated directly to production costs. There is also reduction in the wastage rate resulting from downtime, production of rejects, as well as the costs of accident response. Due to the fact that the operation of the engine does not stop completely after failure detection, foreign experts estimated savings of up to 6% of the annual volume of production of the enterprise.

IOP Conf. Series: Materials Science and Engineering **154** (2016) 012007 doi:10.1088/1757-899X/154/1/012007

References

[1] Gareev R R and Yamaliev V U 2012 Determination of dynamic equipment based on the results of diagnostic measurements of a technical condition *Oil and Gas Business* 10- 3 78-82

[2] Perevezentsev V T 2011 *Diagnosis of gas compressor units with gas turbine engines*

[3] STO GAZPROM 2-3.5-051-2006. *Norms of technological design of gas mains*

[4] Passport of turbine engine DG-90.

[5] GOST R ISO 13373-1-2009 2010 *Condition monitoring and diagnostics of machines. Vibration monitoring of machine condition. Part 1. General methods* (Moscow; Standartinform)

[6] Ishkov A G, Hvorov G A and Yumashev M V 2011 Formation of highly efficient energy-saving innovation technologies in the main gas transport in OAO “Gazprom” *Science and technology in the gas industry* **1**

[7] 2008 *Investigation of thermal processes at pipeline transportation facilities* Edited by Zemenkov Yu D (Tyumen: Vector Buk) 224

IOP Conf. Series: Materials Science and Engineering **154** (2016) 012008 doi:10.1088/1757-899X/154/1/012008

Developing a Method for Operative Diagnostics and Appraisal of Working Capacity of a Combustion Chamber DG-90

A A Razboinikov and V V Vashchilin

Industrial University of Tyumen, 38 Volodarskogo st., Tyumen, 625000, Russia

E-mail: tmn.ers@mail.ru

Abstract. In the paper the problematics of gas transport system, main factors of an urgency of the development are described. Stages of a proposed reconstruction of combustion chamber DG-90 are introduced. Basic elements of the elaborated method for appraisal of risks of an emergency situation occurrence are given. The expected efficiency from implementation of the produced method is described.

1. Introduction

Currently, in the gas industry of the Russian Federation there is an especially sharp problem of complete reequipment of the basic process equipment on compressor stations (CS) – gas compressor units (GPU), as 30 % exhaust the design resource, and about 30 % have worked out more than 75 % of their resource [1]. In the urgency of the given problem a financial factor plays a relevant role as now the cost of transport services, repair and spare parts grows, as well as of power carriers. Thereof, the increasing value is placed on problems linked to the reliability and increase in performance of gas compressor equipment. Identifying the faults at the primary stages of their occurrence results in reduction in expenditures on drive operation, and diminishes the possibility of occurrence of more severe kinds of faults.

A considerable quantity of failures of aggregates with gas turbine units (GTU) consists in a gasdynamic system, monitoring and diagnostics of which use only thermogasdynamic performances (flow rate, temperature, pressure). Calculations based on these parameters help to identify the performance extent (or wear) of separate GPU systems. Their shortage complicates monitoring and diagnostics of many phenomena, for example vibrating combustion, stall, a surging, combustion chamber burn-through, flameout, clogging of flowing parts, etc. [2, 3, 4]. Connection of thermogasdynamic performances and oscillations in a combustion chamber of a GTU is known and proved [5, 6], which allows to increase the quantity of recognition of such failures. Therefore, developing a method for operative diagnostics and appraisal of working capacity of a combustion chamber DG-90 of a gas turbine engine is an urgent problem having a relevant scientific value and practical application.

Accidents and malfunctions during operation arise due to structural and technological errors, material defects, etc. These factors can lead to serious consequences, such as destruction of GTE, which further harms the environment, decrease in reliability, performance and efficiency of the unit. Every year, analysis of the effectiveness of existing gas lines allows not only assessment of the operating state, but also introduction to the industry of requirements such as: improving the quality of pipe metal; improving the quality of insulating coatings of pipes; industrial development of the internal pipe coating technology at the factory; increasing the service life of GTU components and increasing the inter-repair intervals; development of new effective cooling systems of compressed gas; construction of various types of gas storage facilities, including liquefied natural gas storage facilities; creation of special high-performance compressors for pneumatic testing of main gas pipelines; organization of production of special machines for the repair work on the linear part of the gas lines (bucket wheel excavators, cleaning and insulating machines for gas pipes with a diameter of 1020 -

1420 mm); development and implementation of high-performance block-complex systems for gas treatment in the fields; development and launching the production of means of technical diagnostics of the linear part of main gas pipelines; development of reliable independent power supply stations of cathodic protection, etc.

Russian Emergency Situations Ministry predicts the next few years as a period of the peak of technogenic catastrophes in Russia. This situation is caused by the wear of fixed production assets and technological equipment, primarily in the oil and gas industry. The current state of the technological equipment of the oil and gas industry is characterized as critical. To date, 13% of gas lines is operated for over 30 years, 20% - from 20 to 30 years and 34% - from 10 to 20 years [8].

2. Research

To realize the purpose of the research, development of a method for operative diagnostics after conducting retrofits on a gas turbine engine DG-90. It will allow diagnosing the operation in real time and enhancing the odds of early determination of possible faults. After that, operative repairs will be carried in the shortest terms, without leading up to the maximum risk of emergency occurrence.

Because of high pressure and temperatures, a technical solution consists in using a sensor "Sapfir-22EM", model 2170, having pressure measuring limits acceptable for a given retrofit, as well as a margin of error no more than 0.2 %. It is intended for continuous transforming of the value of overpressure, underpressure and differential pressures into the output signal dispatched on a control panel of the compressor station directly in a combustion chamber. Having made an opening in a membrane perpendicular to it, a sensitive member of a pressure unit is mounted on a high steel rod 350 mm in length and 30 mm in diameter, and is tightly fastened on a screwed joint with a sealing ring (figure 1).

Figure 1. The proposed retrofit of a combustion chamber of a gas turbine engine DG-90.

The pressure sensor "Sapfir-22EM" built into the combustion chamber DG-90 will show the operating pressure in a combustion chamber depending on the operating mode, and will signal in the event that the pressure will appear different from the design one. In this case it is diagnosed that a fault has taken place in a combustion chamber, and to exclude part of the main and auxiliary equipment from the list of possible defects. The signal is then fed to a compressor station control panel. That is, the principle of operation is to determine the effects of differential pressure on the membrane of the measuring unit (pressure transducer) and prompts deformation of the sensor and the transmitter deflection resistance. Further there is transformation into the eclectic signal transmitted in the electronic transducer (a PC and a communicator). The advantage is that data is received not only on a stationary computer, but also a portable communicator, and the operator can obtain information in real time even if he is at some distance from the PC (figure 2).

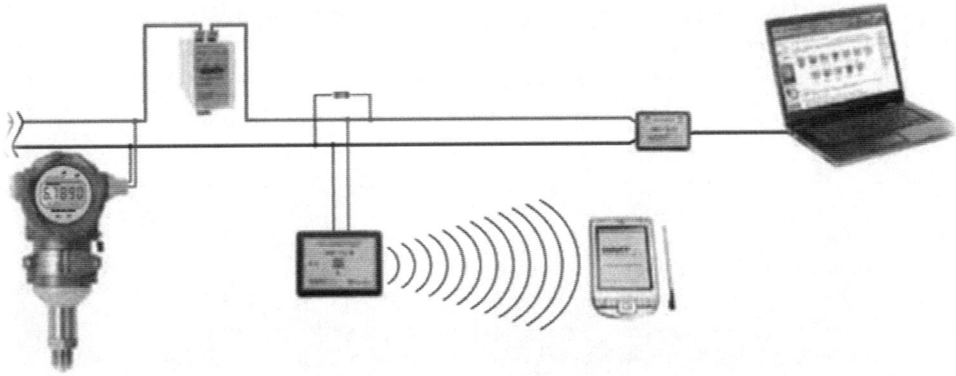

Figure 2. The proposed wiring diagram of the sensor "Sapfir-22EM".

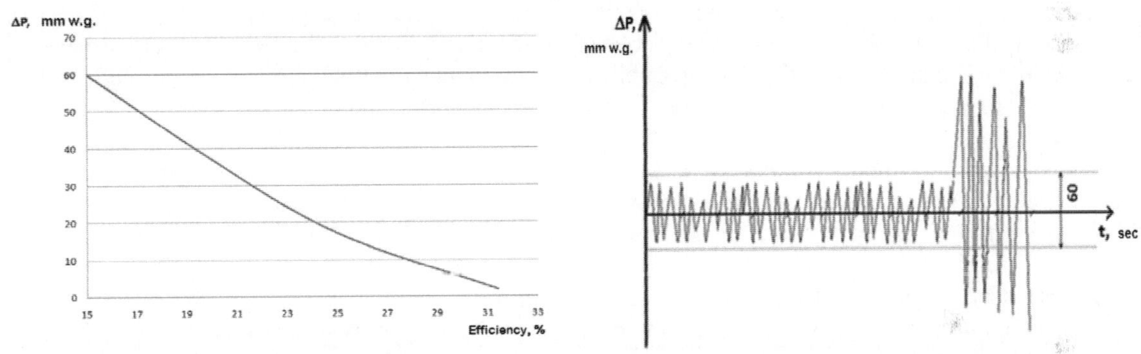

Figure 3. Diagnostic signs of violations of processes on CS DG-90 (mm w.g., efficiency, sec).

Figure 3 shows the approximate dependence of the engine efficiency on the operating pressure swing amplitude of the CS and the pressure pulsations timeline in the combustion chamber as planned, and the area beyond the maximum permissible norm called a surge (it is an abrupt change in a parameter that occurs in less than 5 seconds and holds the value for longer than 15 seconds).

It is important to emphasize that the forecast, diagnostics and monitoring of the GCU state is connected with the recognition of damage leading to gradual failures. The time of the defect development to a critical level is sufficient to predict the moment of the GCU failure. Dividing failures into gradual and sudden is to some extent arbitrary, since any abrupt change of parameters (characterizes the sudden failure) is preceded by a process of gradual change of the defining parameters [7].

In this connection, in the evaluation of efficiency of the combustion chamber a risk matrix has been designed.

The used risk analysis method includes calculating the probability of occurrence of an undesirable event and assessing the implications based on the information collected. A typical risk matrix is shown in Figure 4: horizontally arranged are pressure drop values, and vertically - the angle of deviation. Risks are ranked according to the degree of danger, in this case - low (green), moderate (yellow) and high (red). The low level of the possibility of risk least of all affects the operation of the unit and therefore its degree is minimum. The high level is the one that in the event of occurrence will be havethe greatest value for the degree of danger and can lead to destruction of the equipment, damage to the environment and personnel. The values are a dimensionless empirical measure based on operating experience. The composed matrix is a reference, information basis for rational decision-making in the processing of risks.

ΔP, mm w.g.	10	17	25	37	60
0°	1	2	3	4	5
5°	2	4	6	8	10
17°	3	6	9	12	15
32°	4	8	12	16	20
45°	5	10	15	20	25

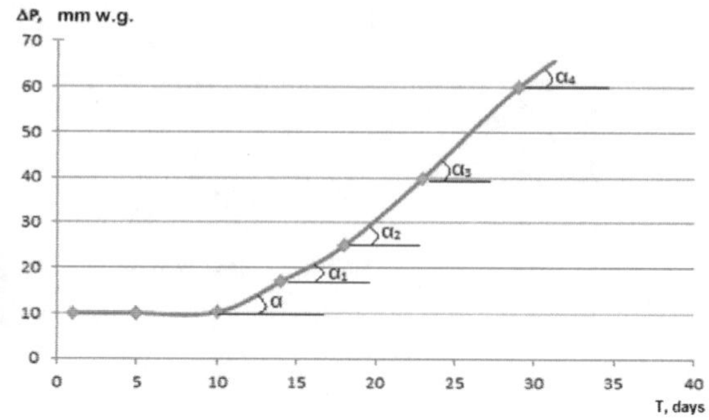

Figure 4. Risk assessment factors of an emergency situation (mm w.g., days)

3. Conclusion

Today, when the risk of accidents and disasters on large and environmentally hazardous facilities is great, one of the first tasks is to anticipate, forecast, prevent and mitigate their effects. Assessment of the accident risk exists to identify the probability (or frequency) and severity of the effects of the accident occurred, the danger to human health, environment and property of the enterprise and is an essential element of its management.

Development and implementation of technical diagnostics, along with the use of new techniques and technologies at the CS, as well as along with the introduction of automated control systems may be regarded as the most important means of improving the reliability and efficiency of the gas-transport equipment. The conducted studies and recommendations on retrofitting the GTD are characterized by a significant expected technical and economic effect:

- the economic advantage is in prolongation of the engine durability without overhauling;
- it will allow to prevent the off-schedule disablement of the aggregate and minimize the fuel gas over-expenditure by inefficient operation of the combustion chamber;
- improving the safety of operations, due to the construction of the risks matrix for accident occurrence, will allow timely fault detection when changing certain parameters, which will help prevent disabling it and further destruction. Currently, under the present conditions of production, transportation and processing of minerals, the developed method for operative diagnostics will help to maintain a reliable operation of the GTU.

References

[1] Kalekin V S and Tokarev V V 2013 Fundamentals of technical diagnostics of oil and gas equipment
[2] STO Gazprom 2-3.5-051-2006 Norms of technological design of gas mains
[3] Vertepov A G 2013 Energy savings at compressor stations by the use of parametric methods of diagnostics of gas pumping units Reference manual
[4] 2010 *Operation of pumping and power equipment at pipeline transportation facilities* Reference manual ed by Zemenkov Yu D (Tyumen; Oil and gas university)
[5] SNiP 2.05.06-85* 1985 *Main pipelines* Design norms (Moscow; Stroyizdat)
[6] Grachev D V 2009 *Assessment of the operational parameters of the gas-dynamic system of gas* compressor *units according to vibrosignals* (Saratov) 181
[7] Sundukov A E 2011 *Improving methods to identify diagnostic features of the technical state of GTE* based *on the broadband vibration spectrum*
[8] 2009 *Occupational Safety and Health* (Siberian University Press) 344

IOP Conf. Series: Materials Science and Engineering **154** (2016) 012009 doi:10.1088/1757-899X/154/1/012009

A Mathematical Model of Gas-Turbine Pump Complex

V A Shpilevoy, S M Chekardovsky and A G Zakirazkov[1]

Industrial University of Tyumen, 38 Volodarskogo st., Tyumen, 625000, Russia

[1]E-mail: zakir1404@mail.ru

Abstract. The articles analyzes the state of an extensive network of main oil pipelines of Tyumen region on the basis of statistical data, and also suggest ways of improving the efficiency of energy-saving policy on the main transport oil. Various types of main oil pipelines pump drives were examined. It was determined that now there is no strict analytical dependence between main operating properties of the power turbine of gas turbine engine. At the same time it is necessary to determine the operating parameters using a turbine at GTPU, interconnection between power and speed frequency, as well as the feasibility of using a particular mode. Analysis of foreign experience, the state of domestic enterprises supplying the country with gas turbines, features of the further development of transport of hydrocarbon resources allows us to conclude the feasibility of supplying the oil transportation industry of our country with pumping units based on gas turbine drive.

1. Introduction

Issues related to energy consumption economy more and more often become the center of world attention. Relevance of energy saving is caused by the fact that traditional sources of energy are limited, while the main transportation of oil is one of the largest consumers of fuel and energy resources (FER) in the Russian Federation. [1] Mainline pumping units, in turn, are energy-intensive machines, so their effective operation is a very important task of operating organizations.

The main part of energy for domestic consumption in the pipeline transportation of oil is electricity. GOST (All Union State standard) for oil transfer pumping units, developed in the USSR, provided the main electric drive of main pumps, which was due to the progressive policies of the vast country in the field of development of the unified system of power supply. Russian standard technical documentation of oil transport companies, also does not leave choice to a project engineer concerning the main drive units.

The existing system of oil and gas pipelines in the Russian Federation has developed in the 60s-80s of the last century, and the greatest development of structures occured due to the need to transport large volumes of oil and gas from Western Siberia to the central regions of the country and for export.

The state policy of Russian export gas pipelines determined the need for the creation of two new destinations:

- northern: from the fields in Western Siberia (beginning in the area of Surgut) and the Timan-Pechora oil and gas province to the Barents Sea coast and further export of oil by sea to the countries of North-West Europe and North America;
- eastern: with the prospects of developing the resources of Eastern Siberia and the Far East and organization of oil exports to China, Japan and other countries in the Asia-Pacific region.

Since a large part of the routes of both projects is planned in the northern latitudes, as well as in Siberia, there is experience in design and construction of such pipeline systems in appropriate climatic conditions in Russia and it can be complemented by foreign expertise in some aspects.

Mainline pumping units are powerful energy consumers, so their efficient and economical operation is a very important challenge for operating companies.

IOP Conf. Series: Materials Science and Engineering **154** (2016) 012009 doi:10.1088/1757-899X/154/1/012009

2. Methods

One of the conditions for the dynamic development of the mainline oil transport is to reduce the cost of pumping, an important component of which is energy costs. There are various ways of reducing energy consumption:

- Optimization of oil pumping conditions applying modern methods of pipeline capacity control;
- Reduction of energy losses in operating equipment, transmission lines, distribution networks;
- Introduction of modern methods of reducing the hydraulic resistance of the pipeline;
- Optimization of techniques for selecting the pump units drive, as the electric main pump has no alternative at the stage of drawing up the specifications and technical documentation;

With the developed system of gas supply, oil pumping units with gas turbine drives can compete with existing one. Their use is especially important in areas where there is no unified of power supply system. The choice of the type of OPS drive is affected by a number of fundamental factors. In the first place, it is the existence of powerful sources of energy, as well as technical, energy and economic aspects.

During the initial period of the national oil transportation development a choice of the type of OPS driving pumps was not difficult. In many cases, there was only one source of energy - the energy of combustion of the pumped product. This ensured a reliable and uninterrupted autonomous power supply. This solution is widely used at present for gas turbine driving of compressor stations of main gas pipelines where the source of energy is the energy of the pumped gas combustion, and the automatic control and regulation system ensures perfect operation of all main gas pipelines.

Energy consumption on the oil transportation can be significantly reduced by choosing the optimal type of pump stations drive both in the construction and renovation of the existing oil pipelines, as the electric drive of main pump has no alternative at the stage of drawing up the specifications and technical documentation already. [3]

Analysis of foreign experience, the state of domestic enterprises supplying the country with gas turbines, features of the further development of transport of hydrocarbon resources allows us to conclude the feasibility of supplying the oil transportation industry of our country with pumping units based on gas turbine drive. [2]

Today there is no strict analytical dependence between main performance properties of the power turbine of gas turbine engine (GTE). At the same time there is necessity to determine the operating parameters when using a turbine at gas turbine pumping unit (GTPU), interconnection between power and speed rate, as well as the feasibility of using this or that mode. [4]

In order to determine the dependence between major operating parameters characteristics of various gas turbine engines were examined, as well as experimental data [5], in relative terms, which are ratio of current parameters to nominal ones:

$$N = \frac{N_{tp}}{N_{nom\,tp}} \;;\; \mathrm{n} = \frac{n_{tp}}{n_{nom\,tp}} \;;\; G_t = \frac{G_t}{G_{t\,nom}}, \tag{1.1}$$

where N – work turbine power;

n – work turbine revolutions frequency;

G_t – fuel consumption.

Analysis of GTE parameters [8] showed that for each power value there exist the most appropriate values of the number of turbine revolutions. [6] As a result of the analysis of turbine revolutions at the height of powers, the optimal modes curve was obtained (Figure 1).

IOP Conf. Series: Materials Science and Engineering **154** (2016) 012009 doi:10.1088/1757-899X/154/1/012009

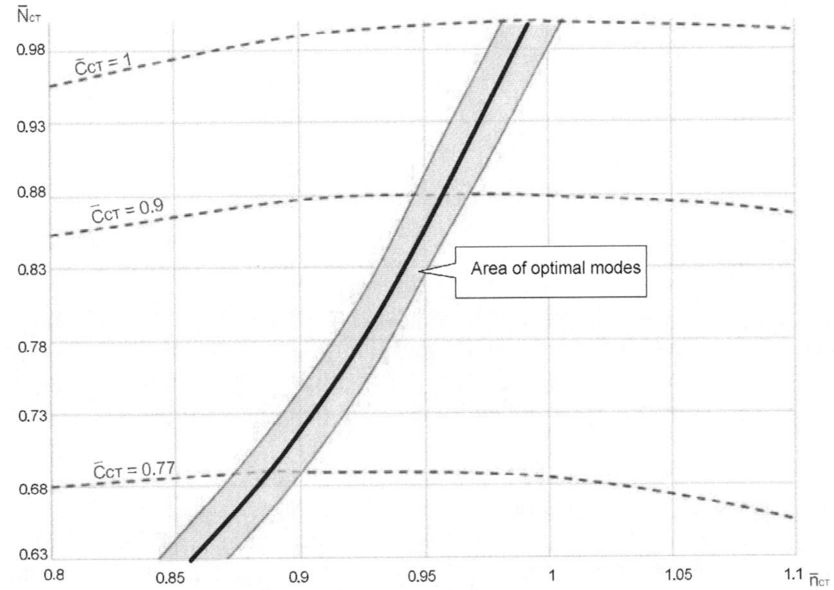

Figure 1. Area of optimal modes of using work turbine.

Considering the efficiency of work turbine of gas turbine engines [7], the optimum engine usage area was determined. In approximating optima curve in the software package «Advanced grapher» , the dependence of power turbine rotor revolutions frequency from power under optimal conditions when R^2=0.99965 was obtained:

$$n = 6.8713N^2 - 9.9611N + 4.1184 \quad (1.2)$$

Taking into account the above, we can conclude that the use of work turbine as a drive of the main pumps can compete with the electric motor in conditions of an uneven supply of oil, and the resulting optimum of mode would save fuel gas at maximum engine power.

A mathematical model of the system, "gas generator, power turbine-pump oil pipeline" as part of the gas-turbine pumping complex (GTPC) was composed (Figure 2).

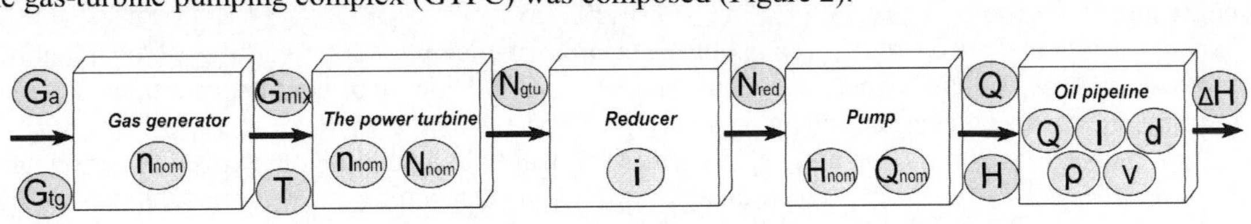

Figure 2. Block diagram of the mathematical model of GTPC.

The model is made in the form of system of equations, identifying the state of GTPC system:

where:

n — work turbine revolutions frequency;

N – work turbine power;

ψ - generalized parameter of the pump head;

ψ_0 – pump head rate at zero flow;

k- number of pumping units;

φ - generalized parameter of pump flow;

β – volume expansion coefficient;

L – length of pipeline;

ν – kinematic viscosity of liquid;

d– pipeline diameter;

Z – the difference in geodesic marks.

The use of gas turbine pumps drive enables the regulation of transfer mode by changing the pump rotor rotating frequencies and introduces an additional option in the management of the pipeline. In this connection, pipeline management is inconceivable without the widespread introduction of automated control systems, requiring developed algorithms, which should be based on the principles of mathematical modeling and system identification oil pipeline - pump - gas turbine drive as a unified energy complex.

Using the model for various tasks will enable taking more effective management decisions and quickly affect the modes of joint work of a pipeline, a pump and a gas turbine drive.

The model is based on the principle of energy conservation. The thermal energy of fuel (Input parameters: G_a - air flow rate; G_{fg} - fuel gas consumption, G_{cmix} - combustible mixture consumption, T - temperature) is converted into mechanical work at the drive shaft (Input parameters: N_{gtu} - engine power, N_{red} - power transmitted to reducer) of the main pump. Then it is converted into hydraulic energy in the pump (Input parameters: Q- pump output, N-pump head) and finally into energy of oil product motion through the pipeline. Nominal parameters of GTPC system are revolution frequency (n_{nom}), power (N_{nom}) gear reduction rate (i), pressure head (H_{nom}) and supply (Q_{nom}) in pump, dimeter (d), the length (l), performance (Q) of the pipeline, as well as density (ρ) and viscosity (ν) of oil product.

One of the main elements of the block diagram characterizing feature of indicated model is a gas turbine drive which can be decomposed into a gas generator and power turbine. The efficiency of the GTE depends on the factors affecting the state of the object.

The aim of the problem solution at optimizing the drive operating mode in an installation is to provide such a ratio between the power and rotation frequency for pump and oil pipeline data, that will minimize fuel consumption. This can be achieved by establishing a certain ratio between n_n and the required pumping capacity, matching rotation frequency of the free turbine rotor and the rotor of the pump by means of a reducer, which is highlighted as a separate element in Fig. 2.

3. Conclusion

The advanced model will enable carrying out a research of modes of interworking of oil pipeline, pump, power turbine and a gas generator, carrying out a rational choice of equipment frame sizes and determining interconnected constructive parameters of the installation and of the pipeline. Such a model will become the basis for solving all kinds of technological problems in engineering and operation of the facility in order to optimize its operation. Depending on the question posed, the solution sequence and its realization program will be modified, and the boundary conditions should reflect the specifics of the problem and minimize the number of possible solutions.

References

[1] Zemenkov Yu D and Kurushina V A 2013 Patterns of innovative and cyclical development of pipeline transport of Russian hydrocarbons *Gorny informational and analytical bulletin (scientific and technical journal)* **S3** 85-98

[2] Zemenkova M Yu, Seroshtanov I V, Kurushina V A, Toropov S Yu and Zemenkov Yu D 2013 Monitoring the reliability of oil and gas facility using the system analysis methods *Territory Neftegaz* **10** 80-86.

IOP Conf. Series: Materials Science and Engineering **154** (2016) 012009 doi:10.1088/1757-899X/154/1/012009

[3] Zemenkov Yu D, Kurushina V A, Vylegzhanina A O, Barmenkova V V and Khairullina N G 2013 *Modern management and Innovative, scientific and technical policy* Monograph (St. Louis)
[4] *Comprehensive program "Energy saving and energy efficiency in Tyumen region" for 2010-2020*
[5] Smirnov A N, Vengerov A A, Dudin S M and Zemenkov Yu D 2010 Monitoring the reliability of industrial and technological processes of data collection and preparation of oil field products *Proceedings of the higher educational institutions. Oil and gas* **3**
[6] Tyrylgin I V, Shpilevoy V A and Zemenkov Yu D 2012 Energy saving and energy efficiency of the economics, recovery, and transport of oil and gas of Russia *Proceedings of the higher educational institutions. Oil and gas* **6**
[7] Zakirzakov A G and Egorov A L 2015 Analysis of the state of the network of main oil pipelines of Tyumen region on the basis of statistical data *Modern problems of science and education*
[8] Gumerov A G, Gumerov R S and Akberdin A M 2001 *Operation of pumping stations equipment* Textbook (Moscow; Nedra-business centers) 475

Modeling the Phase Composition of Gas Condensate in Pipelines

S M Dudin, Yu D Zemenkov and A B Shabarov

Industrial University of Tyumen, 38 Volodarskogo st., Tyumen, 625000, Russia

E-mail: srghome@mail.ru

Abstract. Gas condensate fields demonstrate a number of thermodynamic characteristics to be considered when they are developed, as well as when gas condensate is transported and processed. A complicated phase behavior of the gas condensate system, as well as the dependence of the extracted raw materials on the phase state of the deposit other conditions being equal, is a key aspect. Therefore, when designing gas condensate lines the crucial task is to select the most appropriate methods of calculating thermophysical properties and phase equilibrium of the transported gas condensate. The paper describes a physical-mathematical model of a gas-liquid flow in the gas condensate line. It was developed based on balance equations of conservation of mass, impulse and energy of the transported medium within the framework of a quasi-1D approach. Constitutive relationships are given separately, and practical recommendations on how to apply the research results are provided as well.

1. Introduction

Transportation of raw hydrocarbons (RH) is one of the most energy intensive industry sectors. From the literature it is known that energy costs to pump a unit mass of RH in the form of gas2-3 times exceed energy costs to pump a unit mass of RH in the form of liquid. As far as energy conservation during RH transportation is concerned, an important task is development and application of a method to calculate gas-liquid hydrocarbons flow and heat and mass transfer in process and trunk pipelines during their design and operation.

2. Methods, results and discussion

Methods of calculating thermophysical properties (TPP) and phase equilibria are reviewed inquite a lot of scientific publications and reference books [1, 2, 3, etc.]. However, when it comes to such complex systems as oil, gas condensate, natural gas and their by-products, using even the known techniques is a difficult task. It should also be taken into account that large part of calculation methods is designed for individual hydrocarbons and mixtures of known composition, that is why using them to calculate TPP of complex mixtures of unknown composition is not always possible or can even results in additional, hard-to-estimate errors [1].

To calculate the thermophysical properties of the gas-liquid mixture in the pipeline a method of control volumes is used [2, 4]. The internal volume of the pipeline is divided into a finite number of sufficiently small areas – control volumes V, limited by the inner surface of the pipeline and cross sections S_1 and S_2, located at a distance Δz from each other. With regard to the control volume, the following balance equations are used: of mass, momentum and total energy. Main balance equations of mass, impulse and energy conservation in case of a quasi-1D flow are written as suggested by A.B. Shabarov [2, 4, 5 et al].

The law of conservation of mass of gas-liquid mixture phases in case of applying a quasi-1D approach is:

$$\rho_{2i} v_{2i} S_{2i} = \rho_{1i} v_{1i} S_{1i} - \rho'_i v'_{ni} S'_i + \sum_{\substack{j=1 \\ j \neq i}}^{N} J_{ji} V - \frac{\partial \rho_i}{\partial t} V \cdot \qquad (1)$$

This equation takes into account the flow through the lateral surface, which is characteristic of loss calculations, flows during controlled input and output of the mass in branched pipeline systems and mass exchange between phases.

The mechanical energy balance equation is written as a common form of Bernoulli's equation. For a model with a total phase pressure (i=g, l):

$$\alpha_{kg}\frac{v_{2g}^2}{2}+\frac{\alpha_g P_{2g}}{\rho_g}+gz_2=\alpha_{kg}\frac{v_{1g}^2}{2}+\frac{\alpha_g P_{1g}}{\rho_g}+gz_1+l_{vng}^*-l_{trg}+\tilde{l}_g+l_g^{/}+l_{fg} \qquad (2)$$

$$\alpha_{kl}\frac{v_{2l}^2}{2}+\frac{\alpha_l P_{2l}}{\rho_l}+gz_2=\alpha_{kl}\frac{v_{1l}^2}{2}+\frac{\alpha_l P_{1l}}{\rho_l}+gz_1+l_{vnl}^*-l_{trl}+\tilde{l}_l+l_l^{/}+l_{fl} \qquad (3)$$

where $l_{vni}^*=l_{vn}-gz_1+gz_2$, $l_{vni}=f_{ix}dx$ – specific work of external forces, J/kg, excluding work to overcome gravity; $l_{tri}=\left(\lambda_{tri}\frac{dx}{D}+\xi_{mi}\right)\frac{v_i^2}{2}$ – work expended when moving each of the phases to overcome the forces of friction and local resistance, $\tilde{l}_i=-\frac{\partial v_i}{\partial t}dx$ – specific work of inertial forces; $l_i^{/}=-v_{ni}\frac{\rho_i^{/}S_i^{/}}{\rho_i S_i}(v_{xi}-v_i)$ – work of forces associated with the exchange of an impulse during input and output of the mass through the lateral surface; $l_{fi}=\sum_{\substack{j=1\\j\neq i}}^{N}\frac{1}{\rho_i}\left(P_{jix}-v_i J_{ji}\right)dx$ – specific work of interfacial forces which at small J_{ji} comes down to the work of shear stresses at the phase boundary, α_{ki} – Coriolis coefficients taking into account the uneven distribution of the phase velocity along the radius of the pipe, $\rho_i\approx\frac{2\rho_{1i}\rho_{2i}}{\rho_{1i}+\rho_{2i}}$ – the average density in the section 1-2 on a scheme with a total phase pressure Pi=P. The work of shear stresses at the phase boundary P_{jix} may be presented as $P_{jix}=\tau_{gr}\cdot S_g$, where τ_{gr} – the shear stress at the fluid and gas boundary, S_g – the interface area of the selected volume.

The integral of internal energy conservation for a single-temperature mixture (T_i=T) is written as:

$$\sum_{i=1}^{N}G_{2i}\left(c_iT_2+\upsilon_{g0}\right)=\sum_{i=1}^{N}G_{1i}\left(c_iT_1+\upsilon_{g0}\right)+\sum_{i=1}^{N}G_i^{'}\left(c_iT^{'}+\upsilon_{g0}\right)+\frac{\partial\left(\sum_{i=1}^{N}\rho_i v_i\right)}{\partial t}V+Q_{vn}+N_{tr} \qquad (4)$$

where $G_i^{'}=-\rho_i^{'}v_i^{(n)'}S^{'}$ – the input flow through the lateral surface; Q_{vn} – external thermal power; $N_{tr}=l_{tr}G$ – friction power; $\upsilon_{g0}=l_{gl}+\left(c_l-c_{Vg}\right)\left(T-T_0\right)+\frac{P_0}{\rho_l^{(0)}}-\frac{P_0}{\rho_g^{(0)}}$ – a constant found from the normalization conditions during phase transitions $l\leftrightarrow g$ [4].

Closing relations of the physical-mathematical model of the gas-liquid mixture flow in the condensate line

The dependence of the gas condensate mixture density on pressure and temperature.

Given the gas condensate line's upstream initial conditions (the mass flow rate of a component – G_i, the true density of the gas condensate – ρsm_0, temperature– T_0, pressure – p_0, mass fractions of a component — Nm_i) let us find the density of a group of hydrocarbons hexane+ at the inlet of the rated pipeline section by:

$$\rho_{C6+}=\frac{Nm_n}{\frac{1}{\rho sm_0}-\sum_{i=1}^{n-1}\frac{Nm_i}{M_i}\upsilon_i} \qquad (5)$$

where υ_i – the molar volume of other components calculated for both gas and liquid by a single Peng-Robinson equation of state [6].

Having determined the density of a group of hydrocarbons hexane+ by(5) we find the boiling temperature TbC6+, molar weight MC6+, critical pressure and temperature TcC6+, pcC6+ and acentricity factor ωC6+ using techniques from[1]. Using these data, mass fractions z_i of the RH components are calculated. Therefore, a group of hydrocarbons hexane+ is included in the gas condensate mixture as a single pseudo-component. Having calculated the density of the mixture by substituting values for individual components and a pseudo-component ρsm=661.7 kg/m3 in the equation of state and comparing its value to the givenρsm0=662.0 kg/m3, we prove the adequacy of this approach, as relative deviation of the calculated ρsm value from the given one is -0.05%.

The viscosity of the gas condensate mixture.

The viscosity of individual hydrocarbons is calculated using a database from [3]. The viscosity of a group of hydrocarbons hexane+ is determinedusing techniques from [1]. The viscosity of the entire mixture was calculatedat temperature of 20°Cand pressure of 1 atm by the formula:

$$\mu sm_0 = \frac{\sum_{i=1}^{N} z_i \sqrt{M_i}\,\mu ind_i}{\sum_{i=1}^{N} z_i \sqrt{M_i}} \quad (6)$$

Recalculation of viscosity on thermobaric conditions in different sections of the pipeline was carried out according to the formula:

$$\mu sm = 10^{\lg\left(\mu sm_0\right)+1.4503\cdot 10^{-3}\cdot\left(\frac{P}{10}-0.101325\right)\cdot\left(5.2054\cdot \mu sm_0^{0.278}+0.0239\right)-3}. \quad (7)$$

The isobaric heat capacity of the gas condensate mixture.

The isobaric heat capacity of the individual components in the mixture is calculated using a database from [3]. The isobaric heat capacity of a group of hydrocarbons hexane+ is determined using the Watson-Nelson formula. The heat capacity of the entire mixture was calculated according to the additivity rule:

$$Cpsm = \sum_{i=1}^{N} z_i \frac{Cp_i}{M_i} \quad (8)$$

The thermal conductivity of the gas condensate mixture.

The thermal conductivity of the gas condensate mixture was calculated using a database and recommendations from [3], and the molar volumes of the mixture components were calculated using a single Peng-Robinson equation of state.

The soil heat transfer.

The soil heat transfer is found by [7]:

$$\alpha_{gr} = \begin{cases} \dfrac{2\cdot\lambda_{gr}}{D\cdot\ln\left(\dfrac{2\cdot H}{D}+\sqrt{\left(\dfrac{2\cdot H}{D}\right)^2-1}\right)}, & if\ \dfrac{H}{D} > 3 (\text{Forchheimer - Vlasov}) \\ \dfrac{2\cdot\lambda_{gr}}{D\cdot\ln\left(4\dfrac{H_{pr}}{D}+\dfrac{1}{Nu_{gr}}\right)}, & if\ \dfrac{H}{D} < 3 (\text{Arson-Kutateladze}) \end{cases} \quad (9)$$

Where H – the depth of the pipeline occurrence in the soil, m, H_{pr} is found by:

$$H_{pr} = H + H_e; \qquad H_e = H_{sn}\frac{\lambda_{gr}}{\lambda_{sn}}, \quad (10)$$

and also λ_{sn} and λ_{gr} – the thermal conductivity coefficients for snow and soil:

$$\lambda_{sn} = 0.465\frac{W}{m\cdot K}, \quad \lambda_{gr} = 1.4\frac{W}{m\cdot K},$$

Nu_{gr} – the Nusselt number for soil:

$$Nu_{gr} = \frac{\alpha_{vozd}\cdot D}{\lambda_{gr}}, \tag{11}$$

where α_{vozd} – the heat exchange coefficient for air:

$$\alpha_{vozd} = 11,6\frac{W}{m^2\cdot K}$$

The equation of state for the components of the gas condensate system.

In this paper, a single cubic Peng-Robinson equation of state (PR) was used as a phase state equation. This equation is of a Van-der-Waals type. It is commonly used to solve the tasks of designing, developing and operating hydrocarbon fields, as well as when simulating chemical processes. It is a simple form of a four-coefficient equation of state:

$$p = \frac{RT}{\upsilon - b} - \frac{a}{(\upsilon + c)(\upsilon + d)} \tag{12}$$

Where b, c, d – coefficients, constant for a given substance; the coefficient a depends on temperature ($a = a_c\varphi(T)$, where a_c - a constant, φ – a function of temperature equal to one at the critical temperature).

The coefficient structure of the equation of state is determined as follows [3]:

$$\begin{aligned} a_c &= \alpha R^2T_c^2 / p_c, \\ b &= \beta RT_c / p_c, \\ c &= \sigma RT_c / p_c, \\ d &= \delta RT_c / p_c, \end{aligned} \tag{13}$$

Where

$$\begin{aligned} \alpha &= \Omega_c^3 \\ \beta &= Z_c^* + \Omega_c - 1 \\ \sigma &= -Z_c^* + \Omega_c\left[0.5 + (\Omega_c - 0.75)^{1/2}\right] \\ \delta &= -Z_c^* + \Omega_c\left[0.5 - (\Omega_c - 0.75)^{1/2}\right] \end{aligned} \tag{14}$$

Here Zc, Ωc–independent parameters of the equation of state. Their values together with $\varphi(T)$ completely determine the equation of state of a pure substance.

Determining the composition and quantitative relation of equilibrium vapor and liquid phases.

The distribution coefficient or equilibrium constant of the ith component of the hydrocarbon mixture Ki is the relation of its molar fraction in a vapor phase yi to its molar fraction in a liquid phase xi.

$$x_i = \frac{z_i}{V(K_i - 1) + 1}, \tag{15}$$

$$y_i = \frac{z_i K_i}{V(K_i - 1) + 1}. \tag{16}$$

Equations (12) and (15) are called equations of phase concentrations of the mixture components. They allow us to determine the molar fractions of components in vapour and liquid phases of the mixture composition zi at given values of distribution coefficients Ki and molar fraction of the vapor phaseV.

Calculating the actual volumetric gas content.

By definition, the actual volumetric gas content is equal to:

$$\alpha_g = \frac{V_g}{V_{sm}} = \frac{V_g}{V_g + V_L} = \frac{1}{1 + \dfrac{V_L}{V_g}}. \qquad (17)$$

If gas and liquid volumes are found as a product of the molar volume of gas and liquid (υ_g, υ_L – calculated by the equation of state given the known phase compositions) and the number of moles of the mixture in the vapor and liquid state, respectively (N_V, N_L),

$$\alpha_g = \frac{1}{1 + \dfrac{\upsilon_L}{\upsilon_g}\dfrac{N_L}{N_V}} = \frac{1}{1 + \dfrac{\upsilon_L}{\upsilon_g}\left(\dfrac{1}{V} - 1\right)} \qquad (18)$$

Determining the two-phase gas-liquid medium flow in the condensate line.

At each calculation stage we checked the phase behavior of the gas condensate mixture by the temperature and pressure conditions in the given section of the condensate line. If the mixture was in a single-phase liquid state (V≤0 at equilibrium values of Ki), the system (2, 3) had only one equation, as α_g was assumed to be equal to zero, and α_l — equal to one. If at any calculation stage parameter V was 0<V<1, then by (15) and (16) equilibrium compositions of gas – yi and liquid – xi phases were calculated. After that, according to a single Peng-Robinson equation of state, we calculated the density, and then viscosity of the gas and liquid phases and the surface tension of the liquid phase. According to the known molar ratio between the gas and liquid, mass flow rates of RH in the form of gas - Gg and liquid - GL were calculated. These values were used later to determine the flow regime of the gas-liquid mixture by a known procedure [5].

3. Summary

A physical-mathematical model of a hydrocarbon mixture (HCM) flow in the gas condensate line considering phase transitions and changes in the phase composition was developed. A numerical procedure for determining "liquid-gas" balance equations for each component of the mixture was proposed using a data base on thermophysical properties of the components, including a single Peng-Robinson equation of state. Based on the developed model a technique and calculation algorithm are composed to determine the phase composition in arbitrary pipeline sections and to find the valid values of mass concentration of light hydrocarbons in the inlet condensate line section on condition that a single-phase flow regime along the entire length of the pipeline is provided.

References

[1] Grigoryev B A, Gerasimov A A and Lanchakov G A 2007 *Thermophysical properties and phase equilibria of gas condensates and their fractions* (MEI, Moscow) 344p

[2] Antipyev V N, Zemenkov Yu D, Shabarov A B, et al 2002 *Technical and parametric diagnostics in pipeline systems* (Vector Buk, Tyumen) 432p

[3] Reid R, Prausnitz J and Sherwood T 1982 *Properties of Gases and Liquids: A Reference Guide* Transl.ed. B I Sokolov (Chimia, Leningrad) 592p

[4] Shabarov A B and Saranchin N V 2009 Methods of calculating equilibrium parameters of liquid hydrocarbon mixture in pipelines *Bulletin of Tyumen State University* 6 112-19

[5] Shabarov A B, Saranchin N V and Kutrunov V N 2005 Automated calculation of the flow of oil and gas mixture in the pipeline *Modernization of education in the context of globalization: Round Table "Education through science and innovation"* 126-29

[6] Peng D Y and Robinson D B 1976 A new two-constant equation of state *Ind. Eng. Chem. Fundam.* **15** 59-64

[7] Antonova E O, Bakhmat G V, Ivanov I A and Stepanov O V 1999 *Heat transfer in pipeline transport of oil and gas* (Nedra, St.Petersburg) 228p
[8] Mamadaliev, R.A., Kuskov, V.N., Popova A.A., Valuev D.V., Alloying elements transition into the weld metal when using an inventor power source, J. IOP Conference Series: Materials Science and Engineering. 127 (2016) p. 1-7
[9] Valuev, D.V., Malushin, N.N., Valueva A.V., Dariev, R.S. Mamadaliev R.A. Manufacturing component parts of mining equipment with application of hardening technologies, J. IOP Conference Series: Materials Science and Engineering. 127 (2016) p. 1-7

IOP Conf. Series: Materials Science and Engineering **154** (2016) 012011 doi:10.1088/1757-899X/154/1/012011

Developing Mathematical Provisions for Assessment of Liquid Hydrocarbon Emissions in Emergency Situations

M Yu Zemenkova, Yu D Zemenkov and V D Shantarin

Industrial University of Tyumen, 38 Volodarskogo st., Tyumen, 625000, Russia

E-mail: muzemenkova@mail.ru

Abstract. The paper reviews the development of methodology for calculation of hydrocarbon emissions during seepage and evaporation to monitor the reliability and safety of hydrocarbon storage and transportation. The authors have analyzed existing methods, models and techniques for assessing the amount of evaporated oil. Models used for predicting the material balance of multicomponent two-phase systems have been discussed. The results of modeling the open-air hydrocarbon evaporation from an oil spill are provided and exemplified by an emergency pit. Dependences and systems of differential equations have been obtained to assess parameters of mass transfer from the open surface of a liquid multicomponent mixture.

1. Introduction

Effective development of accident prevention techniques depends largely on goodness-of-fit a priori statistics and their treatment. Damage caused by the loss of pipeline integrity is determined, first of all, by the amount of flowing oil, its physical and chemical properties and the conditions that characterize the accident site, the volume of emissions. Knowing the characteristics and quantity of oil spilled, we can determine the strategy for accident response, necessary effort and resources, as well as prevent deterioration of the situation related to the spread of contamination.

Common methods of characterizing volatility of petroleum and petroleum products can be divided into two groups: experimental (direct and indirect) and theoretical. Direct methods include measuring the fluid volume before and after evaporation or calculating the vapor volume (mass). Direct determination of hydrocarbon emissions is extremely difficult, and in some cases impossible, especially for short-term storage encountered in emergency situations. For example, when determining emissions by measuring the oil level h_{B3} a basic rule of measuring technique is violated, according to which a small value must be obtained by direct measuring, rather than subtracting two inaccurate values.

a) offshore [1] b) onshore [2]

Figure1. Petroleum evaporation during spills.

In the long run, indirect methods may become more commonly used, such as determining losses by a change in density, refractive index, hydrocarbon composition, saturated vapor pressure, etc. Calculations based on comparing samples are labor-intensive, technologically and practically hard-to-apply. For example, analysis of experimental data on evaporative petroleum losses obtained by various methods shows that their results, as a rule, are inconsistent due to large errors of the measurements and the inability to comply with the high requirements for sampling. It may be noted that to calculate emissions and losses of liquid hydrocarbons that are on the daylight surface adequate sampling is practically impossible.

2. Models and methods

According to classical concepts of the kinetic molecular theory of gases, diffusion is the process of distributing a substance from one medium to another. The process of liquid transformation into gas (vapor), occurring only on a free liquid surface, is referred to as evaporation. In both cases, distribution of the substance in the medium is due to a concentration gradient of the vapor. At a constant temperature, the rate of liquid W evaporation is directly proportional to the evaporation area S and the difference between the saturated vapor pressure p_S and pi actually set at a given time and inversely proportional to the pressure in the system p_{am}

$$W = AS\frac{p_S - p_i}{p_{am}}, \tag{1}$$

where A – the coefficient of proportionality.

From (1) we can see that at the initial moment evaporation occurs at an infinite velocity, at its limit equal to the translational velocity of molecules, then it slows down and atp_i=p_S stops. Evaporation from an open surface at p_i=0 happens at a constant and maximal rate. The rate of gas space saturation largely depends on the state of the environment into which evaporation takes place. If there is an uneven temperature distribution in the system or external forces, such as electrical, act upon the system there occur thermal diffusion, pressure diffusion, electrodiffusion, etc., resulting in an uneven concentration distribution.

According to laws of C. Maxwell and J. Stefan, the mass amount dG of a substance, passing by diffusion per time $d\tau$ through the area S, perpendicular to the axis Y, is directly proportional to the diffusion coefficient D and the density gradient (concentration) $d\rho/dy$ in the direction of the diffusion path, i.e. concentration decreases. This is indicated by the sign (-) in (2):

$$dG = -D\frac{p}{p - p_p}\frac{d\rho}{dy}d\tau dS, \tag{2}$$

where D – the diffusion coefficient, m ^{2}/s; p_p– the partial vapor pressure; p – the total vapor pressure.

The known diffusion rate can be determined by Fick's law:

or via the pressure gradient

$$\omega = \frac{dM'}{d\tau} = -D\frac{dM}{dy} = -D\,grad\ C,\ \ \omega = \frac{dM}{d\tau} = -D\frac{d\rho}{dy} = -D\,grad\,\rho. \tag{3}$$

where M' – the vapor amount per unit area, M'=M/S.

We know of a number of such equations derived on the basis of the classical kinetic theory of evaporation and diffusion, however, having been studied experimentally, they showed large discrepancies between the calculated and experimental data. Forexample, in works covering experimental studies of the influence of airflow on evaporation of water at rest in an open container, the scatter of results amounts to 400%. Experimental studies conducted by M. N. Baranaev, M. V. Lourie, S. N. Sklyarenko, A. G. Kolesnikov, N. M. Mihailov, N. Carier and others produced dependences in which the velocity index ranged from 0.3 to 1.0. Theoretical and experimental studies of the losses of petroleum and petroleum products from evaporation in surface and underground tanks were carried out in the works of Abuzova F.F., Bunchuk V.A., Novoselov V.F., Konstantinov N.N., Fathiev N.M., Dushin V. A. et al.

Sklyarenko S.N. and Baranaev M.N. note that the reason for discrepancies in the values are due to the difference in flow conditions in which the evaporation takes place. Increasing the speed of the airflow from minor (where evaporation is determined by the rate of vapor molecular diffusion) to turbulent (where convective currents take precedence in evaporation) leads to a varying evaporation rate of the substance. Thus, the transfer of vapor from the liquid surface to the environment is caused by pure molecular diffusion, as well as convection. It follows that when characterizing mass transfer we must consider two types of evaporation: *static* and *dynamic*.

According to A. V. Nesterenko, thickness of the boundary layer can be expressed by the vapor densityρ, i.e., $\delta = \rho D$. However, δ also depends on the nature of the vapor-air medium flow near the surface. Evaporation processes are caused not only by the vapor concentration, but also by different effects, such as temperature, molecular movement of vapor and air in the boundary layer, etc. During evaporation of substances with a molecular weight substantially less than the molecular weight of air, convection currents will be directed upwards, and for substances with a molecular weight greater than the molecular weight of air - downwards. By this convection currents complicate evaporation.

A small vapor concentration C_S, corresponding to saturation, will occur by the evaporation surface in a relatively thin, so-called boundary layer δ. After the boundary layer, the vapor concentration decreases with increasing distance from the evaporation surface. As various thermodynamic conditions are possible on the surface of the substance and in the environment, it is almost impossible to accurately determine the boundary layer thickness δ.

It becomes clear why the evaporation rates observed experimentally exceed those obtained by calculation using the formula of "free" diffusion, for example

$$\vartheta_X \frac{\partial C}{\partial x} + \vartheta_y \frac{\partial C}{\partial y} + \vartheta_Z \frac{\partial C}{\partial z} = D\left(\frac{\partial^2 C}{\partial x^2} + \frac{\partial^2 C}{\partial y^2} + \frac{\partial^2 C}{\partial z^2} \right) \quad (4)$$

Integrating this differential diffusion equation with partial derivatives in the space with complex boundaries gives many different solutions. In most cases, such as for a stationary process and a three-dimensional diffusion field, it is almost impossible to find a solution satisfying the differential diffusion equation and the single-valuedness condition. Simplification, however, distorts the mechanism of the phenomenon.

Indeed, evaporation causes a loss of petroleum products mainly of light ends, whereby fluid "becomes heavier" changing mass transfer in different ways. To calculate evaporative hydrocarbon emissions that can reach 40% of the weight, of particular importance is the diffusion coefficient D (*molecular, turbulent*)

Data on changes in D of petroleum at various levels of evaporative losses in Russian and foreign literature are clearly insufficient. In practice, diffusion coefficient is determined by the computational-experimental method ofA. S. Irisov using a special evaporator at the liquid-gas phase ratio of 4:1. A number of diffusion coefficient values have been produced in Ufa State Petroleum Technological University byF. F. Abuzova, V. F. Novoselov, V. N. Chernikin and V. A. Martyashova primarily for commercial petroleum products. When temperature changes within the range of 10-40°C, the coefficient changes within a narrow range (25% of the mean value 20.5%). According to the authors, even for gasoline of the same grade this coefficient depends on the source petroleum and processing methods applied at the refinery.

From the analysis of works it follows that the diffusion coefficient decreases with increasing molecular weight of hydrocarbons and increases significantly with increasing temperature. The difference in the diffusion coefficients of various hydrocarbon vapors diminishes with increasing molecular weight. In case of equal molecular weight, arenes have the largest values of D, and cyclanes – the smallest. Alkane hydrocarbons occupy an intermediate position.

Also known is the effective diffusion coefficient D_e, which is used for formal coordination of the experimental data with the theoretical predictions. D_e describes intensification of petroleum vapor transport due to evaporation.

When, for instance, a reinforced concrete storage tank ZhBR–10000 containing petroleum at a higher temperature is emptied, D_e 0-80 times exceeds the molecular diffusion coefficient, and when it is filled – by 40%. Experimental studies of the molecular diffusion coefficients, depending on the temperature and vapor pressure, for gasoline, Samotlor and Ust-Balyk oil, etc. allowed us to determine the value of evaporative losses with an error of 25 - 60% increase.

Loktev V.M. noted that when gasoline grades B-70 and AI-93 are evaporating diffusion coefficients of vapors in the air change significantly. He also proved that diffusion coefficients of gasoline of the same grade but measured in different years of production vary in value.

Based on extensive research, Irisov A.S. determined that when *Re* increases the influence of *D* is not proportional to the evaporation rate but raised to the power m=2/3, i.e. $D^{2/3}$, in other words, becomes smaller.

As far as oil is concerned, differences in the hydrocarbon composition are more significant, resulting in a drastic difference in the diffusion coefficients. Irisov A.S. claims in his works that the decisive factor in determining the rate of evaporation of petroleum products is saturated vapor pressure, and the diffusion coefficient has only an additional effect. Spalding D.B. expresses a similar view after studying diffusion of various mixtures of gases with air. The author notes that the value of the diffusion coefficient is not very sensitive to changes in the composition for binary mixtures, even when compared to the influence of viscosity or molecular weight.

Thus, given the very limited data on the diffusion coefficient of petroleum and petroleum products and a considerable complexity of its determination when calculating oil evaporation losses from an open surface, it is necessary to carry out calculations with great care.

From the analysis it follows that no matter whether the calculated ratios of mass transfer are based on experiment or mathematical analysis, they contain mass transfer coefficients from the open surface of multicomponent hydrocarbon fluids and a large number of dependent factors, which prevents the experimental data from being adequately summarized.

Published data (within their accuracy) on evaporation of multicomponent substances do not allow discovering a clear influence of the mixture heterogeneity on the mass transfer coefficient and the rate of evaporation from the open surface. The methods used for calculating losses based on the determination of evaporation coefficients and mass transfer from the surface of petroleum and petroleum products describe the process of mass transfer only in closed vessels.

A data bank of coefficients of diffusion, mass transfer, thermal conductivity, and so on is required created on the basis of extensive experiments under a wide range of conditions. To date, this type of extensive research has not been carried out.

An alternative solution to the problem of calculating evaporative emissions may be adapting methods based on the equations of state of multicomponent mixtures to the given conditions.

3. Mathematical modeling

According to Raoult's law for the mixture of liquid hydrocarbons, the saturated vapor pressure p_S depends upon the saturated vapor pressure of the individual components and the molar concentrations. It is equal to the sum of pressures of the mixture components that they would have if each of them occupied the entire volume of the mixture at a given temperature, i.e. the sum of partial pressures.

From the combined Dalton-Raoult law, we can draw an important conclusion - in equilibrium, the partial pressure of any component of the mixture in the vapor phase is equal to the partial pressure of that component in the liquid, from which

$$p_{Si}/p_S = y_i/x_i = k_{pi} = const$$

or

$$p_{Si} = k_{pi}p_S\ uy_i = k_{pi}x_I, \tag{5}$$

where p_{Si} is the saturated vapor pressure of a pure component; y_i and x_i is the molar concentration of a component in the vapor and liquid phases.

Coefficient k_{pi} is known as *the equilibrium constant* (or phase equilibrium constant), which under conditions of the equilibrium state (5) allows, for example, finding the gas phase composition by a known fluid composition, and vice versa. If $k_{pi}>1$, then this component will have a higher concentration in the vapor phase than in the liquid, and vice versa.

Besides the phase equilibrium equation (5), during the evaluation of the phase state we can use (6) called the flash evaporation start equation and characterizing the equilibrium system with a small amount of vapor or (7), characterizing the system in the vapor phase with a small amount of liquid.

When forecasting the material balance of multicomponent two-phase systems (or the concentration of components in different phases) k_{Pi} are commonly used when

$$\sum k_{pi} x_{i0} = 1 \tag{6}$$

$$\sum \frac{y_i}{k_{pi}} = 1 \tag{7}$$

pressure and temperature (initial t_{H0} and final t) are given and the molar concentration of the ith component is known, for instance, in the liquid x_{i0} or gaseous y_i state.

For any component, the number of moles in two phases is determined as a sum of N_i^g and N_i^l, i.e. $N_i^0 = N_i^g + N_i^l$. Hence, the molar concentration:

$$x_{i0} = y_i g + x_i l, \tag{8}$$

where g and l – the numbers showing the proportion of the total number of moles in the gaseous and liquid state, respectively ($g+l=1$). In practice, g is calculated using the successive approximation method that is easily implemented by means of a personal computer. It is assumed that $\sum x_i = 1$, which follows from (7).

$$x_i = \frac{x_{i0}}{k_{p_i} + (1 - k_{p_i})\, l}, \qquad y_i = \frac{k_{p_i} x_{i0}}{k_{p_i} + (1 - k_{p_i})\, g}. \tag{9}$$

Considering that $y_i/x_i = 1$, for the ith component concentration in the liquid and gaseous phases we can use the following expressions:

The equilibrium compositions of a two-component system are much easier to determine:

a) for the liquid phase:

$$x_1 = \frac{1 - k_{p2}}{k_{p1} - k_{p2}}; \qquad x_2 = \frac{k_{p1} - 1}{k_{p1} - k_{p2}}; \tag{10}$$

b) for the gaseous phase:

$$y_1 = \frac{k_{p1}(k_{p2} - 1)}{k_{p2} - k_{p1}}; \qquad y_2 = \frac{k_{p2}(1 - k_{p1})}{k_{p2} - k_{p1}}. \tag{11}$$

The processes of open or flash evaporation of liquid multicomponent mixtures are widely used in laboratory and industrial practices and usually involve the use of the iteration method for calculations. To develop a technique for calculation of emissions of liquid multicomponent hydrocarbon mixtures from the open surface we conducted the analysis of existing methods of mathematical modelling of mass transfer. It showed that open evaporation processes are non-stationary and are described by a system of differential algebraic equations. Flash evaporation processes are stationary and are described by a system of ordinary nonlinear algebraic equations.

Let us consider the process of open evaporation of hydrocarbons from an oil spill or, for example, an emergency pit illustrated in Figure 2.The vessel has the shape of a truncated pyramid with an angle of sidewalls inclination α and filled with oil to different heights h_{lev} The amount of initial liquid M can be determined by

$$M = \frac{\rho}{3}\left(S_1 + \sqrt{S_1 S_2} + S_2\right) h_{lev}, \tag{12}$$

where ρ - the density of the petroleum; S_1, S_2 – the areas of the vessel base and the free petroleum surface.

To determine the vapor flow M_u let us build a system of equations of heat and material balances and of the phase state.

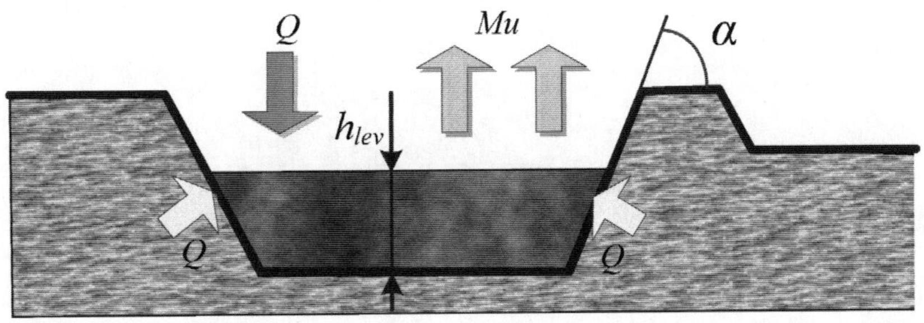

Figure 2. Diagram of an emergency pit.

For the isobaric process, an infinitesimal amount of heat dQ can be represented as a total differential of a certain value J, called enthalpy (formerly known as heat content), i.e.

$$dO = dJ = C_P dt, \tag{13}$$

where C_P - the isobaric heat capacity.

It is known that when temperature changes from 0 to 50°Cthe heat capacity of gasoline varies from 1.93 to 2.28 kJ/kg·K, and that of oil - from 1.80 to 2.52 kJ/kg·K. With sufficient accuracy, the heat capacity of petroleum and petroleum products can be determined analytically by Crego's formula:

where ρ_{293} – the density of the petroleum product, kg/m³.

When determining the isobaric heat capacity of the condensates of the Urengoiskoye field, we can use Mustafayev's formula:

$$C_P = \frac{31,56}{\sqrt{\rho_{293}}}(762 + 3.39T), \tag{14}$$

$$C_P = 3.3909 - 1.6747 \cdot 10^{-3} \rho_T, \text{[kJ/kg·degree]}. \tag{15}$$

A change in the heat capacity depending on the boiling point T_K, of the condensate with the density ρ_K = 650÷720 kg/m³ can be characterized by the following dependence proposed by TSOGU:

$$C_P = 1030.8 + 4.50\, T_K, \quad \text{[J/kg·degree]} \tag{16}$$

for a mixture of NGL with a condensate (ρ_K = 550 ÷ 590 kg/m³):

$$C_P = 1060.4 + 4.796\, T_K, \tag{17}$$

for a stable condensate (ρ_K = 730 ÷ 770 kg/m³):

$$C_P = 887.0 + 3.9460\, T_K, \tag{18}$$

where T_K - the average boiling point of the condensate, к.

The enthalpy is a state parameter of a thermodynamic system. Its change in the isobaric process equals, as is evident from (13), the amount of heat obtained by the system. Of practical interest are changes in the enthalpy rather than its absolute values, i.e. the concept of enthalpy, just as the concept of heat, refers to the process rather than the state. Its physical meaning can be summarized as follows: the enthalpy of any thermodynamic system is the total energy of the system. If overheating of petroleum product vapors is not required we can determine the enthalpy of saturated vapors that, at the given temperature, corresponds to the amount of heat required for heating 1 kg of liquid (petroleum product) from 0 °C to the given temperature and its evaporation at the same temperature. This value is known as the *total heat of evaporation* and equals the enthalpy of the liquid at the boiling point plus the latent heat of evaporation.

Enthalpy is calculated for liquid hydrocarbons and their vapors. For practical calculations of enthalpies of liquid petroleum products at pressures close to atmospheric and different temperatures *t* we can use known nomograms or, for example, dependences developed by Ilembitov R. I.:

$$h = (\frac{4.187}{\sqrt{\rho_4^{20}}})(0.3886t + 0.00045t^2), \tag{19}$$

for petroleum vapors:

$$I = 4.187\begin{bmatrix}(139 - 62.92\rho_4^{20}) + (0.4576 - 1.43\rho_4^{20})t + \\ +(503 - 57.2\rho_4^{20})10^{-6}t^2 - 35(1 - \rho_4^{20})^3\end{bmatrix}, \tag{20}$$

where ρ_4^{20} – the relative density of the substance at 20°C.

The enthalpy of liquid hydrocarbon mixtures is virtually independent of the pressure and can be determined by the value of the enthalpy of individual components using the rule of additivity

$$I = \sum k_{pi}(t)x_i I_i(t) \tag{21}$$

For vapor mixtures, such calculation introduces some error, although in some cases it is feasible, for example, when characterizing the processes of open or flash evaporation of liquid multicomponent mixtures widely used in laboratory and industrial practices

$$h = \sum x_i h_i(t) \tag{22}$$

Thus, considering equations (6)÷(19) mass transfer from the exposed surface of the liquid multicomponent mixtures can be characterized by the following system:

$$\frac{d}{d\tau}M = -M_u; \tag{23}$$

$$\frac{d}{d\tau}x_i = \frac{M_u}{M}\left(1 - k_{pi}\right)x_i; \quad i = \overline{1, n}; \tag{24}$$

$$\frac{d}{d\tau}(Mh) = -M_u I + c_p\left(t - t_{HO}\right)M + \theta; \tag{25}$$

$$\sum k_{pi}x_i = 1 \tag{26}$$

with initial conditions at τ=0

$$M = M_0;\ x_i = x_{i0};\ \sum k_{pi}x_{i0} = 1, \tag{27}$$

where τ - the time of evaporation, s; M – the molar amount of evaporated liquid; M_u – the molar vapor flow; $\{x_i\}$ – the concentration of components in the liquid phase; I - the enthalpy of the vapor mole, J/kg; h - the enthalpy of the evaporated mixture mole, J/kg; $\{c_p(t_{HK}- t_{H0})M + \theta\} = Q$ - the heat flow supplied to the evaporated mixture, J/s; t_{HK}, t_{H0} – the initial and the final temperature of the mixture; θ - the heat flow required to transform M_u into vapor.

The system of ordinary differential equations (23÷26) contains (n+3) equations and (2n+5) variables: k_{pi}, $\{x_i\}$, M, Mu, t_{HK}, Q. Therefore, to carry out accurate calculations of mass transfer it is enough to bring the resulting system in line with the number of variables. If the a priori information on the deterministic values k_{pi}, t_{H0}, and Q is available, solving the system with determining $\{x_i\}$, M, Mu and t_{HK} is not particularly difficult. The authors obtained a solution of the system of differential equations of mass transfer by a Runge-Kutta method, and proposed a technique of determining the constants of phase equilibrium $k_{pi}, \{x_i\}$. The research results produced in TSOGU are taken as a basis for the development of a standard document and are published as monographs and textbooks.

4. Summary

Thus, the developed models and technique for calculating parameters of multicomponent substance mass transfer and flash evaporation processes as applied to petroleum and petroleum product spills with the deterministic composition allow us to assess the quantity of spills and liquid hydrocarbon losses during their evaporation from an open surface.

References

[1] Elimination of emergency oil spills *Online resource* (http://za4istkarvs.ru/likvidatciya-avariynykh-rozlivov-nefti-i-nefteproduktov) Access date 01.02.2016

[2] In the Gulf of Mexico a new oil leak is found *Online resource* (http://www.ecolife.ru/infos/news/1622/) Access date 01.02.2016

[3] Zemenkov Yu D 1985 Evaporation of oil from the exposed surface in case of failure on the main oil pipelines *Thesis for the degree of candidate of technical sciences (Ufa, UNI)* 202

[4] Zemenkov Yu D 1999 Methodological support of expert calculations of leakage and emissions from pipelines when transporting liquid hydrocarbons *Thesis for the degree of Doctor of Technical Sciences (Tyumen, TSOGU)* 360

[5] Zemenkov, Yu.D., V.V. Shalay, M.Yu. Zemenkova. Immediate Analyses and Calculation of Saturated Steam Pressure of Gas Condensates for Transportation Conditions (2015) Procedia Engineering, Volume 113 (2015), pp. 254-258 doi:10.1016/j.proeng.2015.07.330

[6] Walas Stanley M 1998 *Phase equilibria in chemical engineering* Translated from English (Moscow, Mir) **2** 360

[7] Stepanov S, Cheremisin N, Sokolov S, Altunin A and Shabarov A 2012 Computational method for relative permeability curves based on the bernoulli generalized porous media equations *ECMOR 2012 13th European Conference on the Mathematics of Oil Recovery*

[8] Vlasenko V S, Slesarenko V V, Gulkov A N and Lapshin V D 2015 Preparation of arctic oil and gas condensate deposit formation fluid for storage and transport in the form of hydrocarbon hydrate-containing dispersed system *Proceedings of the International Offshore and Polar Engineering Conference* 1863

[9] Soldatenko M V, Sidorov V G and Vasiljev S S 1972 The role of the turbulent vapor diffusion in the kinetics of liquids evaporation from a free surface Russian Journal of Physical Chemistry A 46 2399-401

[10] Chekardovskiy, M.,Chekardovskiy, S.,Ilyukhin, K.,Gladenko, A. Upgraded Algorithm for Calculating the Turbo-Expander of Gas Distribution Stations 2016 MATEC Web of Conferences 73

IOP Conf. Series: Materials Science and Engineering **154** (2016) 012012 doi:10.1088/1757-899X/154/1/012012

Development of Thermophysical Hydrocarbon Wastes Pyrolysis Model (in the Case of Wood)

V D Shantarin, M Yu Zemenkova and Yu D Zemenkov

Industrial University of Tyumen, 38 Volodarskogo st, Tyumen, 625000, Russia
[1]E-mail: muzemenkova@mail.ru

Abstract. The article is devoted to solving environmental problems in the operation in oil and gas industry objects. Reduction of environmental damage by pollution with hydrocarbons can be achieved by disposing oil-contaminated hydrocarbon wastes, using high-temperature pyrolysis process. Authors proposed a recycling method by which in the output there generates the maximum amount of syngas, which, in its turn, is an expensive resource

1. Introduction

Considering the problems in the oil and gas sector and reducing the geo-ecological environmental damage, it is possible to take a comprehensive approach to the problem by disposing oil-contaminated hydrocarbon wastes using high-temperature pyrolysis process. By means of this method of recycling hydrocarbon waste, in the output there generates maximum amount of syngas, excluding high-cost process of obtaining it from natural gas. Contributing to efficient use of resources, methyl alcohol having great demand in industry is obtained from this industry wastes.

2. Models and methods

Existing technologies of producing syngas are unprofitable because of big energy costs on syngas generation and purification, which makes them capital-intensive and applicable only for extremely large-scale production. If a major producer of methanol generates gas from the main pipeline, he becomes entirely dependent on its constantly rising prices. It is known that the field enterprise in Yakutia and the Far North have undeveloped transport network. Creating a module mobile compact facility, located directly in gas recovery areas and easy transportable, would allow recycle waste products directly in the field and to reduce the environmental load. The solution of a technical task, providing improvement of technologies of pyrolysis method of oil sludge recycling using a synthesis reactor, which allows obtaining methanol from hydrocarbon wastes, was experimentally reasoned.
In experiment, birchwood was used.

Volume:

$$V = 5.65 \cdot 10^2 \cdot 3.75 \cdot 10^{-3} \cdot 15.50 \cdot 10^{-3} = 3.28 \cdot 10^{-4} m^3 \tag{1}$$

Mass:

$$m^w = 212 \cdot 10^{-3} kg \, ; m = 198 \cdot 10^{-3} kg \tag{2}$$

Density:

$$\rho^w = \frac{m^w}{V} = \frac{212 \cdot 10^{-3}}{3.28 \cdot 10^{-4}} = 646.3 kg / m^3 \tag{3}$$

$$\rho = \frac{198 \cdot 10^{-3}}{3.28 \cdot 10^{-4}} = 603.7 kg / m^3 \tag{4}$$

Humidity:

$$W^p = \frac{\rho^w - \rho}{\rho} = 0.07(7\%) \quad (5)$$

Material balance (mass conservation law)

$$m(T)_z^w = m(C)_{pyr} + m(C_xH_yO_zNe)_{pyr} + m(H_2O)_{pyr} + m(\Gamma\Gamma)_{pyr} + m(Z)_{pyr}, \quad (6)$$

where $m(T)_z^w$ – wood fuel mass, kg;

$m(C)_{pyr}$ – mass of generating charcoal, kg;

$m(C_xH_yO_zNe)_{pyr}$ – total mass of generating organic product as a liquid phase (resin mixture, acetic acid, methanol, etc.);

$m(CG)_{pyr}$ – mass of generating combustion gas, kg;

$m(H_2O)_{pyr}$ – mass of generating water, kg;

$m(Z)_{pyr}$ – mass of generating ash, kg.

Chemical reaction of wood pyrolysis in a general form:

$$C_aH_bO_cN_d(rig) \xrightarrow{T=500C} \alpha_1C(rig) + \alpha_2C_xH_yO_zNe(liq) + \alpha_3H_2O(liq) + \alpha_4CO_2(g) + \alpha_5CO(g) + \alpha_6CH_4(g) + \frac{d}{2}N_2(g), \quad (7)$$

where $C_aH_bO_cN_d$ – chemical formula of the atomic composition of wood;

$m(T)_z^w = 0.05kg$ – mass of wood fuel with a relative humidity $w^p = 0.07$ – and ash content $A^p = 0.01kg$;

Mass of humidity in fuel:

$$m(W)_T = m(T)_z^w \cdot W^p = 0.05 \cdot 0.07 = 0.0035kg \quad (8)$$

Mass of ash in fuel:

$$m(Z)_T = m(T)_z^w \cdot A^p = 0.05 \cdot 0.01 = 0.0005kg \quad (9)$$

Mass of fuel in terms of absolutely dry, ashless:

$$m(T) = m(T)_z^w (1 - A^p - W^p) = 0.05 \cdot (1 - 0.01 - 0.07) = 0.0460kg \quad (10)$$

Elemental composition of wood will take as:

Carbon $(C) - \phi(C)_T = 0.5075$; (11)

Oxygen $(O) - \phi(O)_T = 0.4253$; (12)

Hydrogen $(H) - \phi(H)_T = 0.0612$; (13)

Nitrogen $(N) - \phi(N)_T) = 0.0006$. (14)

Mass costs of elemental carbon, ambient oxygen and fuel nitrogen on pyrolysis:

$$m(C)_T = \phi(C) \cdot m(T) = 0.075 \cdot 0.046 = 0.0233kg; \quad (15)$$

$$m(O)_T = \phi(O)_T \cdot m(T) = 0.4253 \cdot 0.046 = 0.0196kg; \quad (16)$$

$$m(H)_T = \phi(H)_T \cdot m(T) = 0.0612 \cdot 0.046 = 0.0028kg; \quad (17)$$

$$m(N)_T = \phi(N)_T = 0.006 \cdot 0.0460 = 0.00002kg. \quad (18)$$

Experimental fuel carbon gasification completeness rate:

$$\varsigma(C)^{g} = 0.09 \tag{19}$$

Experimental rate of charring completeness on fuel carbon:

$$\varsigma(C)_{T} = 0.63 \tag{20}$$

From here rate of completeness of forming a liquid organic product (resin, acetic acid, methyl alcohol) from the fuel carbon:

$$\varsigma(C)_{org} = 1 - \varsigma(C)^{g} - \varsigma(C)_{T} = 1 - 0.09 - 0.63 = 0.28 \tag{21}$$

Mass of gasified fuel carbon:

$$m(C)_{T}^{g} = \varsigma(C)^{g} \cdot m(C)_{T} = 0.09 \cdot 0.023 = 0.0021 kg \tag{22}$$

Mass of generating product of wood coal from fuel:

$$m(C)_{pyr} = \varsigma(C)_{T} - m(C) = 0.63 \cdot 0.023 = 0.0147 kg \tag{23}$$

Mass of fuel carbon spent on formation of a liquid organic product:

$$m(C)_{opz} = \varsigma(C)_{opz} \cdot m(C)_{T} = 0.28 \cdot 0.023 = 0.0065 kg \tag{24}$$

Mass of the free nitrogen gas generating from fuel (N_2):

$$m(N_2)^{g} = m(N)_{T} = 0 \tag{25}$$

According to the experimental studies, it was found that in the pyrolysis process at a temperature of +450-500°C, free gas hydrogen (H_2) is almost not formed (H_2).

Experimental rate of chemical water formation from fuel in terms of absolutely dry at pyrolysis:

$$\psi(H_2O)_{T} = 0.28 \tag{26}$$

Mass of generating "chemical" water from completely dry fuel:

$$m(H_2O)_{T}^{X} = \psi(H_2O)_{T} \cdot m(T) = 0.28 \cdot 0.046 = 0.0129 kg \tag{27}$$

Mass of elemental hydrogen of fuel that went into interaction with elemental oxygen of fuel to form "chemical" water:

$$m(H)_{T}^{X} = \frac{2m(H_2O)_{T}^{X} \cdot A(H)}{M(H_2O)} = \frac{2 \cdot 0.0129 \cdot 1}{18} = 0.0014 kg, \tag{28}$$

where

$A(H) = 1 kg / kmol$ – atomic mass of hydrogen,

$M(H_2O) = 18 kg / kmol$ – molar mass of water.

Mass of elemental oxygen of fuel interacting with elemental hydrogen of fuel to produce "chemical" water:

$$m(O)_{T}^{X} = m(H_2O) - m(H)_{T}^{X} = 0.0129 - 0.0014 = 0.0115 kg \tag{29}$$

Mass of hydrogen of fuel generating methane gas and other organic products:

$$m(H)_{org} = m(H)_T - m(H)_T^X = 0.0028 - 0.0014 = 0.0014kg \quad (30)$$

Mass of oxygen of fuel forming carbon dioxide, carbon monoxide and organic products:

$$m(O)_{pyr} = m(O)_T - m(O)_X^T = 0.0196 - 0.0115 = 0.0081kg \quad (31)$$

Mass of water generated in wood pyrolysis products:

$$m(H_2O)_{pyr} = m(W)_T + m(H_2O)_T^X = 0.0035 + 0.0129 = 0.0164kg \quad (32)$$

Experimental coefficient of mass ratios of generated carbon dioxide and monoxide in the pyrolysis of secondary wood resources:

$$k(\frac{CO_2}{CO}) = \frac{m(CO_2)^g}{m(CO)^g} = 3.0 \quad (33)$$

Assume that in the analysis of the resulting carbon dioxide CO2, by means of adsorptive absorption tubes by gravimetry (weighing the tubes before and after the analysis on an analytical balance), its mass is as follows:

$$m(CO_2)_T^g = 0.0043kg \quad (34)$$

Mass of generated carbon monoxide:

$$m(CO)_T^g = \frac{m(CO_2)^g}{3} = \frac{0.0043}{3} = 0.0014kg \quad (35)$$

Mass of carbon fuel spent on the formation of carbon dioxide:

$$m(C)_{CO_2} = \frac{m(CO_2)_T^g \cdot A(C)}{M(CO_2)} = \frac{0.014 \cdot 12}{44} = 0.0012kg \quad , \quad (36)$$

where $A(C) = 0.012$ – atomic mass of carbon,
$M(CO_2)$ – molar mass of carbon dioxide.
Mass of carbon fuel spent on the formation of carbon monoxide:

$$m(C)_{CO} = \frac{m(CO)_T^g \cdot A(C)}{M(CO)} = \frac{0.0014 \cdot 12}{28} = 0.0006kg \quad (37)$$

where $M(CO) = 28kg / кмоль$ – molar mass of carbon monoxide.
Mass of carbon fuel spent on the formation of methane gas:

$$m(C)_{CH_4} = m(C)_T^g - m(C)_{CO_2} - m(C)_{CO} = 0.0021 - 0.012 - 0.0006 = 0.0003kg \quad (38)$$

Mass of generated methane gas (*CH₄*):

$$m(CH_4) = \frac{m(C)_{CH_4} \cdot M(CH_4)}{A(C)} = \frac{0.0003 \cdot 16}{12} = 0.0004kg, \quad (39)$$

where $M\left(CH_4\right)=0.016kg/kmol$ – molar mass of methane.

Mass of generating combustible gas in terms of absolutely dry gas:

$$m\left(\Gamma\Gamma\right)=m\left(N_2\right)^g+m\left(CO_2\right)^g+m\left(CH_4\right)^g=0.027+0.044+0.014+0.025=0.0061kg \quad (40)$$

The volume of the generated gaseous phase in form of combustible gas (CG) under normal conditions in terms of absolutely dry gas:

$$V\left(GG\right)=V\left(N_2\right)_T^g+V\left(CO_2\right)_T^g+V\left(CH_4\right)_T^g=$$

$$22.4\cdot\frac{m\left(N_2\right)_T^g}{M\left(N_2\right)}+\frac{m\left(CO_2\right)_T^g}{M\left(CO_2\right)}+\frac{m\left(CO\right)_T^g}{M\left(CO\right)}+\frac{m\left(CH_4\right)_T^g}{M\left(CH_4\right)}=0.004m^3 \quad (41)$$

$$\rho\left(GG\right)=1.525\frac{kg}{m^3} \quad (42)$$

The number of kilomoles of generating vapour:

$$n=\frac{V}{22.41}=\frac{0.0040}{22.4}=0.0002kmol \quad (43)$$

Mole fraction of nitrogen gas in generated combustion gas:

$$\mu\left(N_2\right)=\frac{m\left(N_2\right)_T^g}{M\left(N_2\right)\cdot m\left(GG\right)}=\frac{0.00003}{0.0002}=0.005 \quad (44)$$

Mole fraction of methane gas in generated combustion gas:

$$M\left(CH_4\right)^g=\frac{m\left(CO\right)_T^g}{M\left(CO\right)\cdot n\left(GG\right)}=\frac{0,0004}{16\cdot 0.0002}=0.125 \quad (45)$$

Mole fraction of carbon monoxide in the generated combustion gas:

$$\mu(CO)^\Gamma=\frac{m(CO)_T^n}{M(CO)\cdot n(GG)}=\frac{0.0014}{28\cdot 0.0002}=0.250 \quad (46)$$

Mole fraction of carbon dioxide in the generated combustion gas:

$$\mu\left(CO_2\right)^g=1-\mu\left(N_2\right)^g-\mu\left(CH_4\right)^g-\mu\left(CO\right)^g=1-0.005-0.125-0.250=0.620 \quad (47)$$

The total mass of generated organic products as a liquid phase:

$$m\left(C_xH_yO_zNe\right)_{pyr}=m(resin)+m(Methanol)+m(Acetic\ acid)=$$

$$m\left(T\right)_z^w-m\left(H_2O\right)_{pyr}-m\left(CO_2\right)_T^g-m(CO)_T^g-m\left(N_2\right)_T^g-m\left(CH_4\right)-m\left(C\right)_{pyr}-m\left(Z\right)_T= \quad (48)$$

$$=0.050-0.0164-0.0043-0.0014-0-0.0004-0.0147-0.0005=0.0123kg$$

The percentage-mass yield of pyrolysis products from the fuel mass with humidity $w=0.07$ and ash-content $A^p=0.01$.

1. Charcoal: $$\frac{m(C)_{pyr}}{m(T)_Z^W}\cdot 100\% = \frac{0.0147\cdot 100\%}{0.05} = 29.4\% \qquad (49)$$

2. Combustible gas in terms of absolutely dry:

$$\frac{m(GG)^g\cdot 100\%}{m(T)_Z^W} = \frac{0.0061\cdot 100\%}{0.05} = 12.2\% \qquad (50)$$

3. The total yield of liquid phase of organic products:

$$\frac{m(C_xH_yO_zNe)\cdot 100\%}{m(T)_Z^W} = \frac{0.0123\cdot 100\%}{0.05} = 24{,}6\% \qquad (51)$$

4. Water: $$\frac{m(H_2O)_{pyr}}{m(T)_Z^W}\cdot 100\% = \frac{0.0164\cdot 1005}{0{,}05} = 32.8\% \qquad (52)$$

5. Ash: $$\frac{m(Z)_T}{m(T)_Z^W}\cdot 100\% = \frac{0.0005}{0.05}\cdot 100\% = 1\% \qquad (53)$$

Total: 100%

Table 1. The percentage-mass yield of pyrolysis products from the fuel mass.

№	*Name of product of wood pyrolysis*	*The percentage-mass yield, %*
1.	Charcoal	29.4
2.	Combustible gas	12.20
3.	The total yield of liquid phase of organic products	24.60
4.	Water	32.8
5.	Wood ash	1

3. Conclusion

Thus, authors have for the first time proved a possibility of this technology and its applicability for oil and gas branch and an and mass parameters are determined. Comparison of experiment results and calculation researches allows us to say that hydrocarbon wastes, disposed by high-temperature pyrolysis have huge resource potential. It allows justifying efficiency of the proposed technology from the standpoint of geoecology. High-temperature pyrolysis technology includes recycling (reuse of gaseous products of pyrolysis), environmental impact on atmospheric air (no harmful emissions), water resources (not being contaminated) and soil cover (polygon areas are being reduced). The effect was determined, that during thermal processing of hydrocarbon wastes, quantitative correlation dependence between the mass of recyclable wastes, generated pyrolysis gas volume and the time of recycling is observed. This dependence proves the great potential of high-temperature pyrolysis recycling method.

References

[1] Antipyev Y N, Nevolin A P and Zemenkov Yu D 1981 Operation of intermediate pumping stations on transport of gas saturated oils *NEFT. KHOZ,* **10** 46

[2] Zemenkova M Yu, Shalay V V and Zemenkov Yu D 2015 Expert systems of multivariable predictive control of oil and gas facilities reliability *Procedia Engineering* **113** 312-5

[3] Mamadaliev R A, Kuskov V N, Zemenkov Yu D and Popova A A 2015 Influence of high-concentrated heat sources on alloying elements transition into the weld metal *Applied Mechanics and Materials* **770** 19-22

IOP Conf. Series: Materials Science and Engineering **154** (2016) 012012 doi:10.1088/1757-899X/154/1/012012

[4] Ctivation and catalytic reactions of alkanes / Lane with English; under the editorship of K. Hill. M.: World, 1992.
[5] General toxicology. Under the editorship of Loit A. About. SPb.: ELBI-SPb., 2006.
[6] Polya D. Combinatory calculations for groups, counts and chemical compounds (Russian)//Enumerative tasks of the combinatory analysis. M.: World, 1979. P. 36-138.
[7] Pereushanu V. Production and use of hydrocarbons. M.: Chemistry, 1987.
[8] Fizer L., Fizer M. Organic chemistry. Advanced course. M, 1966. V.1. 680 pages.
[9] Haynes A. Methods of oxidation of organic compounds. Alkanes, alkenes, alkynes and arenas. M.: World, 1988.
[10] Chemical encyclopedia/ Hl. edition I. L. Knunyants. M.: Big Russian encyclopedia, 1992. V.3: Copper sulfides Polymeric dyes. 640 pages.

IOP Conf. Series: Materials Science and Engineering **154** (2016) 012013 doi:10.1088/1757-899X/154/1/012013

Prospects of Applying Vibration-Resistant Pressure Gauges in the Oil and Gas Industry

S P Pirogov, D A Cherentsov and B A Gulyaev

Industrial University of Tyumen, 38 Volodarskogo st., Tyumen, 625000, Russia

E-mail: cherentsovda@bk.ru

Abstract. The article presents justification for improving vibration protection of pressure gauges used in the oil and gas industry. A mathematical model of manometric tubular spring oscillations in a viscous medium is viewed. By the developed model, the authors have determined the impact of manometric spring geometric characteristics and damping fluid viscosity on oscillation attenuation parameters, as well as provided evaluation of the impact of the cross-sectional shape on the oscillation attenuation rate.

1. Introduction

The application and development of pressure gauge constructions, aimed at increasing vibration resistance when working in harsh conditions in the oil and gas industry, are governed by a number of regulatory documents:

- "Regulations on industrial safety of hazardous production facilities where pressurized equipment is used " of 25.03.2014;
- Federal Law "On industrial safety of hazardous production facilities" of 21.07.97 №116-FZ (ed. of 04.03.2013);

"Framework of the federal system of monitoring the critical facilities and (or) potentially hazardous infrastructure facilities of the Russian Federation and dangerous goods" approved by the RF government order of 27.08.2005 №1314-r.

Harsh operational conditions lead to increased measurement error, wear of operating device and, as a result, cause breakdown in production process. At present, vibration damping by a fluid is an effective method of improving protection against vibration. Hereafter, evaluation of the impact of various characteristics on improving vibration protection of vibration-resistant pressure gauges is given.

2. Development of a mathematical model and comparison with test values

A sensing element – manometric tubular spring (MTS) – is presented as a curved bar. The motion of an infinitesimal element (figure 1) is composed of the longitudinal u and cross w components. The motion resistance force in the fluid is the distributed load q. Forces from the internal pressure on upper and lower parts of the cross-section of the infinitesimal element, at pressure P, are presented as distributed load inside the tube.

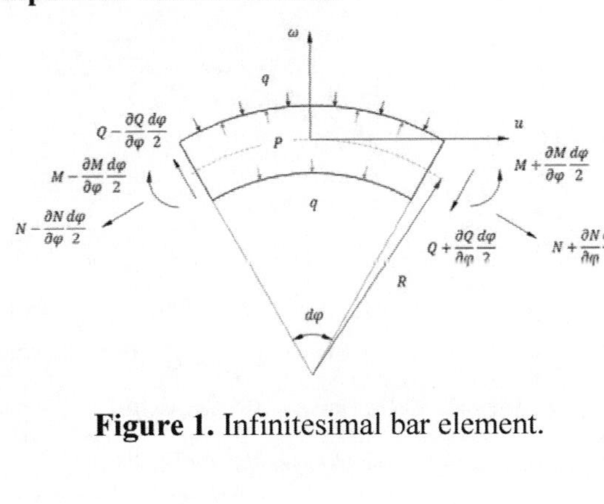

Figure 1. Infinitesimal bar element.

The system of differential equations of MTS motion is [1, 2]:

$$m(\varphi)\frac{\partial^2 w}{\partial t^2}+(1+\frac{b}{R})\beta\frac{\partial w}{\partial t}-\frac{\partial^2}{\partial\varphi^2}\{H_R(\frac{\partial u}{\partial\varphi}-\frac{\partial^2 w}{\partial\varphi^2})\}+\{D_R(\frac{\partial u}{\partial\varphi}+w)\}=\frac{4ab}{R}P,$$

$$m(\varphi)\frac{\partial^2 u}{\partial t^2}-\frac{\partial}{\partial\varphi}\{H_R(\frac{\partial u}{\partial\varphi}-\frac{\partial^2 w}{\partial\varphi^2})\}-\frac{\partial}{\partial\varphi}\{D_R(\frac{\partial u}{\partial\varphi}+w)\}=0. \quad (1)$$

where $m(\varphi)$– the mass of the single tube length ($m(\varphi)=\rho S(\varphi)$); β – the resistance coefficient of the damping fluid; a – the major semi-axis of the cross-section; b – the minor semi-axis of the cross-section; R – the radius of the central axis curvature; $\partial\varphi$ – the angle of the infinitesimal element cut out from the curved bar; $H_R=\frac{EJ(\varphi)K_k(\varphi)}{(1-\mu^2)R^4}$; $D_R=\frac{ES(\varphi)}{(1-\mu^2)R^2}$.

At the base of the manometric spring (at the section where the spring is rigidly fixed) - φ=0 longitudinal and cross motion, as well as the rotation angle of the spring cross-section equal zero. At the loose end φ=γ shear and tensile forces, as well as bending moment equal zero, therefore, we obtain boundary conditions:

Essential boundary conditions at φ=0:

$$U(0)=0;\ w(0)=0;\frac{\partial w}{\partial\varphi}(0)=0$$

Natural boundary conditions at φ=γ:

$$\left(\frac{\partial u}{\partial\varphi}(y)-\frac{\partial^2 w}{\partial\varphi^2}(y)\right)=0,\ \left(\frac{\partial u}{\partial\varphi}(y)+w(y)\right)=0\ ,\left(\frac{\partial^2 u}{\partial\varphi^2}(y)-\frac{\partial^2 w}{\partial\varphi^2}(y)\right)=0.$$

Correlation of the results produced by this model and the test results is given in figures 2-3. The comparison has been conducted on 10 steel samples with various geometric characteristics, glycerol being a damping fluid [3].

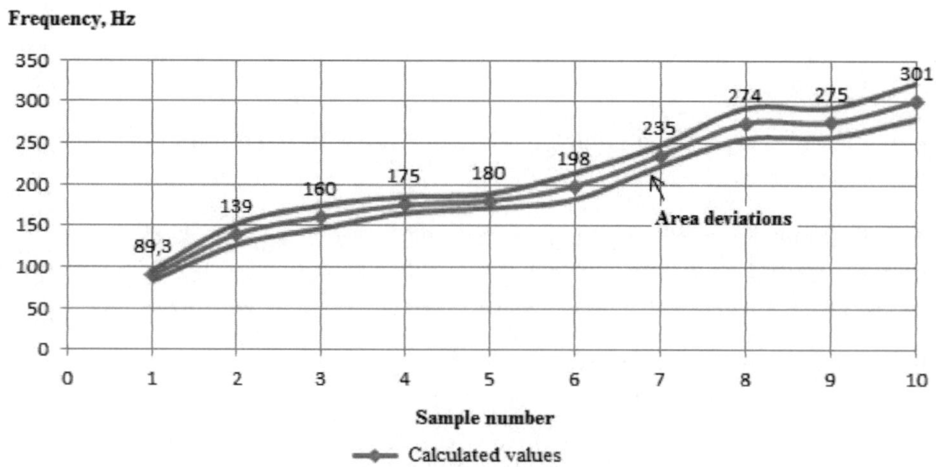

Figure 2. Comparison of calculated and test values of oscillation attenuation frequencies.

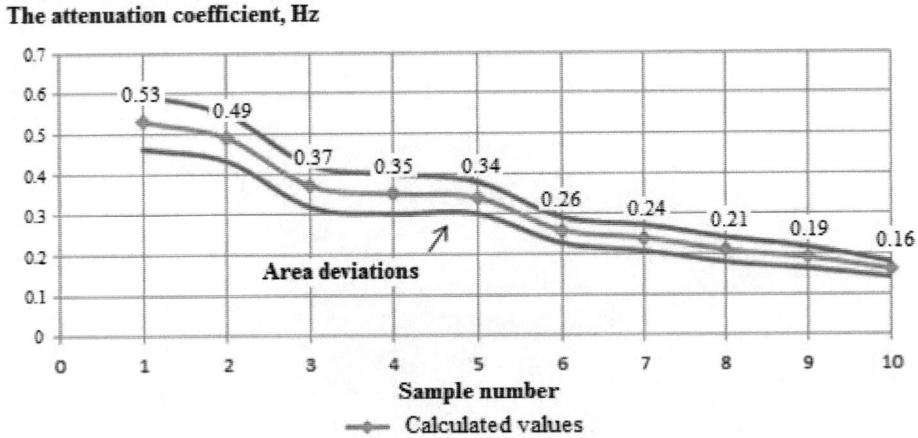

Figure 3. Comparison of calculated and test values of attenuation coefficient.

The comparison of theoretical and test values has demonstrated that deviations of frequencies do not exceed 10%, and those of the attenuation coefficient do not exceed 15%. This proves a reasonable accuracy of calculations, thus this model can be applied to calculate vibration-resistant pressure gauges.

3. Results and discussion

Using the developed model we have studied various vibration-resistant pressure gauges constructions [4], with the impact of MTS geometric characteristics and damping fluid viscosity on the attenuation coefficient shown in figures 4-7.

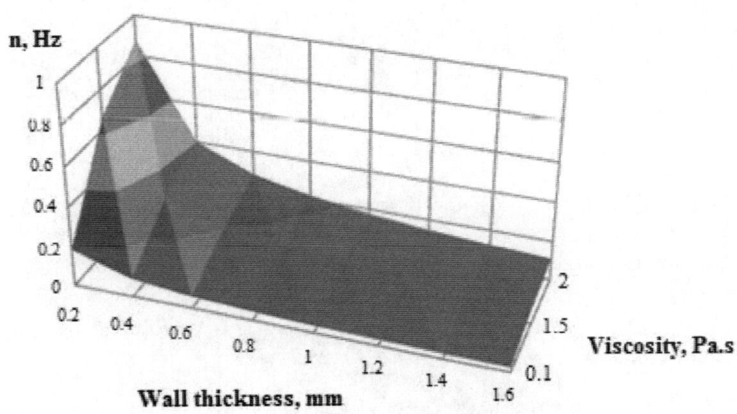

Figure 4. Impact of the MTS wall thickness and damping fluid viscosity on the attenuation coefficient.

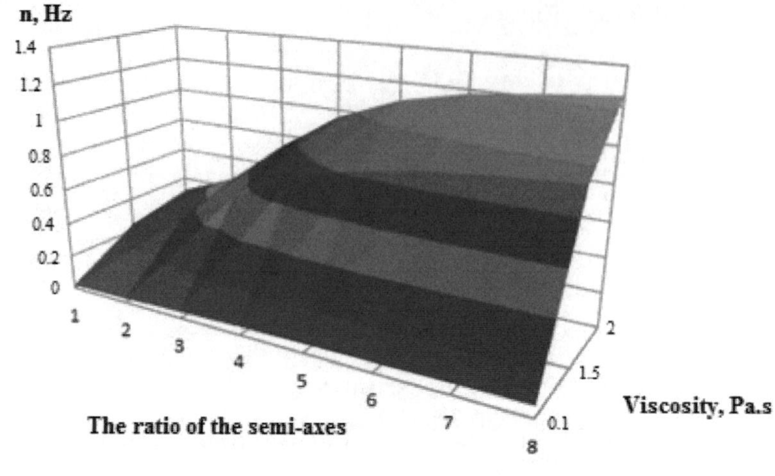

Figure 5. Impact of the ratio of the MTS semi-axes and damping fluid viscosity on the attenuation coefficient.

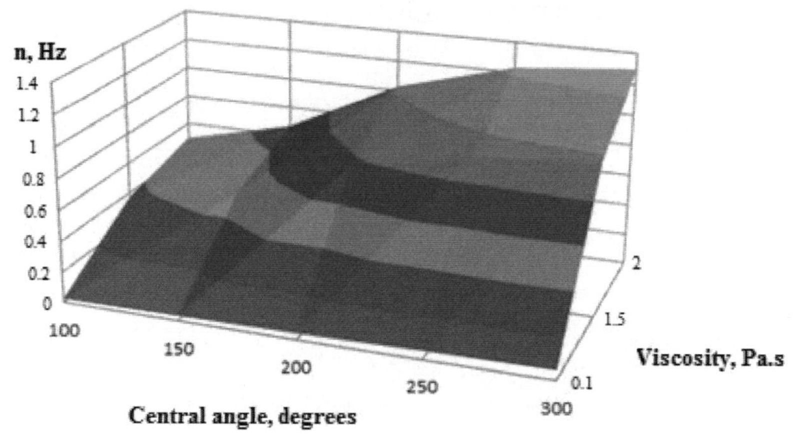

Figure 6. Impact of the MTS central angle and damping fluid viscosity on the attenuation coefficient.

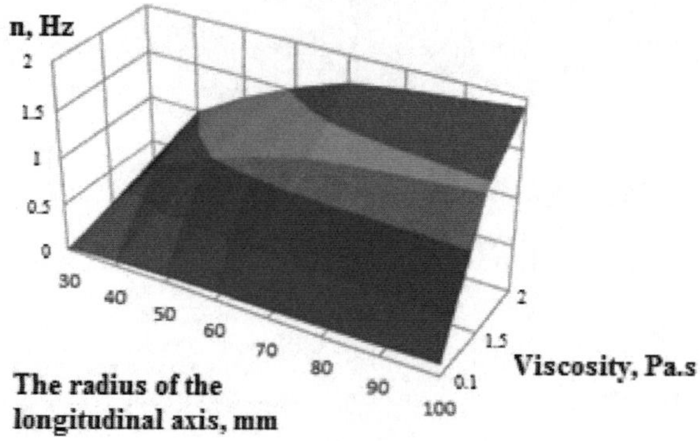

Figure 7. Impact of the radius of the MTS longitudinal axis and damping fluid viscosity on the attenuation coefficient.

The impact of MTS geometric characteristics and damping fluid viscosity on the oscillation attenuation frequency is shown in figures 8-11.

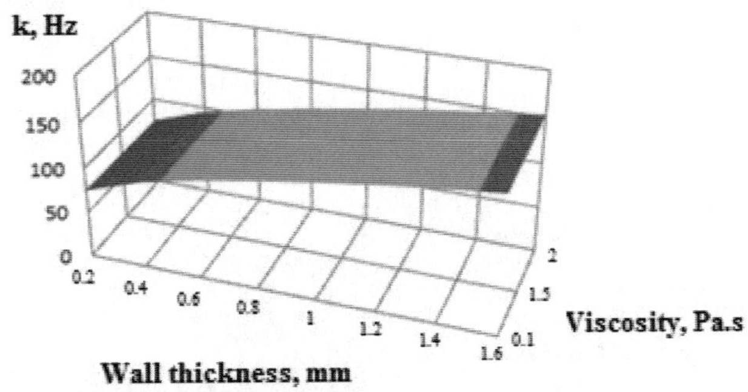

Figure 8. Impact of the MTS wall thickness and damping fluid viscosity on the oscillation attenuation frequency.

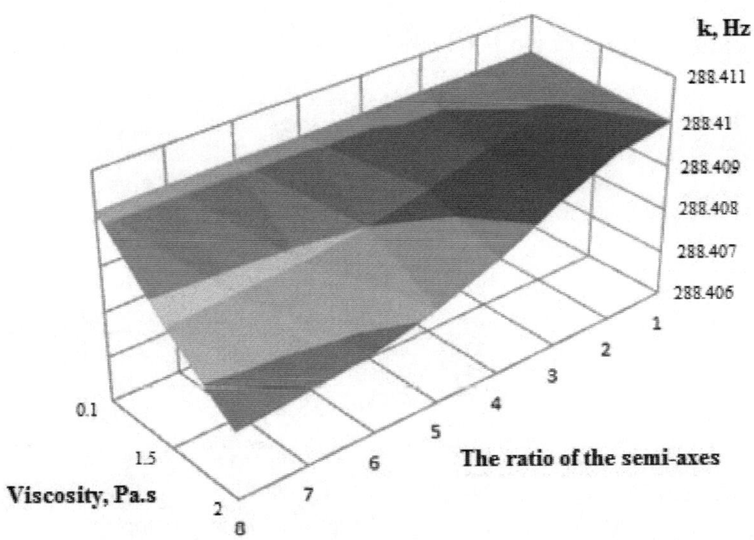

Figure 9. Impact of the ratio of the MTS semi-axes and damping fluid viscosity on the oscillation attenuation frequency.

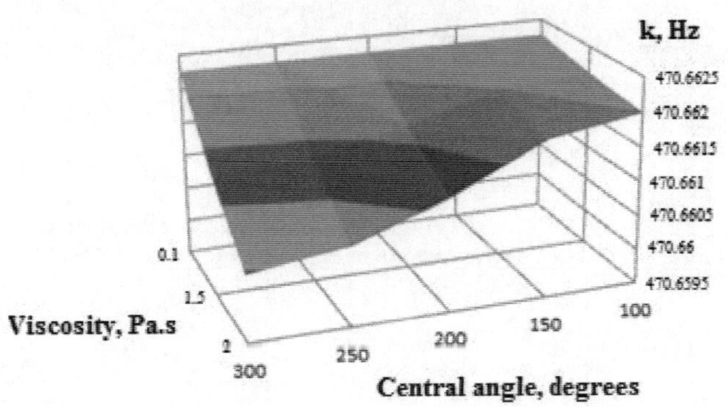

Figure 10. Impact of the MTS central angle and damping fluid viscosity on the oscillation attenuation frequency.

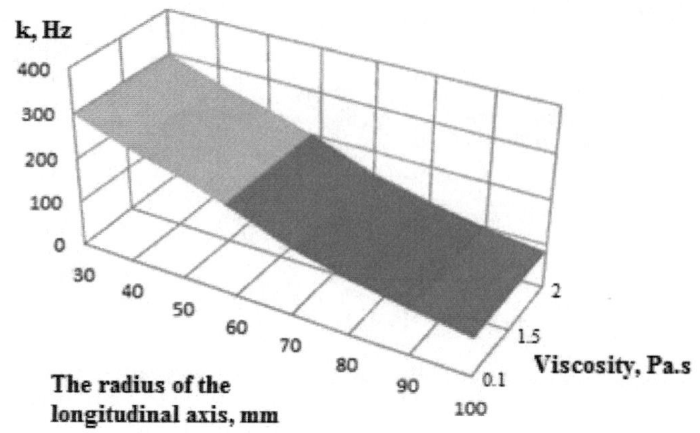

Figure 11. Impact of the radius of the MTS longitudinal axis and damping fluid viscosity on the oscillation attenuation frequency.

The study of the impact of the MTS sectional shape on the oscillation attenuation rate is shown in figure 12.

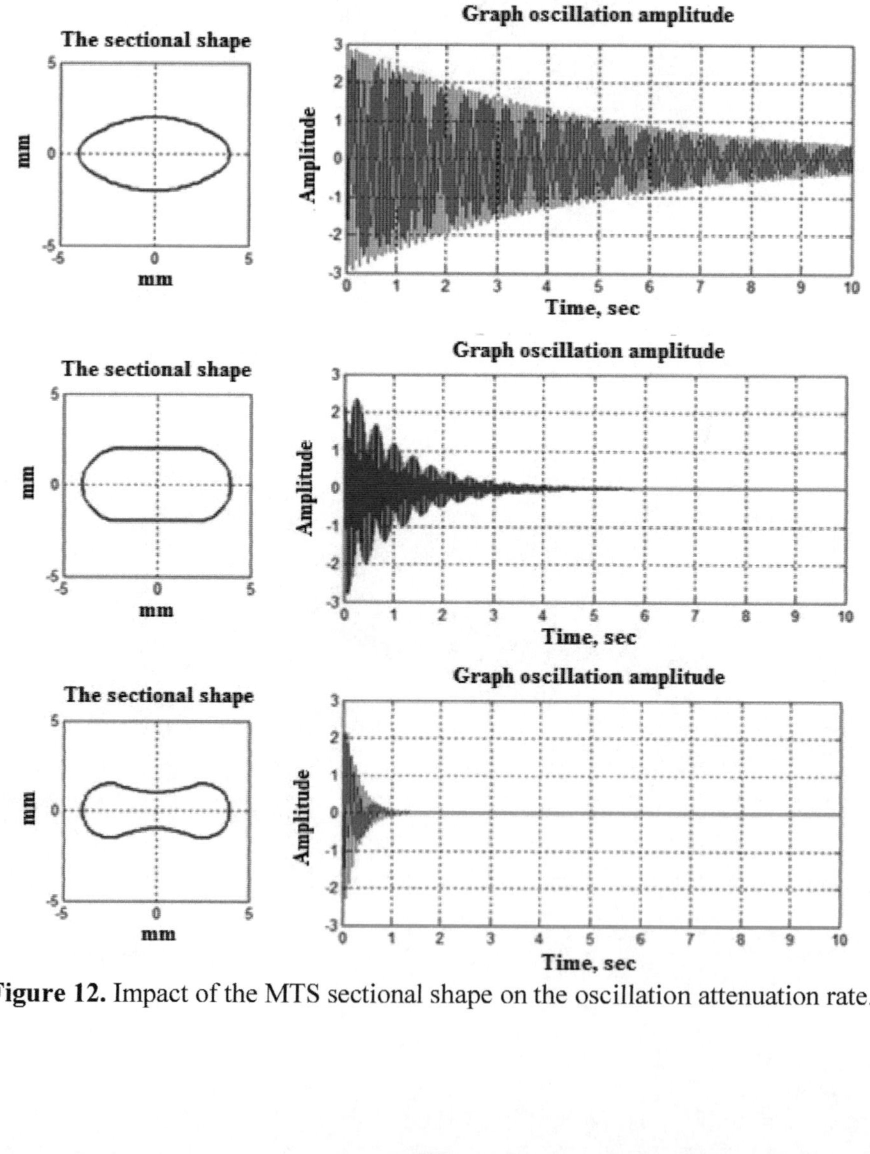

Figure 12. Impact of the MTS sectional shape on the oscillation attenuation rate.

4. Conclusions

For pressure gauges working within the resonance zone vibration damping by a fluid significantly reduces the time for oscillation attenuation. The study of the impact of the MTS geometric characteristics and damping fluid viscosity on the oscillation attenuation parameters has demonstrated that increasing the wall thickness leads to increasing the oscillation attenuation frequency and decreasing the attenuation coefficient, and increasing the ratio of the semi-axes, central angle и radius of the longitudinal axis leads to decreasing the oscillation attenuation frequencies and increasing the attenuation coefficient. The oscillation attenuation frequencies depend on the geometric characteristics of tubular springs. The attenuation coefficient is determined by dynamic viscosity of the damping fluid and, to a lesser extent, depends on the tubular spring geometric characteristics. A significant impact on the attenuation rate is exerted by the cross-sectional shape, tubes with the elliptical cross-section showing the smallest rate, and with the 8-shaped – the greatest.

References

[1] Cherentsov D A, Pirogov S P, Dorofeev S M and Chuba A Yu 2015 Calculating the oscillation attenuation parameters of manometric tubular springs *Tyumen State University Herald* **1-1(1)** 136-45

[2] Zemenkov Yu D, Shalay V V and Zemenkova M Yu 2015 Expert systems of multivariable predictive control of oil and gas facilities reliability *Procedia Engineering* **113** 312-5

[3] Kurushina V and Zemenkov Y 2014 Innovative cyclical development of the Russian pipeline system *WIT Transactions on Ecology and the Environment* **190-2** 881-8

[4] Zemenkov Yu D, Shalay V V and Zemenkova M Yu 2015 Immediate analyses and calculation of saturated steam pressure of gas condensates for transportation conditions *Procedia Engineering* **113** 254-8

IOP Conf. Series: Materials Science and Engineering **154** (2016) 012014 doi:10.1088/1757-899X/154/1/012014

Calculation the Margin of Safety and Durability According to Change Thickness the Pipeline

S Y Toropov[1], S M Chekardovskiy[1], A E Brand[1] and V M Stroenko[2]

[1] Industrial University of Tyumen, 38 Volodarskogo st., Tyumen, 625000, Russia
[2] PJSC SIBUR Holding, 16/1 Krzhizhanovskogo st., Moscow, 117997, Russia

E-mail: Allbert-@mail.ru

Abstract. In the article the analysis and to mathematically calculated the margin of safety and durability according to change thickness of pipeline. The proposed conditions on the application of the obtained results.

1. Introduction
The security level of the technological line section of the main pipeline (MP) directly depends on the following parameters: level and character of the intense deformed state, pipeline geometry, chemical and structural composition [1].

In the conditions of a large number of short-term changes of pressure in MP is observed increase level of intensity in a pipeline wall that it leads to reduction of an operational resource.

In the course of product transport at the stationary mode on MP with the internal pressure P_1 in the wall of MP appears circumferential stresses, which lead to changes of internal geometry of the pipeline.

At emergence of the non-stationary mode, which is characterized by increase of pressure from P_1 to P_2, appear ring tensions leading to reduction in reserve of strength and durability. By the achievement of considerable indicators of tension in a wall of MP appears a bend, which is followed by the displacement of the axis of the pipeline and the soil massif, loss of stability. As a result appears an emergency situation in the form of break (Figure 1) [2].

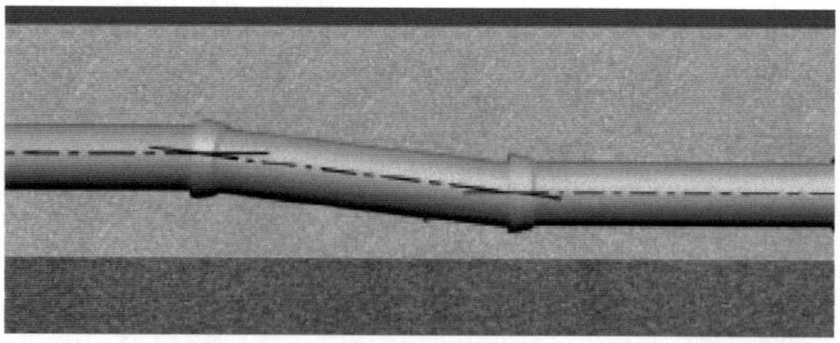

Figure 1. Appearance of bending moments.

The research objective is to mathematically calculate the margin of safety and durability according to change thickness the pipeline.

2. Research
We will apply the differential equation of a curved surface:

$$\frac{\partial^4 w}{\partial x^4}+\frac{\partial^4 w}{\partial x^2 \partial y^2}+\frac{\partial^4 w}{\partial x^4}=\frac{q}{D}, \tag{1}$$

where q – the loading distributed on an inner surface of a wall of the pipeline
w – the deflection of the pipeline wall
D –flexural rigidity

We will apply the membrane theory of thin-walled casings and we will write down the differential equation of a bend of a pipeline wall with a cylindrical casing (form) [3]:

$$\frac{\partial^4 w}{\partial x^4} + 4\beta w = \frac{q}{D}, \qquad (2)$$

where β – mechanical-geometrical coefficient

We will transform the differential equation of a bend of a wall and we will write down a final form of the common decision of a bend:

$$w = e^{\beta x}(C_1 \sin\beta x + C_2 \sin\beta x) + e^{-\beta x}(C_3 \sin\beta x + C_4 \sin\beta x) + f(x), \qquad (3)$$

a fast-rising function a fast damped function

In the conditions of the solving of a task on a bend of the pipeline long the L is important a fast damped function characterizing a small section on length of L with the operating maximum bend. We will consider appearance of the maximum deflection of a pipeline wall at the non-stationary mode at figures 2 and 3.

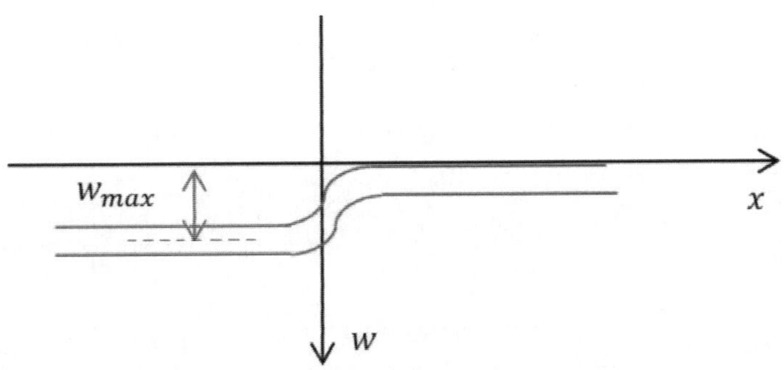

Figure 2. Appearance of the maximum deflection of a pipeline.

Taking into account a projection $w_{max} = \dfrac{P_2 \cdot R^2}{\delta \cdot E}$

$$w = e^{-\beta x}(C_3 \sin\beta x + C_4 \sin\beta x) + f(x), \qquad (4)$$

where $f(x) = \dfrac{\Delta P \cdot R^2}{\delta \cdot E}$

by C_3, C_4, $X = 0 \rightarrow \dfrac{\Delta P \cdot R^2}{\delta \cdot E}$ where $\Delta P = P_2 - P_1$

Calculated the bending moment of a MT:

$$M_x = -D\frac{d^2 w}{dx^2}, \qquad (5)$$

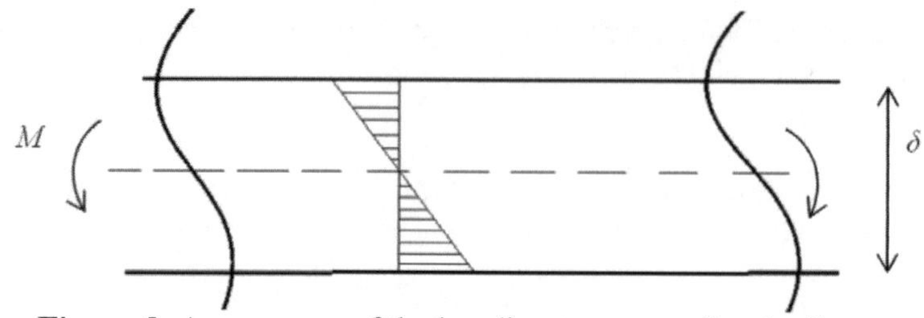

Figure 3. Appearance of the bending moment of a pipeline.

$$\sigma_{\max.} = \pm\frac{6(P_2 - P_1)\cdot R^2}{\delta^3 \cdot E}\cdot 2\left(\sqrt[4]{\frac{3(1-\nu^2)}{(R\delta)^2}}\right)^2 \cdot \frac{\delta^3 \cdot E}{12(1-\nu)^2} = \pm\frac{(P_2 - P_1)\cdot R}{\delta}\cdot\sqrt{\frac{3}{1-\nu^2}}, \qquad (6)$$

Taking into account action of internal pressure and a bend will applicable the fourth (power) theory of durability, which is often used in practice and coincides with experiences for plastic materials, for the purpose of determination of equivalent tension [4].

$$\sigma_{com.\max} = \sigma_{\max.}\cdot\sqrt{\left[1 + 4.9\cdot\frac{P_2 - P_1}{P_1}\cdot\left(1 + 1.73\cdot\frac{P_2 - P_1}{P_1}\right)\right]} = \sigma_{\max}\cdot K, \qquad (7)$$

For definition of durability of N_{dur} by changing of pressure from P_1 to P_2 we will use Manson's formula for low-cyclic destruction.

$$E_a = \frac{1}{2}\left(\ln\frac{1}{1-\psi}\right)^{0.6}\cdot n_g^{-0.6} + 0.75\cdot\frac{\sigma_b - \sigma_m}{E}\cdot n_g^{-0.12}, \qquad (8)$$

where ε_a - amplitude of variable deformations; Ψ - transverse contraction of the material; σ_b – limit of strength of metal; σ_m – the average tension of a cycle; N_{dur} – the number of cycles before destruction.

$$n_g = \left[\frac{3.5\cdot\left[n_b - 0.5\left(1 + \frac{1}{K}\right)\right]}{1 - \frac{1}{K}}\right]^{8.33}, \qquad (9)$$

Will hold the count key indicators by the initial conditions: P_1= 4 MPa, D = 700 mm, σ_b = 520 MPa, n_g = 30 000, Steel - 17G1C at tablet 1 and figures 3-5.

Table 1. Key indicators MT.

$\frac{P_2 - P_1}{P_1}$	δ	n_b	n_g
0.1	7	2.114	3.4E+06
0.1	8	2.423	2.2E+07
0.1	9	2.734	1.0E+08
0.1	10	3.046	3.9E+08
0.1	11	3.361	1.2E+09
0.1	12	3.677	3.3E+09

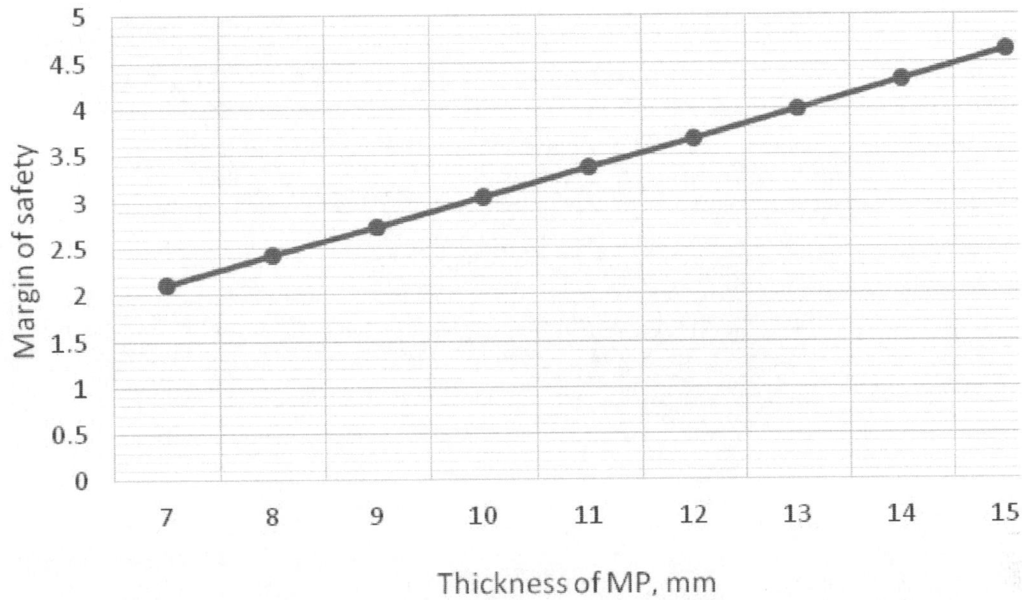

Figure 4. The dependence of the margin of safety from the thickness of MP.

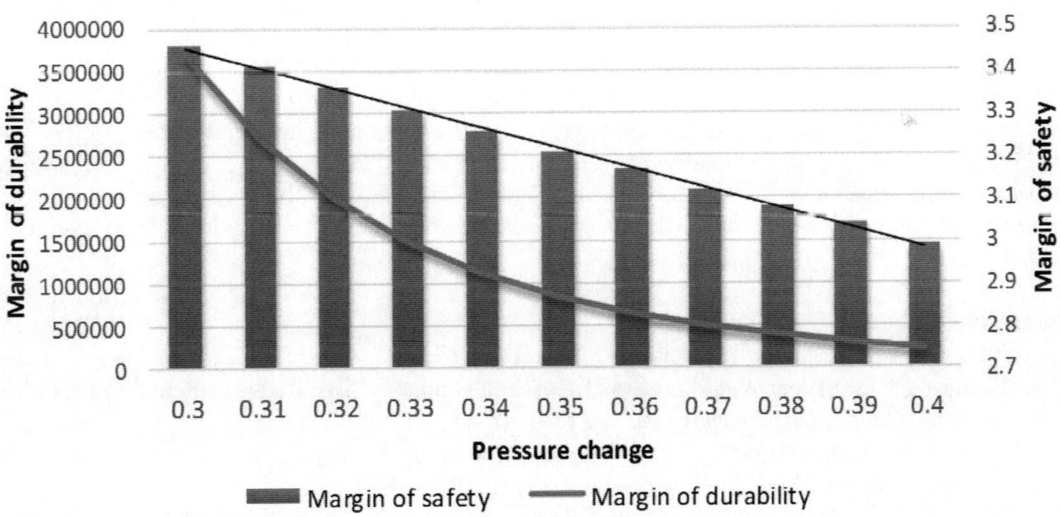

Figure 5. The dependence of the margin of safety and margin of durability from the pressure change.

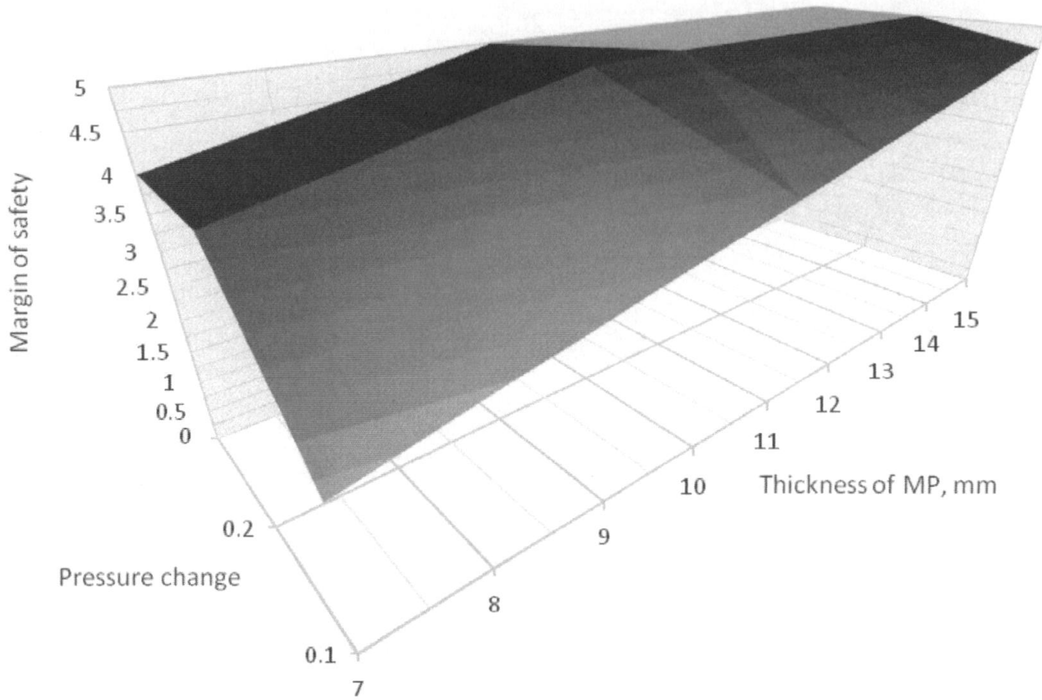

Figure 6. The dependence of the margin of safety from the thickness of MP and pressure change.

4. Conclusion

It is shown that increasing in thickness of the pipeline wall at 1 mm on average increases margin of safety from 6 to 18%, and the durability from 3 to 10 times, and these indicators decrease with increase P_2. Thus it is mathematically proved that increase of the thickness of MP wall increases durability and increases margin of safety of MP. Design institutes and the construction organizations for increase of safety at operation of MP, the transporting wide fractions of hydrocarbons, can use the obtained data.

References

[1] Pasquill F. Atmospheric Diffusion. 1974 *J. Wiley* **17** 429

[2] Lisanov M V, Savina A V, Samuseva E A and Sumskoy S I 2010 Accidents on offshore oil and gas facilities *Oil and Gas Journal Russia* **5 (39)** 48-53

[3] Sklavounos S and Rigas F 2006 Estimation of safety distances in the vicinity of fuel gas pipeline. *J. of Loss Prevention in the process industries*, **19** 24–31

[4] Lisanov M V, Grazhdankin A I, Pchelnikov V A, Savin A V and Sumy S I 2006 Risk analysis of accidents on oil pipeline systems BTS and MN Druzhba *J. Labour Safety in industry* **1** 34-40

[5] Zemenkova, M., Shalay, V., Zemenkov, Y., Kurushina, E. 2016 Improving the Efficiency of Administrative Decision-Making when Monitoring Reliability and Safety of Oil and Gas Equipment *MATEC Web of Conferences* **73 Volume** **73**, 11 August 2016, Article number 07001 Code 123284; DOI: 10.1051/matecconf/20167307001

[6] Zemenkova, M.Y.,Shipovalov, A.N.,Zemenkov, Y.D. 2016 Mathematic Modeling of Complex Hydraulic Machinery Systems When Evaluating Reliability Using Graph Theory *IOP Conference Series: Materials Science and Engineering* **127 (1)** DOI: 10.1088/1757-899X/127/1/012056

Study of Elastic Sensing Elements for Vibration-Resistant Pressure Gauges

S P Pirogov[1], **A Yu Chuba**[2] **and D A Cherentsov**[1]

[1] Industrial University of Tyumen, 38 Volodarskogo st., Tyumen, 625000, Russia
[2]State Agricultural University of Northern Zauralye, 7 Respubliky st., Tyumen, 625001, Russia

E-mail: cherentsovda@bk.ru

Abstract. The paper provides substantiation of the necessity to determine the natural oscillations frequencies of manometric tubular springs. A mathematical model of manometric tubular springs is overviewed. Based on the given model a computer program was developed that was incorporated into a software suit for manometric tubular spring calculations. The results of numerical experiments in the software suit allowed us to identify the impact of geometric parameters on the natural oscillation frequency.

1. Introduction

Pressure gauges often operate under vibration and tubular springs, primary elastic sensing elements, oscillate affecting the accuracy of gauges. Therefore, when designing manometric springs much attention should be given to dynamic calculations, resistance calculations in particular. Natural oscillation frequency is an important characteristic of vibration resistance, so key factors influencing this parameter must be identified. When designing manometric springs it is necessary to select such geometric parameters of tubes that will eventually ensure the required spring properties.

2. Manometric tubular spring designs

Manometric springs utilize the ability of a hollow thin-walled tube with a noncircular section to undergo deformation under pressure. A spring is a circular-bent tube with a noncircular cross-section that has one or two mutually perpendicular axes and is placed in such a way that one of its axes is a continuation of the curvature radius of the locus of the cross-sections' centers of gravity (figure1).

We will call this locus the central axis of the tube. One end of the spring is fixed in a holder with an opening through which pressure is transmitted into the inner cavity of the spring. The other end, called free, is tightly closed with a tip and is connected to the mechanism of the pressure gauge. Under the action of pressure the spring rebounds, and its free end performs a travel λ. The spring geometric parameters defining its dimensions are the curvature radius of the longitudinal axis – R, the central aperture angle – γ, the wall thickness – h, the major semi-axis – a and the minor semi-axis of the cross-section – b.

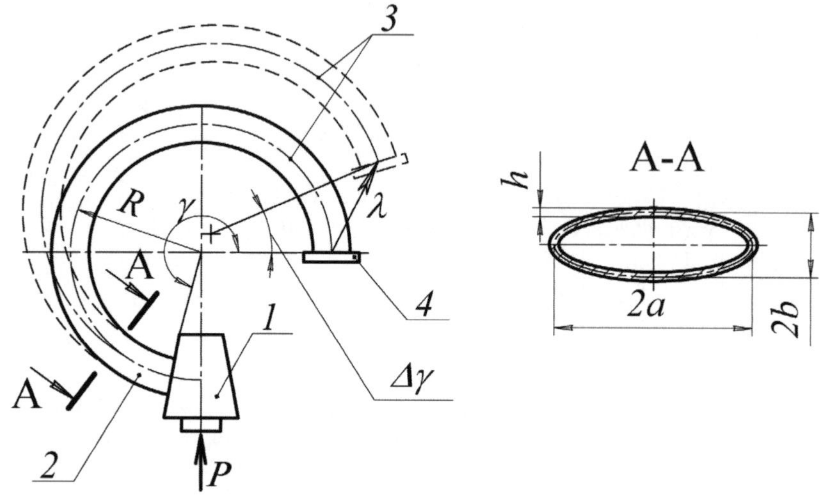

Figure 1. A manometric tubular spring.
1 – holder; 2 – tube; 3 – central axis; 4 – tip

There are known spring designs with a variable cross-section along the longitudinal axis. Both the size and shape of the cross-section can be variable. Springs with varying geometric section parameters have better performance compared to springs with a constant cross-section. Despite their advantages, springs with a variable cross-section are not widely available. The main reason for holding back the introduction of spring designs with variable cross-section has been the lack of methods for determining the dynamic characteristics, including the frequency of natural oscillations. Therefore, an urgent task to determine the frequency of natural oscillations of manometric tubular springs was set and solved.

To improve the manometric springs performance a number of spring designs were proposed with a variable cross-section along the central axis and wall thickness. According to the patent [1], in order to increase the natural oscillation frequency a manometric spring was proposed with a variable section wall thickness decreasing from the base to its free end, with the cross-section dimensions reducing from the base to the free end as well (figure 2).

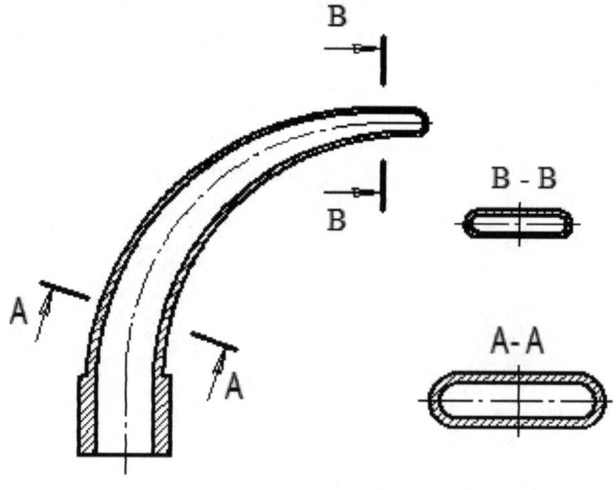

Figure 2. A conical manometric tubular spring.

To increase the sensitivity, relative stiffness and strength of manometric springs operating in a force mode and under vibrations, a spring is proposed with a variable length section [2] gradually changing from elliptical at the free end to 8-shaped at the fixed one. A similar design with longitudinal ridges (corrugations) is proposed [3].

A manometric spring design is also proposed characterized by enhanced durability and reliable performance [4]. It features an asymmetrical cross-section with respect to its major axis which is formed smoothly varying along the longitudinal spring axis, при этом the outer contour of the cross section is formed smoothly varying from flat oval at the free end to 8-shaped at the fixed one, and the inner contour of the cross section is formed smoothly varying from elliptical at the free end of the spring to flat oval at the fixed end. To simplify the variable section designs and extend the functionality a spring is proposed [5] that has a section variable along the length and consisting of several interconnected tubes each of which has a constant wall thickness and semi-axes cross-section ratio. The wall thickness and size of the tube section semi-axes increase from the free end of the spring to the fixed one.

A composite spring design is also known made with a section variable along the length, characterized in that it consists of interconnected tubes each of which has a constant sectional shape along the length and wall thickness. As they approach the fixed end of the spring, the tubes sections that constitute the spring change from elliptical to flat oval and then from flat oval to 8-shaped, with the wall thickness of the tubes also increasing from the free end to the fixed one.

Also a design is patented with composite spring inserts made with a cross-section variable along the length, characterized in that it consists of several interconnected tubes each of which has a constant wall thickness and semi-axes cross-section ratio. Interconnection of the tubes is done by elastic inserts. The size and wall thickness of the tube section semi-axes increase from the free end of the spring to the fixed one.

3. Developing a mathematical model

Let us present a dynamic model of the Bourdon tube in the form of a thin-walled curved bar oscillating in the central axis curvature plane. Figure 3 shows a curved rod and an infinitesimal element cut out of it. Travel of the center of gravity of the cross-section with the coordinate φ can be decomposed into radial w and circumferential u components.

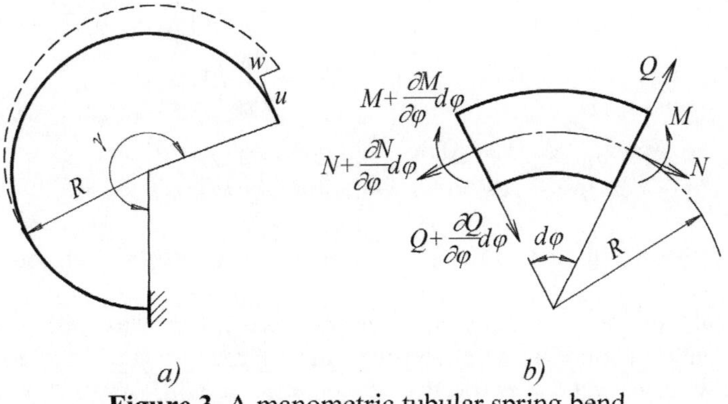

Figure 3. A manometric tubular spring bend.
a) a curved rod; *b)* a rod element

According to the principle of D'Alembert let us build the equations of motion of a tube element $Rd\varphi$, projecting forces applied to the element (considering the inertia force) on the normal and tangent.

$$-\frac{N}{R}-\frac{1}{R}\frac{\partial^2}{\partial\varphi^2}\left(\frac{M}{R}\right)=m_i(\varphi)\cdot\frac{\partial^2 w}{\partial t^2}$$
$$\frac{1}{R}\frac{\partial N}{\partial\varphi}-\frac{1}{R^2}\frac{\partial M}{\partial\varphi}=m_i(\varphi)\cdot\frac{\partial^2 u}{\partial t^2}, \quad (1)$$

where N – axial force; $N = D\varepsilon$,

D – tensile sectional stiffness; $D=\frac{ES(\varphi)}{1-\mu^2}$;

ε – elongation of the longitudinal axis of $\varepsilon=\frac{\partial u}{R\partial\varphi}+\frac{w}{R}=\frac{1}{R}\left(\frac{\partial u}{\partial\varphi}+w\right)$;

$S(\varphi)$ – cross-sectional area of the tube, depending on the angular coordinate φ of this section;

$\frac{\partial M}{\partial\varphi}=Q$ – shear force;

$m_i(\varphi)$ – mass per unit length of the tube (cross-sectional mass with the coordinate φ);

M – bending moment in the cross section of the tube $M_{è}=-B\chi=-\frac{B}{R^2}\left(\frac{\partial u}{\partial\varphi}-\frac{\partial^2 w}{\partial\varphi^2}\right)$

B – tube bending stiffness; for the cross section of the manometric tube $B=\frac{E\cdot J(\varphi)\cdot K_k(\varphi)}{(1-\mu^2)}$,

E – elasticity modulus of the pipe material;

$J(\varphi)$ – moment of inertia depending on the section angular coordinate φ;

$K_k(\varphi)$ – Karman constant, depending on the section angular coordinate φ;

μ – Poisson's ratio.

Boundary conditions. In the section of hard spring fixing ($\varphi=0$) tangent, normal movements and rotation angle of the tube cross-section are equal to zero, and at the opposite end ($\varphi=\gamma$) bending moment, shear, tensile forces become zero, which leads to the following boundary conditions:

$$\text{At } \varphi=0: u(0)=0;\ w(0)=0;\ \frac{\partial w}{\partial\varphi}(0)=0,$$
$$\text{At } \varphi=\gamma: M(\gamma)=0;\ Q(\gamma)=0;\ N(\gamma)=0. \quad (2)$$

To solve the system of equations a Bubnov - Galerkin method was used.

4. A computer program for calculating the natural oscillation frequencies of manometric tubular springs

In order to make calculations automatic a program "PKRMTP" was compiled in the MATLAB environment that includes a subroutine to determine the Karman constant and allows one to calculate several natural oscillation frequencies of the manometric tubular spring. With this program, we calculated the natural oscillation frequencies. The calculation results allow us to estimate the influence of the geometric parameters of tubular springs on their natural oscillation frequencies.

5. Conclusions

1. It has been established that increasing the wall thickness and ratios of the lateral section curvature radius to the minor semi-axis increases the natural oscillations frequency, and increasing the central axis curvature radius, central angle and ratios of the minor semi-axis to the major one decreases the frequency.

2. It has been established that manometric springs with a variable section changing from 8-shaped to flat oval, as well as springs composed of conical tubes, have natural oscillation frequencies 20-40% higher than those of springs with a constant section, therefore, it is reasonable to use them in vibration-resistant pressure gauges.

References

[1] Patent 325522 USSR, Cl. G 01 L 7/04. *Manometric tubular spring* / SA Sibeykin, VA Zhukov, KG Shein. - №1237570/18 – 10; Filed 26.04.68; Published 07.01.72; Bulletin №3.
[2] Patent 2093805 RF, MPK 7, G 01 L 7/04. *Manometric tubular spring* / SP Pirogov, TG Ponomareva. – № 96105858/28; Filed 26.03.96; Published 20.10.97.
[3] Patent 2216001 RF, MPK 7, G 01 L 7/04. *Manometric tubular spring (variants)* / SP Pirogov, NP Mityagin, NN Ustinov, SS Samakalev. – № 2001122276/28; Filed 08.08.2001; Published 10.11.2003; Bulletin №31.
[4] Patent 2215274 RF, MPK 7, G 01 L 7/04. *Manometric tubular spring* / SS Samakalev, SP Pirogov. – № 2002127092/28; Filed 10.10.2002; Published 27.10.2003; Bulletin №30.
[5] Patent 2111465 RF, MPK 7, G 01 L 7/04. Manometric tubular spring / SP Pirogov, TG Ponomareva, AA Volzhakov. – № 96121675/28; Filed 10.11.96; Published 20.05.98.
[6] Zemenkov Yu D, Shalay V V and Zemenkova M Yu 2015 Expert systems of multivariable predictive control of oil and gas facilities reliability *Procedia Engineering* **113** 312-5.
[7] Kurushina V and Zemenkov Y 2014 Innovative cyclical development of the Russian pipeline system *WIT Transactions on Ecology and the Environment* **190-2** 881-8
[8] Zemenkov Yu D, Shalay V V and Zemenkova M Yu 2015 Immediate analyses and calculation of saturated steam pressure of gas condensates for transportation conditions *Procedia Engineering* **113** 254-8.

IOP Conf. Series: Materials Science and Engineering **154** (2016) 012016 doi:10.1088/1757-899X/154/1/012016

Development of Nature Protection Technologies of Hydrocarbon Wastes Disposal on the Basis of High-Temperature Pyrolysis

V D Shantarin, M Yu Zemenkova[1] and Yu D Zemenkov

Industrial University of Tyumen, 38 Volodarskogo st., Tyumen, 625000, Russia

[1]E-mail: muzemenkova@mail.ru

Abstract. The research shows the thermal balance of low-temperature pyrolysis of birch sawdust with the possibility of further development of nature protection technology of hydrocarbon wastes disposal with secondary useful products production. The actual problem was solved by preventing environmental pollution by greenhouse gases using pyrolysis process as a method of disposal of hydrocarbon wastes with secondary useful products production. The objective of paper is to study features of the processes of thermal processing of wastes and development of environmentally sound technology of disposal C-containing wastes, contributing to the implementation of the pollution prevention concept.

1. Introduction

For the preservation and development of civilization it is necessary to move to the noosphere - the sphere of mind (V I Vernadsky). The future of the planet hangs by a thread and, to prevent this, humankind has decided to establish firm control over emissions into the atmosphere of six greenhouse gases: carbon dioxide, methane, nitrous oxide, hydrofluorocarbons, perfluorocarbons, sulfur hexafluoride (the Kyoto Protocol, 1997).

Environmental problems have excited the world community long time ago. In 1992, the United Nations Framework Convention on Climate Change was adopted by 185 states in order to unite efforts to prevent dangerous climate change and contribute to the stabilization of greenhouse gases i.e. provide a low-carbon development path.

On January 24 in Davos, UN Secretary General Ban Ki-moon spoke at the plenary session of the World Economic Forum, dedicated to the fight against climate change and promoting sustainable development and economic growth (Published on 01/24/2015). He stated that 2015 should become the year of global actions. Ban Ki-moon appealed of achievement a low-carbon economic development in order to counter the threats of global warming. Unfortunately, this is a long way of development and at the moment, development of civilization is impossible without carbon.

In 1972 in Stockholm (Sweden), the UN environmental conference, relating to the relationship between economic development and degradation of the environment, was held for the first time. After the Conference, the Government established the United Nations Environment Program (UNEP), which remains the leading institution in the world on environmental issues.

On 3-14 June, 1992 in Rio de Janeiro (Brazil, Rio-20) was held the UN Conference on Environment and Development. At the meeting, a declaration was adopted, which stated that "for achieving sustainable development, environmental protection shall constitute an integral part of the development process and it cannot be considered in isolation from it." The declaration includes 27 principles that define the rights and obligations of countries to ensure the development and welfare of the people. At a conference in Rio de Janeiro, «Agenda 21» was adopted - a program of how to make development sustainable from a social, economic and environmental points of view.

Recently, the 21st UN Conference on Climate Change was held. It ended in December 2015 in Paris with the adoption of a new agreement. In the negotiations participated 195 countries and have set emission reduction commitments, taking into account all the challenges and opportunities. In the negotiations, 195 countries participated that set emission reduction commitments, taking into account all the challenges and opportunities.

Development of nature protection technology of hydrocarbon wastes disposal with producing secondary useful products is an urgent task of preventing environmental pollution, using the pyrolysis process as a method of disposal hydrocarbon wastes with producing secondary useful products, contributing to the implementation of the concept of pollution prevention.

2. Models and methods

The amount of heat, generated by the complete combustion of the entire mass of generated organic product in the liquid phase [1-3] (resin, acetic acid, methanol, etc.).

$$Q\left(T_z^w\right)=Q\left(C\right)_{pyr}+Q\left(C_xH_yO_zNe\right)_{pyr}-Q\left(H_2O\right)+Q\left(GG\right)+Q_{p,} \quad (1)$$

where $Q\left(C\right)_{pyr}$ is the amount of heat released by full combustion of the entire mass of the resulting charcoal, kJ;

$Q\left(C_xH_yO_zNe\right)_{pyr}$ is amount of heat released by complete combustion of the entire mass of generated organic product as a liquid phase (resin, acetic acid, methanol, etc.), kJ;

$Q(H_2O)$ - the amount of heat absorbed by evaporation of moisture from the fuel and resulting "chemical" water, kJ;

$Q\left(GG\right)$ - the amount of heat released by full combustion of the entire mass of the resulting combustible gas, kJ;

Q_P - the amount of heat generated in the pyrolysis of the entire mass of wood fuel, kJ.

The calorific value of wood fuel, calculated using the formula of Mendeleyev:

$$Q((T)_W^Z)_H^P=4.19(8100\phi(C)_T+24600\phi(H)_T-600\cdot W^P)=18647.55kJ/kg. \quad (2)$$

The lowest working calorific value of generated combustible gas:

$$Q\left(GG\right)_H^P=4.19(2580\cdot\mu\left(H_2\right)^G\oplus3045\mu(CO)^G+8517\mu(CH_4)^G\ \ )=7650.5kJ/kg. \quad (3)$$

Let us assume the lowest calorific value of liquid phase to be:

$$Q\left(C_xH_yO_zNe,m\right)_H^P=32500kJ/kg. \quad (4)$$

Assume the calorific value of the charcoal as following:

$$Q\left(C\right)_p^H=34000kJ/kg \quad (5)$$

The amount of heat generated during the pyrolysis of wood:

$$\begin{aligned}Q_p&=Q\left(T\right)_W^Z-Q(C)_{np}-Q(C_xH_yO_zNe,m)-Q(GG)^T=m(T_W^Z)_n^P-m(C)_{np}Q(C)_H^P-\\&-m(C_xH_yO_zNe,m)Q(C_xH_yO_zNe)_H^P-V(GG)^GQ(GG)_H^G=0.050\cdot18647.55-0.0147\cdot\\&\cdot34000-0.0123\cdot32500-0.004\cdot7650.5=2.2kJ,\end{aligned} \quad (6)$$

where $Q(T_t^w)$ - the amount of heat released by the complete combustion of the entire mass of fuel wood, taken for pyrolysis, kJ.

According to the developed model, let us consider how is the calorific value of generated combustible gas changing, depending on the wood pyrolysis temperature. It is known that in the processes of gasification of wood fuel in the gas generator with the inverse process of combustion, wood resin and organic products are not formed, since the temperature 700-750°C, because they are almost entirely gasified to generator gas. Then, the ratio of fullness of fuel carbon gasification will be equal to:

$$\varsigma^*(C)^G = 1-\varsigma(C)_{pyr} = 1-0.63 = 0.37. \tag{7}$$

Mass of gasified carbon fuel is:

$$m^*(C)_T^G = \varsigma^*(C)^G \cdot m(C)_T = 0.37 \cdot 0.0233 = 0.0086kg. \tag{8}$$

Mass of generated carbon dioxide in combustion gas is equal to:

$$m^*(CO_2)_T^G = \frac{m(CO_2)_T^G \cdot \varsigma^*(C)^G}{\varsigma(C)^G} = \frac{0.0043 \cdot 0.37}{0.09} = 0.0177kJ. \tag{9}$$

Mass of generated carbon monoxide in the combustion gas is equal to:

$$m^*(CO)_T^G = \frac{m^*(CO_2)_T^G}{3} = \frac{0.0177}{3} = 0.0159kg. \tag{10}$$

Mass of carbon fuel spent on generation of carbon dioxide is equal to:

$$m^*(C)_{CO_2} = \frac{m^*(CO_2)_T^G \cdot A(C)}{M(CO_2)} = \frac{0.0177 \cdot 12}{44} = 0.0048kg. \tag{11}$$

Mass of carbon fuel spent on the generation of carbon monoxide:

$$m^*(C)_{CO} = \frac{m^*(CO)_T^G \cdot A(C)}{m(CO)} = \frac{0.0059 \cdot 12}{28} = 0.0025kJ. \tag{12}$$

Mass of carbon fuel spent on methane gas generation:

$$m^*(CH_4)_T^G = \frac{m^*(C)_{CH_4} \cdot M(CH_4)}{A(C)} = \frac{0.0013 \cdot 16}{12} = 0.0017kJ. \tag{13}$$

The amount of kilomoles of generated nitrogen gas (N2) as a part of the fuel gas:

$$n^*(N_2)_T^G \frac{m(N_2)_T}{M(N_2)} = \frac{0.00003}{28} = 10^{-6}kmole \tag{14}$$

The number of kilomoles of generated carbon dioxide as a part of combustible gas:

$$n(CO_2)_T^G = \frac{m^*(CO_2)_T^G}{M(CO_2)}, = \frac{0.0177}{44} = 0.0004kmole \tag{15}$$

The number of kilomoles of generated carbon monoxide as a part of combustible gas:

$$n^*(CO)_T^G = \frac{m^*(CO)_T^G}{M(CO)} = \frac{0.0059}{28} = 0.0002kmole \tag{16}$$

The number of kilomoles of generated methane gas as a part of combustible gas:

$$n^*(CH_4)_T^G = \frac{m(CH_4)_T^G}{M(CH_4)} = \frac{0.0017}{16} = 0.0001kmole \tag{17}$$

The number of kilomoles of generated combustible gas in terms of the absolutely dry one:

$$\begin{aligned} n^*(GT)^\Gamma &= n^*(N_2)_T^G + n^*(CO_2)_T^G + n^*(CO)_T^G + n^*(CH_4)_T^G = \\ &= 10^{-6} + 0.00004 + 0.0002 + 0.0001 = 0.0007kmole \end{aligned} \tag{18}$$

Mass of generating combustible gas:

$$m^*(GG)^G = m^*(CO_2)_T^G + +m^*(CO)_T^G + m^*(N_2)_T^G + m^*(CH_4)_T^G = 0.253kJ \tag{19}$$

The amount of generated gas at the initial conditions:

$$V^*(GG)^G = 22.4n^*(GG)^G 22.4 \cdot 0.0157m^3 \tag{20}$$

Density of absolutely dry combustible gas:

$$\rho^*(GG)^{\Gamma} = \frac{m^*(GG)^G}{V^*(GG)^G} = \frac{0,0253}{0,0157} = 1.6115 kg/m^3 \quad (21)$$

The mole fraction of gases CO2, CO, CH4, N2 in the combustion gas:

$$\mu^*(CO_2)^G = \frac{n^*(CO_2)_T^G}{n^*(GG)^G} = \frac{0.0004}{0.0007} = 0.571; \quad (22)$$

$$\mu^*(CO)^G = \frac{n^*(CO)_T^G}{n^*(GG)^G} = \frac{0.0002}{0.0007} = 0.286; \quad (23)$$

$$\mu^*(CH_4)^G = \frac{n^*(CH_4)_T^G}{n^*(GG)^G} = \frac{0.0001}{0.0007} = 0.143; \quad (24)$$

$$\mu^*(N_2) = \frac{n^*(N_2)^G}{n^*(GG)^G} = \frac{10^{-6}}{0.0007} = 0.001. \quad (25)$$

The lowest working calorific value of the gas:

$$Q^*(GG)_H^P = 4.19(2580 \cdot \mu(H_2)^G + 3045 \cdot \mu(CO)^G + 8517 \cdot \mu(CH_4)^G = 8752.1 kJ/m^3. \quad (26)$$

Specific lowest working calorific value of gas:

$$Q_{y\partial}^*(GG)_H^P = \frac{Q^*(GG)_H^P}{p^*(GG)^G} = \frac{8752.1}{1.6115} = 5431.0 kJ/kg. \quad (27)$$

Let us consider the pyrolysis process at a temperature of + 1000 ° C, at which the reaction of water gas begins to flow, i.e. the interaction of carbon fuel with water vapor:

$$C_{(solid)} + H_2O_{(vap)} \rightarrow CO_{(T)} + H_{2(G)}. \quad (28)$$

Then the mass of carbon fuel spent on the reaction of water gas, is equal to:

$$m'(C)^{WG} = \frac{m(H_2O)_{pyr} \cdot A(C)}{M(H_2O)} = \frac{0.0164 \cdot 12}{18} = 0.0109 kg. \quad (29)$$

Mass of ungasified fuel carbon:

$$\dot{m}(C)_{pyr} = m(C)_T - m^*(C)_T^G - m'(C)^{WG} - 0.0233 - 0.0086 - 0.0109 = 0.0038 kJ. \quad (30)$$

Mass of generated carbon monoxide and hydrogen gas by a water gas reaction:

$$m'(CO)^{WG} = \frac{m'(C)^{WG} \cdot M(CO)}{A(C)} = \frac{0.0109 \cdot 28}{12} = 0.0254 kg; \quad (31)$$

$$m'(H_2)^{WG} = \frac{m'(C)^{WG} \cdot M(H_2)}{A(C)} = \frac{0.0109 \cdot 2}{12} = 0.0018 kg. \quad (32)$$

Mass of generating hydrogen gas as a part of combustible gas:

$$\dot{m}(H_2)_T^G = m'(H_2)^{WG} = 0.0018 kg \quad (33)$$

Mass of generating nitrogen gas as a part of combustible gas:

$$\dot{m}(N_2)_T^G = m(N)_T = 0.00003 kg \quad (34)$$

Mass of generated carbon monoxide as a part of combustible gas:

$$\dot{m}(CO)_T^G = m^*(CO)_T^G + m'(CO)^{WG} = 0.0059 + 0.0254 = 0.0313 kJ. \quad (35)$$

Mass of generated methane gas as a part of combustible gas:

$$\dot{m}(CH_4)_T^G = m^*(CH_4)_T^G = 0.0017 kg. \quad (36)$$

Mass of generated carbon dioxide as a part of combustible gas:

$$\dot{m}(CO_2)_T^G = m^*(CO_2)_T^G = 0.0177 kg. \quad (37)$$

Mass of generated combustible gas:

$$\dot{m}(GG) = m(T)_z^w - \dot{m}(C)_{pyr} - m(Z)_T = 0.05 - 0.0038 - 0.0005 = 0.0457 kg. \quad (38)$$

The number of kilomoles of generated hydrogen as a part of the combustible gas:

$$\dot{n}(H_2)^G = \frac{\dot{m}(H_2)_T^G}{M(H_2)} = \frac{0.0018}{2} = 0.0009 kmole \ . \quad (39)$$

The number of kilomoles of generated nitrogen gas as a part of the combustible gas:

$$\dot{n}(N_2)^G = \frac{\dot{m}(N_2)_T^G}{M(N_2)} = \frac{0,00003}{28} = 10^{-6} kmole. \quad (40)$$

The number of kilomoles of generated carbon monoxide as a part of the combustible gas:

$$\dot{n}(CO)^G = \frac{\dot{m}(CO)_T^G}{M(CO)} = \frac{0.0313}{28} = 0,0011 kmole \ . \quad (41)$$

The number of kilomoles of generated carbon dioxide as a part of the combustible gas:

$$\dot{n}(CO_2)^G = \frac{\dot{m}(CO_2)_T^G}{M(CO_2)} = \frac{0.0177}{44} = 0.0004 kmole \ . \quad (42)$$

The number of kilomoles of generated fuel gas in terms of the absolutely dry:

$$\dot{n}(GT)^G = \dot{n}(H_2)^G + \dot{n}(N_2)^G + \dot{n}(CO)^G + \dot{n}(CO_2)^G = 0.0025 kmole \quad (43)$$

The amount of generated combustible gas under normal conditions:

$$\dot{V}(GT) = 22.4 \cdot \dot{n}(GT)^T = 22.4 \cdot 0.0025 = 0.0560 m^3. \quad (44)$$

The density of generated gas:

$$\dot{\rho}(GG) = \frac{\dot{m}(GG)}{\dot{V}(GG)} = \frac{0.0457}{0.0560} = 0.8160 kg/m^3 \ . \quad (45)$$

The mole fractions of gasesH2, N2, CO, CH4, CO2

$$\dot{\mu}(H_2)^G = \frac{\dot{n}(H_2)^G}{\dot{n}(GG)^G} = \frac{0.0009}{0.0025} = 0.3600; \quad (46)$$

$$\dot{\mu}(\ N_2)^G = \frac{\dot{n}(N_2)^G}{\dot{n}(GG)^G} = \frac{10^{-6}}{0.0025} = 0.0004; \quad (47)$$

$$\dot{\mu}(CO)^G = \frac{\dot{n}(CO)^G}{\dot{n}(GG)^G} = \frac{0.0011}{0.0025} = 0.4400; \quad (48)$$

$$\dot{\mu}(CO_2)^G = \frac{\dot{n}(CO_2)^G}{\dot{n}(GG)^G} = \frac{0.0004}{0.0025} = 0.1600; \quad (49)$$

$$\dot{\mu}(CH_4)^G = \frac{\dot{n}(CH_4)^G}{\dot{n}(GG)^G} = \frac{0.0001}{0.0025} = 0.0400 \quad (50)$$

The lowest working calorific value of combustible gas:

$$\dot{Q}(GG)_H^P = 4.19(2580 \cdot \dot{\mu}(H_2)^T + 3045 \cdot \dot{\mu}(CO)^G + 8517 \cdot \dot{\mu}(CH_4)^G) = 10932.9 kJ/m^3 \quad (51)$$

Specific lowest working calorific value of combustible gas:

$$\dot{Q}_{spec.}(GG)_H^P = \frac{\dot{Q}(GG)_H^P}{\dot{\rho}(GG)} = \frac{10392.9}{0.816} = 13398.2 kJ/kg \quad (52)$$

As follows from calculations, with increasing temperature of pyrolysis, composition and calorific value of the combustible gas does not change since all the water has been used up to the water gas reaction, and no additional reagent is supplied in wood pyrolysis area. But the mole fraction of gases changes. The comparative table shows that with the increase of pyrolysis temperature, hydrogen appears at the output that was not found in the low-temperature pyrolysis.

Table 1. Structure and calorific value of the combustible gas depending on the temperature of wood pyrolysis.

Parameters	Temperature, °C		
	500	750	1000
Composition of combustible gas in fractions, CO	0.250	0.286	0.440
CO2	0.620	0.571	0.160
CH4	0.125	0.143	0.040
N2	0.005	0.001	0.0004
H2	0	0	0.360
The calorific value of combustible gas, kJ / kg	7650.5	8752.1	10932.9

Table 1 shows that with increasing of pyrolysis temperature from 500 to 1000 °C calorific value of generated combustible gas increases significantly, and is not changed further.

The section from 0 to 300 ° C means removing moisture from wood and beginning of thermal decomposition of wood.

The section from +300 to + 500 ° C means the wood pyrolysis with generation of charcoal, combustible gas with calorific value of about 5000 kJ / kg, vapors of acetic acid, methanol, resins, and other.

The section from + 500 to 750 °C means the formation of charcoal, the thermal decomposition of organic products of pyrolysis to a combustible gas, and an increase in its calorific value to 5431.0 kJ/kg.

The section from +750 to less than 1000 ° C means a process similar to the process in the "C" point without changing the calorific value of combustible gas and looks like a straight line.

The section from less than 1000 to + 1000 ° C is a transition process on a significant increase in the calorific value of combustible gas to 13398.2 kJ/kg due to the beginning of the water gas flow and expenses of generated charcoal on this reaction.

The section from +1000 to over 1000 °C refers to the process of water gas reaction flow without changing the calorific value of generated combustible gas.

Thus, the developed thermophysical model of pyrolysis of hydrocarbon-containing wastes (on the example of wood) allows speaking about the efficiency of this process and its application for the disposal of hydrocarbon wastes with the expansion of the recycling industry.

IOP Conf. Series: Materials Science and Engineering **154** (2016) 012016 doi:10.1088/1757-899X/154/1/012016

3. Conclusion

The TIU developed technologies that reduce the geo-ecological damage to the environment. There was used an integrated systematic approach to the problem of disposal of oil-contaminated hydrocarbon wastes using high-temperature pyrolysis process.

The effectiveness of the technology that we offer is confirmed by comparing the experimental results and theoretical studies. High-temperature pyrolysis technology includes recycling (reuse of gaseous products of pyrolysis), the environmental impact on the atmosphere is reduced (no emissions), water resources and soil cover are not contaminated, landfill areas are decreasing.

The authors identified the effect, that at the thermal processing of hydrocarbon wastes, quantitative correlation between the mass of recyclable wastes, generated pyrolysis gas volume and the time of recycling is observed, proving the great potential of the high-temperature pyrolysis recycling method.

References

[1] Antipyev Y N, Nevolin A P and Zemenkov Yu D 1981 Operation of intermediate pumping stations on transport of gas saturated oils *NEFT. KHOZ* **10** 46

[2] Zemenkova M Yu, Shalay V V and Zemenkov Yu D 2015 Expert systems of multivariable predictive control of oil and gas facilities reliability *Procedia Engineering* **113** 312-5

[3] Mamadaliev R A, Kuskov V N, Zemenkov Yu D and Popova A A 2015 Influence of high-concentrated heat sources on alloying elements transition into the weld metal *Applied Mechanics and Materials* **770** 19-22

[4] Ctivation and catalytic reactions of alkanes / Lane with English; under the editorship of K. Hill. — M.: World, 1992.

[5] General toxicology / Under the editorship of Loit A. About. — SPb.: ELBI-SPb., 2006.

[6] Petrov Ave. A. Himiya of alkanes. — M.: Science, 1974. — 243 pages.

[7] Polya D. Combinatory calculations for groups, counts and chemical compounds (Russian)//Enumerative tasks of the combinatory analysis. — M.: World, 1979. — P. 36 — 138.

[8] Pereushanu V. Production and use of hydrocarbons. — M.: Chemistry, 1987.

[9] Fizer L., Fizer M. Organic chemistry. Advanced course. — M, 1966. — T. 1. — 680 pages.

[10] Haynes A. Methods of oxidation of organic compounds. Alkanes, alkenes, alkynes and arenas. — M.: World, 1988.

[11] Chemical encyclopedia/Hl. edition I. L. Knunyants. — M.: Big Russian encyclopedia, 1992. — T. 3: Copper sulfides — Polymeric dyes. — 640 pages.

IOP Conf. Series: Materials Science and Engineering **154** (2016) 012017 doi:10.1088/1757-899X/154/1/012017

Diagnosis of the Computer-Controlled Milling Machine, Definition of the Working Errors and Input Corrections on the Basis of Mathematical Model

A I Starikov[1, a], R Yu Nekrasov[2], O J Teploukhov[3], I V Soloviev[4], K A Narikov[5, b]

[1,2,3,4]Tyumen Industrial University, Russia
[5] West Kazakhstan agrarian-technical universiti named Zhangir khan, Kazakhstan

E-mail: [a] hapsai@rambler.ru; [b] syncler@mail.ru

Abstract. Manufactures, machinery and equipment improve of constructively as science advances and technology, and requirements are improving of quality and longevity. That is, the requirements for surface quality and precision manufacturing, oil and gas equipment parts are constantly increasing. Production of oil and gas engineering products on modern machine tools with computer numerical control - is a complex synthesis of technical and electrical equipment parts, as well as the processing procedure. Technical machine part wears during operation and in the electrical part are accumulated mathematical errors. Thus, the above-mentioned disadvantages of any of the following parts of metalworking equipment affect the manufacturing process of products in general, and as a result lead to the flaw.

Introduction

With a view to provide the necessary precision and quality of manufactured products, machine tools, by convention to fractionalize of accuracy by achieving sizes. At the same time, in consequence of equipment deterioration is transferred from a precise group in the less precise group. For this purpose, large machine-building enterprises use centralized diagnostic and repair services serving turning lathes, milling machines and grinder machines respectively.

The purpose of this study is the definition of errors that occur during the convoluted surface processing of oil and gas products on CNC milling machines.

To achieve this purpose it is necessary to solve following problems:

1Perform a diagnostics operation of executive work tools of the milling machine with the impact force and without it.

2. According to information diagnostic operation, determine the errors caused by process system, and form deviation and measurement of machined detail surfaces from the ideal model.

3. Identify ways of elimination arising errors without repairs.

At this stage, we made diagnostic operation of executive work tools of the milling machine without forced action. The device for diagnostics in detail processing is developed for the following experiment.

1. Materials and technique of research

Milling machine diagnostic identifies imprecisions of finish and the accuracy of the workpiece, the straightness of the slide ways, the flatness, the levelness and plumbness of actuated parts. Diagnosis can be performed by methods of functional and testing diagnostics.

Testing of geometrical accuracy involves in verifying capability of the real precision specified parameters to the parameters from milling machine check, depending on the accuracy class (N – normal grade of accuracy, P – heightened grade of accuracy , B – high grade of accuracy, A – extended high grade of accuracy, C – super-accurate grade of accuracy), which are defined by GOST "Standards of accuracy". However, these normative documents require that the geometric machine errors are systematic and completely transferred to the workpiece.

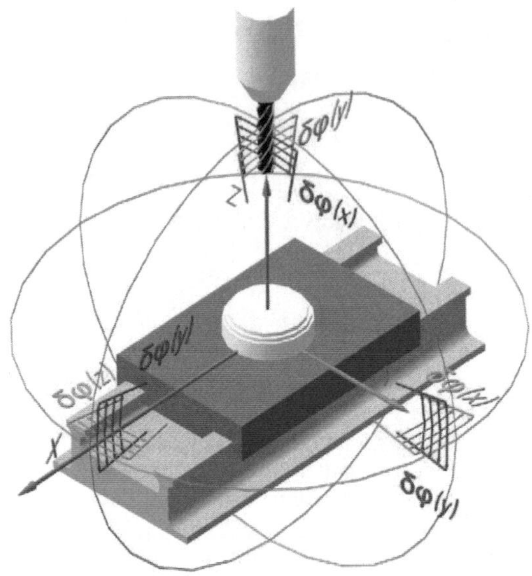

Figure 1. Errors arising from the movement of forming machine components.

Diagnostics related to determining positioning error of the shape-generation by cutting tool and moving the shape-generation machine units. These errors include: precision motion (radial and axial runout of the spindle, the straightness of working bodies movement, etc.); accuracy of mutual situation and movement (perpendicularity of longitudinal and transverse movement of the table, perpendicularity of the spindle axis to the lathe table, etc.); positioning accuracy (moving to the predetermined position); the accuracy of the surfaces defining the relative position of the workpiece and tool; the accuracy of the guide surfaces.

On figure 1 is shown chart for accuracy appraisals of the two main types by shape-generation machine units with progressive advance and circumduction. At the first moving inaccuracy may occur imprecisions of linear position δx in the direction of travel and error δy δz, characterizing deviation from straightness in mutually perpendicular directions.

As a result, there are variations in the angular movement of the table, $\delta\varphi$ (x) with respect to X-axis errors associated with revolution bodies. Uncertainties about the transverse axis $\delta\varphi$ (y) and the vertical axis Acp (z) are mainly associated with the shape and position error of the guide with the obligatory account of the rigidity of the lathe.

With the second movement arises angular error $\delta\varphi$ (z), related to the accuracy of positioning and angular error $\delta\varphi$ (x) and $\delta\varphi$ (y), determining bias when axes X and Y are moving.

On the diagnostic, the numerical values of tolerance depend on the type, size and class of machine accuracy. Imprecisions are measured for turning machine by normal accuracy, which largely influence over machining accuracy (radial and axial run-out of the spindle, positioning accuracy), at an average deviation should be within the range of 10-30 microns. At the same time, it uses a variety of tools and tooling metrology. In order to obtain results of measurements completely characterized the accuracy of the machine necessarily make measurements of its strength and heat deformation.

For tolerances of geometric accuracy of lathes limit values determine in accordance with GOST 17734-88. According to GOST 17734-88 vertical milling machines are subjected to the following types of inspections:

- estimation of radial runout of the outer centering spindle neck;
- axial runout of the spindle;

- perpendicularity of the spindle rotation axis and the working surface of the table;
- radial runout of the axis of the conical spindle bore;
- flatness of the working surface of the table;
- parallelness of table surface of longitudinal slide;
- parallelness of table surface to guideway console etc.

Table 1 shows the limit values of tolerance on the listed types of inspection of normal accuracy lathes and diagnosed vertical milling machine VDL-500.

These diagnostic methods (as above) do not include errors caused by the dynamic load, temperature and elastic deformations. A more objective picture of existent inaccuracy can be obtained after processing the product by using the circular interpolation. Measurements are made on the inner and the outer cylindrical surface treated. Using the proposed method of measurement, with the benefits that working error of workpiece surface, that is not the geometric accuracy of the machine, without disassembling the machine to appraise its technical condition, identify the individual components and spare parts that are defective. Due to this, the repair will be carried out purposefully to address the specific defects. This will ensure the stability of production, will predict the time of trouble-free operation of the equipment, pre-ordering the necessary components for repair.

Table 1. - Comparative analysis of limit values of permissible variation.

Type of verification	Tolerance for the class of accuracy of lathes, micron	
	N – normal grade of accuracy	VDL-500 (milling machine)
magnitude of radial run out of the outer centering spindle neck	10	6
axial run out of the spindle	10	6
perpendicularity of the spindle rotation axis and the working surface of the table	20	12
radial runout of the conical spindle bore axis	12	8
flatness of the working surface of the table	25	16
parallelness of table surface of longitudinal slide	20	12
parallelness of table surface to guideway console	20	12

To determine the milling machine accuracy VDL-500 based on the power and thermal load, using detail - a disc (Steel 40), are based on three-jaw chuck mounted on the lathe table. This form workpiece shows fully potential manufacturing errors. The shank milling cutter Ø16mm (R6M5) was used as a cutting tool.

Processing details (Figure 2) was carried out on the outer and inner cylindrical surface with the dimensions given in Figure 3.

Figure 2. - Picture of ready detail.

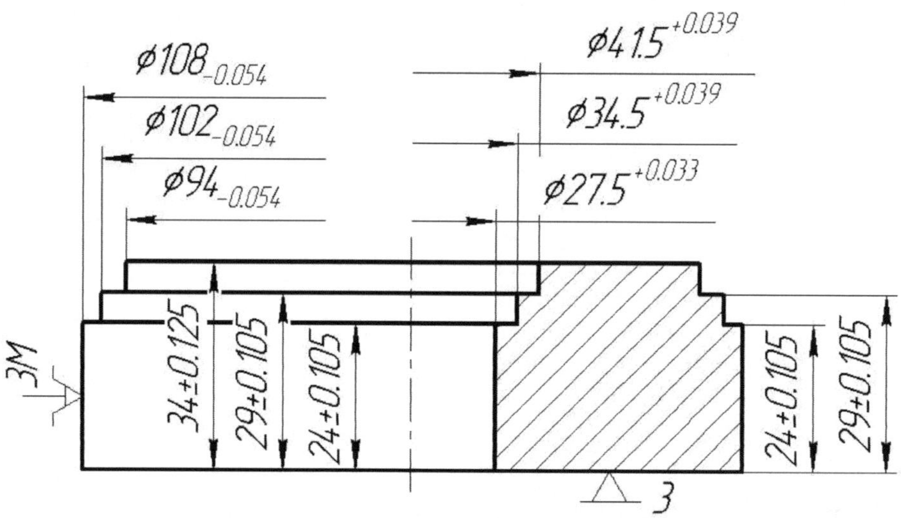

Figure 3. - Dimensional parameters of the measured detail.

2. Results and discussion

After machining of the prototype on a milling machine, measurements were made on the inner and outer surface contact method on the control-measuring machine CNC EOS 5-4-4, direct probe with thread M2, ruby ball 2 mm, length: 20 mm, ERE 14 mm stainless steel rod 1.4 mm in diameter. For the interpretation of these measurements has been used the software Metrologic group.

The measurement results are presented in the following figures:

- 8 and 9 - processing of interior cylindrical surface (hole), B = 8 mm, B = 4 mm, respectively (B - milling width);
- 10 and 11 - processing of external (outword) of the cylindrical surface, B = 8 mm, B = 4 mm, respectively.

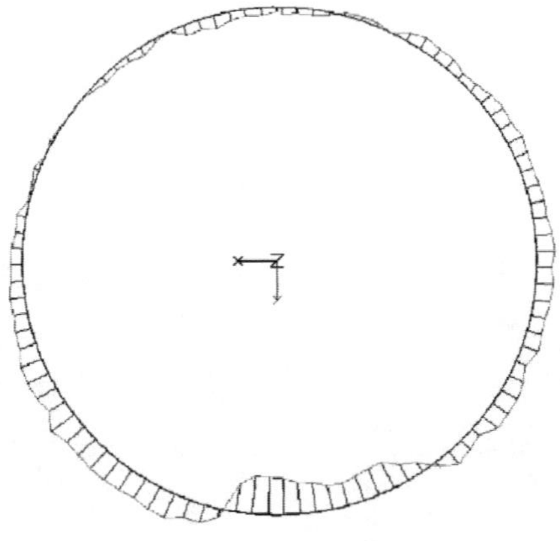

RESULTS

REFER.	NOMINAL	ACTUAL	TOL-	TOL+	DEV.	OUT OF TOL.
SCT1	Сечение 0 Pts [0 < LT ; LT < 0 < UT ; 0 > UL]					
MAX	0.000		-0.050	0.050		
МИН	0.000		-0.050	0.050		
SAvg						
UAvg						
F.F.						
SCT2	Сечение 100 Pts [0 < LT ; LT < 96 < UT ; 4 > UL]					
MAX	0.000	0.062	-0.050	0.050	0.062	0.012
МИН	0.000	-0.046	-0.050	0.050	-0.046	
SAvg		-0.008				
UAvg		0.023				
F.F.		0.109				
SCT2.1	Поверх. Pnt На Поверхность @144-513					
N.D.	0.000	0.007	-0.050	0.050	0.007	
X	0.005	0.005	-0.050	0.050	-0.000	
Y	-20.796	-20.789	-0.050	0.050	0.007	
Z	-1.501	-1.501	-0.050	0.050	0.000	

Figure 4. - The interior surface ½ of diameter.

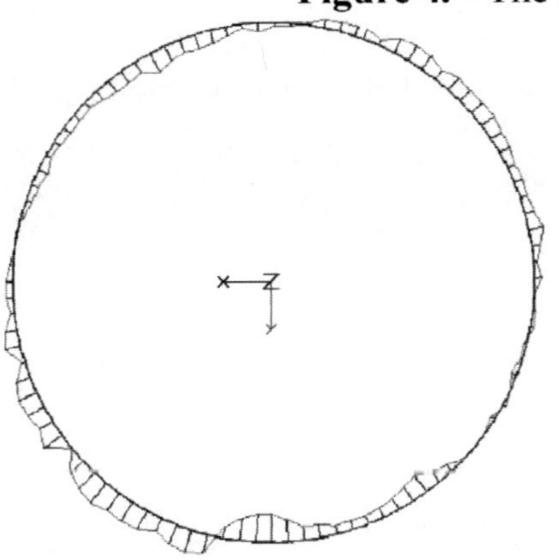

RESULTS

REFER.	NOMINAL	ACTUAL	TOL-	TOL+	DEV.	OUT OF TOL.
SCT1	Сечение 100 Pts [0 < LT ; LT < 100 < UT ; 0 > UL]					
MAX	0.000	0.037	-0.050	0.050	0.037	
МИН	0.000	-0.034	-0.050	0.050	-0.034	
SAvg		-0.002				
UAvg		0.015				
F.F.		0.072				
SCT1.1	Поверх. Pnt На Поверхность @144-479					
N.D.	0.000	-0.014	-0.050	0.050	-0.014	
X	-5.588	-5.593	-0.050	0.050	-0.004	
Y	-16.345	-16.358	-0.050	0.050	-0.013	
Z	-7.428	-7.428	-0.050	0.050	0.000	
SCT1.2	Поверх. Pnt На Поверхность @144-479					
N.D.	0.000	-0.012	-0.050	0.050	-0.012	
X	-4.561	-4.564	-0.050	0.050	-0.003	
Y	-16.660	-16.672	-0.050	0.050	-0.012	
Z	-7.417	-7.417	-0.050	0.050	0.000	

Figure 5. - The interior surface ¼ of diameter.

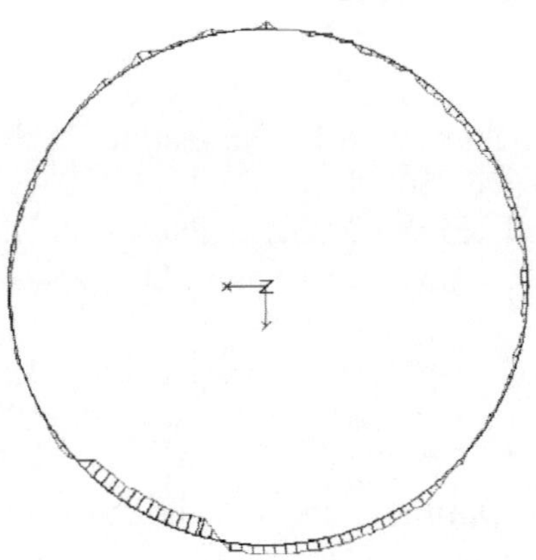

RESULTS

REFER.	NOMINAL	ACTUAL	TOL-	TOL+	DEV.	OUT OF TOL.
SCT1	Сечение 150 Pts [55 < LT ; LT < 42 < UT ; 53 > UL]					
MAX	0.000	0.039	-0.050	0.050	0.039	
МИН	0.000	-0.067	-0.050	0.050	-0.067	-0.017
SAvg		-0.004				
UAvg		0.018				
F.F.		0.106				
SCT1.1	Поверх. Pnt На Поверхность @144-223					
N.D.	0.000	-0.065	-0.050	0.050	-0.065	-0.015
X	13.068	13.050	-0.050	0.050	-0.018	
Y	45.206	45.144	-0.050	0.050	-0.062	-0.012
Z	-1.802	-1.802	-0.050	0.050	0.000	
SCT1.2	Поверх. Pnt На Поверхность @144-223					
N.D.	0.000	-0.066	-0.050	0.050	-0.066	-0.016
X	14.960	14.939	-0.050	0.050	-0.021	
Y	44.616	44.554	-0.050	0.050	-0.062	-0.012
Z	-1.798	-1.798	-0.050	0.050	0.000	

Figure 6. - The exterior surface ½ of diameter.

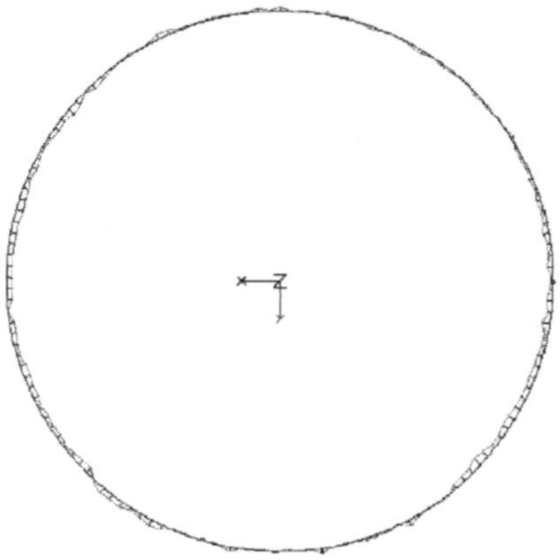

RESULTS							
REFER.		NOMINAL	ACTUAL	TOL-	TOL+	DEV.	OUT OF TOL.
SCT1		Сечение 150 Pts [0 < LT ; LT < 150 < UT ; 0 > UL]					
	MAX	0.000	0.026	-0.050	0.050	0.026	
	МИН	0.000	-0.033	-0.050	0.050	-0.033	
	SAvg		-0.006				
	UAvg		0.012				
	F.F.		0.059				
SCT1.1		Поверх. Pnt На Поверхность @144-371					
	N.D.	0.000	-0.009	-0.050	0.050	-0.009	
	X	14.060	14.058	-0.050	0.050	-0.002	
	Y	49.007	48.999	-0.050	0.050	-0.008	
	Z	-7.264	-7.264	-0.050	0.050	0.000	
SCT1.2		Поверх. Pnt На Поверхность @144-371					
	N.D.	0.000	-0.002	-0.050	0.050	-0.002	
	X	16.093	16.093	-0.050	0.050	-0.001	
	Y	48.377	48.376	-0.050	0.050	-0.002	
	Z	-7.256	-7.256	-0.050	0.050	0.000	

Figure 7. - The exterior surface ¼ of diameter

Due to the circular interpolation was ascertained that the form accuracy of surfaces (interior and exterior) varies depending on the position of the machined surface. The closer the surface to the center of the part, the more error handling. Based on the decryption of polar diagrams obtained results allow to state:

- arising inaccuracy exceeds the circularity form tolerance;
- in moments of cutting accord of motion data, as a consequence of its elastic deformation.

According to GOST 17734-88 has been made machine diagnostics VDL-500, a comparative analysis of its results are presented in Table 1.

Conclusion

1. From the results of polar diagram was seen that by reducing the contact of the working surface of the cutting tool, the resulting size errors due to the reduced contact stress will decrease. Also was educed that in the place of cutting tool is arised upgrade of dimensional allowance.
2. When changing the motion path of the tool perpendicular to tangent (outer surface) and helical (internal surface) is maintaining the size within a tolerance

REFERENCES

[1] 1. Producing powder by rotary grinding / V.G. Shalamov, S.D. Smetanin, D.A. Savel'ev // Russian Engineering Research. -2013. - T. 33, №3. - c.133 - 135.

[2] 2. Kinematics rotary milling in the production of cell chips / V.G. Shalamov, S.D. Smetanin // British Journal of Science, Education andCulture. - London: London University Press, 2014. - №,2(6). - c. 35-42.

[3] O. S. Lomova, Mathematical modeling of structural changes in the surfaces of the workpieces at thermal perturbations in the process of sanding, Omsk scientific bulletin. 2(120) (2013), pp. 95-98.

[4] S. I. Petrushin, I. M. Bobrovich, M. A. Korchuganova, Optimal design of the shape of the cutting blade tools, Tomsk, 1999.

[5] S. I. Takhman, Patterns of wear and forecast basis the wear resistance of tools from the standard hard alloys. Mechanics and physics of processes on the surface and in the contact of solid bodies, parts of technological equipment, S. I. Tahman. 3(2010), pp. 64-72.

[6] V. G. Shalamov, Mathematical modeling in machining of metals, Chelyabinsk, 2007.

[7] A. S. Yanushkin, S. O. Safonov, D. V. Lobanov, Improvement of technological processes of machine-building productions, Bratsk, 2006.

IOP Conf. Series: Materials Science and Engineering **154** (2016) 012017 doi:10.1088/1757-899X/154/1/012017

[8] A. I. Afonasov, A. A. Lasukov, Elementary Chip Formation in Metal Cutting, Russian Engineering Research. 3 (2014), pp. 152–155.
[9] U S Putilova, Yu I Nekrasov, A A Lasukov Loading of the manufacturing systems elements in the process of unsteady mode cutting and the models of their arrangement deviations, Applied Mechanics and Materials 682 (2014) 192-195.
[10] R Yu Nekrasov, A I Starikov, I V Soloviev Simulation of technological systems for diagnosis and management machining with CNC Applied Mechanics and Materials 770 (2015) 617-621
[11] R Yu Nekrasov, U S Putilova, A I Starikov, D A Kharitonov Diagnostics of cut-layer deformation and rational tool loading in numerically contolled lathes, Russian engineering research 34 (2014) 826-828.
[12] A A Lasukov, A A Mokhovikov Influence of modified layer of tool on stress - Strain state of cutting wedge, Proceedings (2012)
[13] R Yu Nekrasov, A I Starikov, A A Lasukov Entering the operative correcion machining processes CNC Materials Science and Engineering 91 (2015).
[14] A A Lasukov Selection of machining conditions in terms of the temperature dependence of chip formation, Russian Engineering Research 9 (2015) 679–681.

Improvement of Information Processing Quality in Preproduction Engineering Efficiency Assessment

L N Bakanovskaya and I A Chekardovskaya

Industrial University of Tyumen, 38 Volodarskogo st., Tyumen, 625000, Russia

E-mail: mila27@tyuiu.ru

Abstract: The article has considered the method of obtaining, analyzing and processing of expert information using modern software environment of MATLAB computer mathematics that will allow increasing the efficiency of decisions about the expected economic performances (the desired profitability and the desired output of products).

1. Introduction

Currently, the necessity of using modern methods and algorithms of computer information processing [1-4], effective in preproduction engineering in automatic mode and in calculation of sensitivity to production items, is undebatable. This is especially necessary when making "managerial decisions" on optimization of production strategies, production costs reduction [5], revenue increase, and reduction of production cycle duration, etc. In designing products, the methods of obtaining, analyzing and processing of information (the degree of prototypes possession [6]) tend to remain the same; in other words, solving the problem of model data processing is the "bottleneck" limiting further development of product design automation and efficiency improvement in the development of new economically profitable algorithms of production. The results of both analytical and descriptive prototyping largely depend on proper organization and planning of scientific works, collection, analysis, visualization and interpretation of the obtained data. The consistency and efficiency of such studies are of particular relevance in connection with the development of the common concept of production. Thus there is an urgent need to develop automated applications to the existing subsystems (e.g. [7]), capable of processing the available and incoming data in online user mode, as well as allowing to design predictable systems under the expert-technologists' assumptions (hypotheses) proved by formalized statistical methods. Expert systems have already shown their efficiency in many fields of human activities. In this regard, the task of development of expert modules (artificial intelligence systems) for economic issues of garment production is urgent.

2. Research

Optimization-analytic tasks almost always have ambiguous, "fuzzy" decision. The technologist formulates a hypothesis about the desired profitability of production or planned production on the basis of the existing awareness on the sales market, and then proves it with experiments and observation. The use of various tasks of linear programming and discrete optimization, in particular, when making decisions, is quite common [8]. There are many mathematical criteria to prove or rule out the hypotheses put forward by the expert-technologist, among which is the production cost reduction at the calculated sensitivity. When proving a hypothesis and defining important features the expert faces the task of creation on their basis of the description and evaluation system, suitable for use in practice by the technologist and the Economic Department of the company. But the use of expert systems specialists is not always available at the enterprises of the sewing industry. A program, simple in mastering and management, will provide for a solution to this task without additional costs.

Engineers' automated workstations (AWS) should possess compatibility with newly created software systems for their use in scientific research and real optimization practice. Currently, there is a certain number of methods for expert systems synthesis, developed on the basis of linear programming methods, discrete optimization, decision tree, clustering, multiple linear regression, as well as with the help of genetic algorithms, neural networks and fuzzy logic. There is a significant number of unified software products for statistics, data analysis and expert systems synthesis (e.g., MS Exel, "Statistica for Windows", MAPLE, MATLAB and many others), having a large application scope. Of the above software tools the MATLAB computer mathematics environment is distinguished as the most developed, primarily due to the fact that it comprises modern algorithms of fuzzy logic means construction and the problems of neural network optimization and training [9]. The MATLAB software environment is an interactive system; it is open and enables implementing complex algorithms of data analysis. In addition, MATLAB has a GUI application (graphical user interface) enabling the creation of flexible applications with a suitable user interface, in particular, experts-technologists oriented. Having many specialized functions, this software can be used to automate the required calculations in almost every field of science and technology.

Based on the above, in the present work the MATLAB computer mathematics environment has been selected as the main development means of the expert evaluation program module. The production cost of a series of garments and the desired output for a specified period (figure 1 a. b) are used as initial data, where *Zi* is the amount of expenses on all counts, *n* is the number of items, V_{opt} – optimal output, *E* - efficiency:

№ n/n	Article the cost on manufacture of the product	Expenses are minmal, thousand of rubles	Expenses are maximum, thousands of rubles	Sensitivity, thousands of rubles / piece	Expenses expected thousands of rubles
1	Wages of workers	8.00	18.00	10.00	-3000.34
2	Development of standad design	1.50	3.00	1.50	-449.75
3	Technology of manufacturing	1.00	2.00	1.00	-299.83
4	Fabric (including accessories)	0.58	2.50	1.92	-577.02
16	Zi	11.08	25.50	14.42	-4326.95
17	n, thousands of piece			1.00	
18	Voptimal, piece				300.83
19	Effectiveness, %				56.55%

a)

№ n/n	Article the cost on manufacture of the product	Expenses are minmal, thousand of rubles	Expenses are maximum, thousands of rubles	Sensitivity, thousands of rubles / piece	Expenses expected thousands of rubles
1	Wages of workers	8.00	18.00	10.00	-2931.95
2	Development of standad design	1.50	3.00	1.50	-439.49
3	Technology of manufacturing	1.00	2.00	1.00	-292.99
4	Fabric (including accessories)	0.80	2.50	1.70	-498.99
16	Zi	11.30	25.50	14.20	-4163.42
17	n, thousands of piece			1.00	
18	Voptimal, piece				293.99
19	Effectiveness, %				55.69%

b)

Figure 1. MS EXCEL work window with initial data

In the work [1] the choice of initial data for the main means of software module development has been justified and their usefulness proved without losses for the production; which enables to achieve not only the breakeven point but also monitoring the manufacturing process throughout the entire life cycle of preproduction engineering.

Based on the results of MATLAB calculation, with the imported initial data, a graph (figure 2) of organizational-technical and economic indicators is constructed according to the production efficiency evaluation method developed by the authors [2].

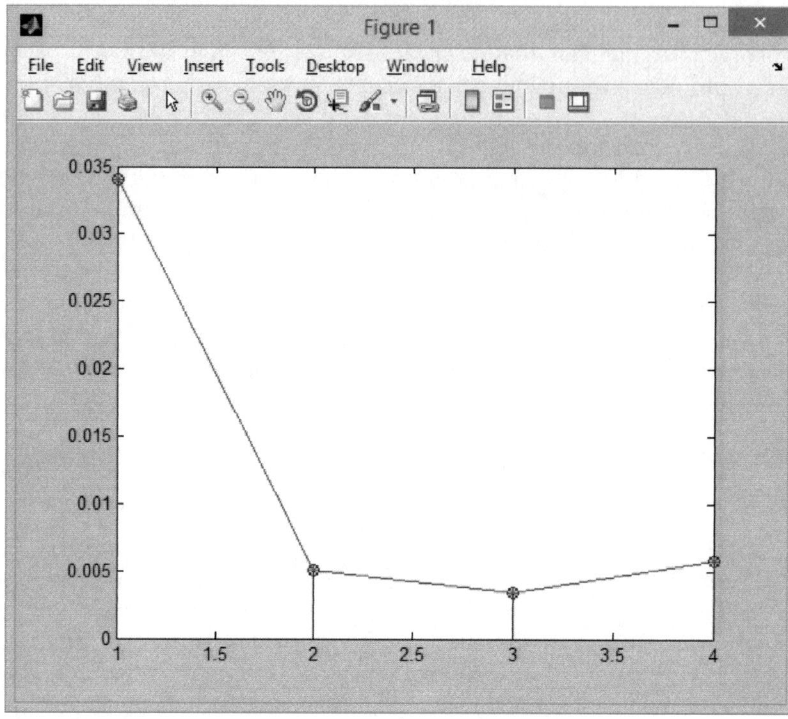

a)

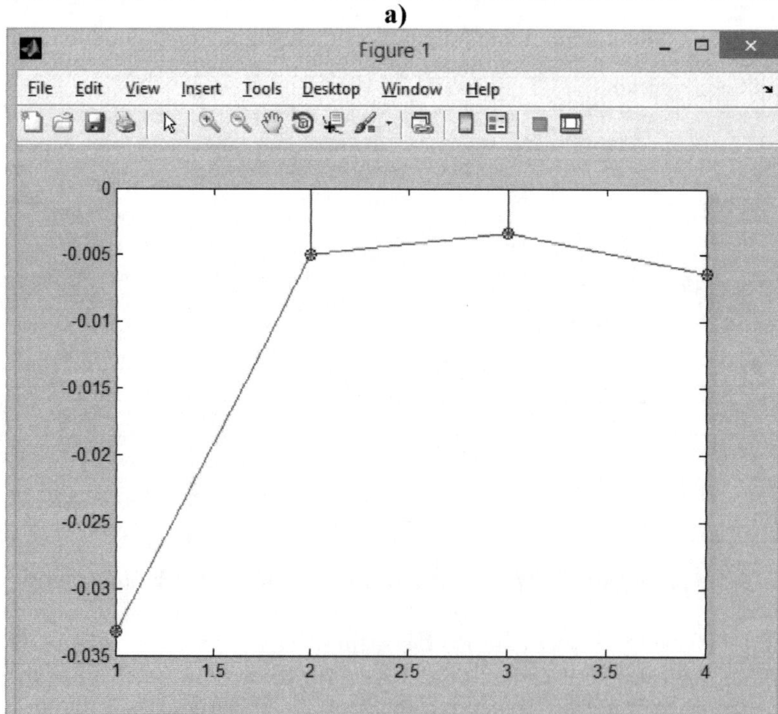

b)

Figure 2. Calculation visualization with graph construction.

Based on the application of the of the most effective calculation methods for [2] manufacturing a series of apparel, the analysis of the operation of LLC "Fashion Studio "Silhouette" (Tyumen) and LLC "Profile" (Tyumen) has been accomplished. The sensitivity for each of the above indicators has been defined, and based on the visualization of graphical and mathematical proofs [1, 3] the production efficiency – expenses sensitivity relationship has been clearly defined. In this case 40-60% of preproduction engineering economic efficiency based on the information processing quality has been implied.

3. Conclusion

Study of the methods of obtaining, analyzing and processing the information about preproduction engineering using one of the developed MATLAB computer mathematics environments has been implemented that will enable prototyping and enhancing production efficiency on the basis of modern scientific research and "managerial decisions" made by the expert-technologist.

The work has also solved the following problems by calculating the production cost of two enterprises (small-scale ones with the output of a series of more than 100 units): the sensitivity of each expense to the produced units has been defined; cost reduction efficiency depending on the expense item at each of the operations has been identified. The efficiency of preproduction engineering increases for both "Fashion Studio "Silhouette" (with cost reduction by 9%) and "Profile" (with cost reduction by 13%). The optimized use of the available resources and the improvement of the information processing quality will provide for the efficiency of preproduction engineering.

References

[1] Kitanina, K. Y., Khromushin, V. A., Dzasokhov A. S., Khromushin O. V. (2015) Features of construction of the expert system on the basis of algebraic models of constructive logic. *Bulletin of new medical technologies. Electronic edition,* 3 (Publication 1-1). http://www.medtsu.tula.ru/VNMT/Bulletin/E2015-3/5209.pdf doi: 10.12737/13073.

[2] Benderskaya E. N., Nikitin K. V. (2015) Using Spiking Recurrent Neural Networks for ECG Analysis. Information and control systems. 1(85-91). doi:10.15217/issn1684-8853.2015.1.85.

[3] Maass, W., Natschla, ger T., Markram, H. (2002) Real-Time Computing without Stable States: A New Framework for Neural Computations Based on Perturbations. *Neural Computation,* 11 (2531–2560). doi:10.1 162/089976602760407955.

[4] Izhikevich, E. M. (2004) Which Model to Use for Cortical Spiking Neurons? *IEEE Transactions on Neural Networks,* 15 N 5. (1063–1070). doi:10.110 9/TNN.2004.832719.

[5] Chekardovskaya I A and Chekardovskiy M N 2015 Golden Square. Methodology of Definition of Economic Performances in the Scope of Oil-and-Gas Transport Production *Business Magazine Neftegaz.RU* **11-12** 72-7

[6] Chekardovskaya I A, Bakanovskaya L N and Toropova I A 2015 Effective system of design production processes *Scientific Review* **5** 272-8

[7] Turenko S K and Ahmadulin R K 2006 The formation of many of the results of allocation when you build a quality criterion based on the formalization of expert assessments *Geology and petroleum potential of the West Siberian basin: proceedings of the Fourth all-Russian scientific-practical conference* **2** 101-3

[8] Bakanovskaya L N, Mokeeva N S and Zaev V A 2011 Application "greedy" algorithm to optimize of priority of launch the models of sewing products into production *Bulletin of computer and information technology* **2** 23-6

[9] Bakanovskaya L N 2015 Revisiting the development of neural network expert systems *Materials of the XI International scientific and practical conference "Conduct of modern science" Economic science* **3** 11-14

The Effect of Dynamic Processes in the System "Pipe-Soil" on the Pipeline Deviation from Design Position

K S Voronin and E V Ogudova

Industrial University of Tyumen, 38 Volodarskogo st, Tyumen, 625000, Russia

[1]E-mail: voronin_tsogu@mail.ru

Abstract. This article explains the changes of the main gas pipeline geometrical shape, i.e. the deviation of its position from the project one, as one of the important factors affecting the operating safety of underground main pipelines. In the study of the processes occurring at interaction between main gas pipeline and foundation soil, a rod model, fixed on one side, with the load applied to, it was examined. Action of elastic foundation across the pipeline is modeled with spring substrate that resists cross motion. As a result, the effect of soil impact on the vibration frequency of the pipe wall, appearing at the stage, preceding its deviation from the design position, was determined. Detection the frequency of these vibrations can make it possible to forecast pipe bending not at the stage of growth but at the stage preceding its appearance.

1. Introduction

An important factor, affecting the operating safety of underground pipelines, is changing the geometric shape of the pipeline, that is, the deviation of its position from the project one. Pipelines upfloating and bending is one of the factors that affect their stress-strain state, which is known to affect significantly the mechanical properties of the pipe.

According to many authors [1, 3, 6, 7], the most important element of this problem is the interaction between underground pipeline and the surrounding soil. This interaction takes place not only on the part of ground due to thermal processes, the humidity, the cohesive forces, but also on the part of the pipeline on the ground due to the current loads caused by temperature and pressure.

2. Methods

During the construction of the pipeline, its stable position is achieved through the use of weighting agents, as a means of ballasting, or direct anchoring. However, in many areas, especially in northern ones, despite the effect of weighting agents, pipelines turn out to be located higher than the project level, up to the formation of vertical and horizontal bends.

From the analysis of the above models, the following general patterns can be identified. The models consider longitudinal motion of the linear section of pipeline with low soil resistance in the perpendicular direction. This area can be defined as dangerous from the standpoint of the arch formation. If the pipe wall is compressed, with increasing longitudinal force (on its origin models vary) on the studied area, there is a loss of stability of rectilinear form above the critical value. The value of the critical force depends on the coefficient of soil bedding and bending stiffness of pipe [5]. Bedding ratio sharply decreases with increase in soil water cut, which corresponds to the seasonal thawing. The interaction with the soil can significantly weaken due to vibration.

Along with soil water cut, buoyancy increases, which changes the high-rise pipe position in the ground. It is considered that the emergence of dangerous areas (with low resistance and high water cut) is caused by a complex combination of climatic and hydrogeological conditions, which complicates the forecasting of arcing. In models studied, arcing forecast is based on calculations of the dangerous area parameters depending on the time and soil characteristics.

Therefore, it can be concluded that despite the difference of models in their forecasting part, based on the calculation of implementing "bending deflection", they are good enough in forecasting the increase of deflection, if it already exists.

The mechanisms that cause increase in length of the linear part of the pipeline, are still not completely understood. Thus, the interaction between weighting agents and a pipeline turns out to be beyond the scope of models. Because before the coming of realization the condition of instability of the pipeline horizontal position, weighting agents should be dropped.

Let us consider a stage, prior to pipeline bending. As a model, we choose rod with length L fixed at one side and with the applied load F. Motion of the rod will be determined by the function φ (x). The curvature of the rod is determined by changing the angle:

$$\frac{d\theta}{dx} = \frac{d}{dx} arcsin\varphi'(x) = \dot{\varphi}\sqrt{1-\varphi'^2} = \frac{d^2\varphi}{dx^2}\sqrt{1-\left(\frac{d\varphi}{dx}\right)^2} \quad (1)$$

where θ is the angle of the rod curvature; φ (x) is the motion of the rod.

Deformation energy of a rod will be determined by the equation:

$$\delta V = \frac{1}{2} EI \left(\frac{d\theta}{dx}\right)^2 \delta x \quad (2)$$

where E is elastic modulus; I is a moment of inertia of the pipe section.

As the unit of length, choose mass value m, then the expression for the kinetic energy will be:

$$\delta\varepsilon = \frac{1}{2} m(\varphi'(x))^2 \delta x \quad (3)$$

Potential energy of a bent rod is given by:

$$V = U - \frac{1}{2} F \int_0^L \varphi'^2 dx \quad (4)$$

where U - potential energy of deformation; F - compressive force.

Flexural mode φ (x) with arbitrary amplitude can be selected as a half of sinusoid:

$$\varphi(x) = w(t) \sin\frac{\pi x}{L} \quad (5)$$

According to a paper [2], at a stage preceding the time of pipeline bending, vibrations should be developed in it, as a result of instability:

$$\omega_0^2 \approx \frac{\pi^4 EI}{2L^3}\left[1-\frac{Sl}{\pi^2 I}\left(\frac{16}{\pi^2 \rho h}P\sum_{m=1}^{\infty}\frac{\cos\alpha_m\xi\sin\varphi}{m\omega_{mn}^2}\sin\omega t\right)^2\right] \quad (6)$$

where L - length of the pipe; P - pressure; p - the mass of the shell per unit area; h - thickness of the wall; ; $\xi = \frac{x}{r}$;x - changing shell radius; φ - angle; , $\alpha_m = \frac{m\pi r}{l}$, l - length of the shell, m - number of half-waves in the direction of a generator of a shell, n - the number of waves around the circumference; ω_{mn} - natural frequency of vibration.

Extraction of frequency of these vibrations can serve as a basis for the development of the pipeline condition monitoring method at a stage preceding its bending. In this case, a geometry deformation is a consequence of the nonlinearity of pressure surges at an unstable gas flow. Reducing the angular vibration frequency (1) defines an approximation of the system to the bifurcation point. The task outlines two balance state paths. In subcritical area, vibrations (ω_0) correspond to the steady state with amplitude (Q_1).

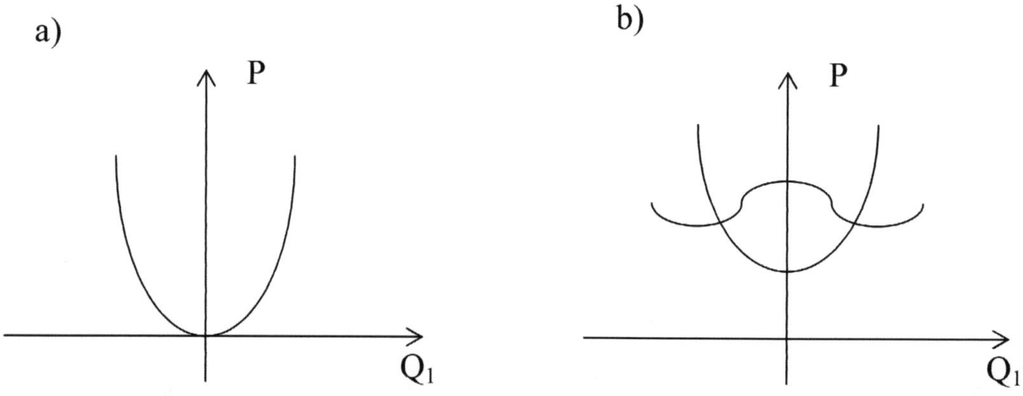

Figure 1. The dynamic behavior of the pipe.
a) far from the bifurcation point, $P < P_c$
b) near the bifurcation point, $P \approx P_c$

Control parameter in this task is $\lambda = P_c - P$. As we approach to the bifurcation point, the control parameter decreases to zero, along with it natural vibrations frequency (ω0) approaches to zero. At the bifurcation point, the potential energy is modified, going from single-well (steady) state (Fig. 1a) to a two-well (Fig. 1b).

The emergence of the critical load P_c corresponds to the disappearance of the potential energy minimum. As a result, the vibration frequency tends to zero. Thus, the transformation of vibration frequency into zero defines the moment of the pipeline bending formation, as a transition to a new stable state.

The main component of the load (P) is a longitudinal motion caused by a dynamic load. Researches [1, 7, 9] calculate the value of the longitudinal displacement at interaction with the ground not at the stage of bending formation, but at its growth when bending has already occurred. It seems that at the stage of bending formation, the interaction with ground will also be an important condition.

Let us consider emergence of a bending mode at longitudinal loading of pipeline with variable dynamic load in terms of interaction with the ground. Since the ground has the binding ability, from all types of interaction we choose its elastic shear resistance. Consider the stationary case, then the movement of u (x) must satisfy the equation of longitudinal-transverse bending:

$$\frac{d^2u}{dx^2} - \frac{\pi D_1 k_n}{ES} = \frac{dP}{dx} \tag{7}$$

where D_1, k_n – coefficients characterizing the elastic properties of the ground [9]; x - distance from the beginning of pipeline section.

The pressure distribution along the length of pipeline we choose as a model function:

$$P(x) = P_0 - a_1 e^{-\alpha_1 x} \tag{8}$$

Define the boundary conditions:

$$u(0) = 0$$
$$\frac{du(L)}{dx} = \frac{1}{ES} F \tag{9}$$

where L - length of pipeline section on which the bending occurred.

Equation (4) must be supplemented by the expression of the internal strain forces, compressing the pipe [1].

$$F = \frac{ES}{2} \int_0^L \left[\left(\frac{dz}{dx}\right)^2 + \left(\frac{dy}{dx}\right)^2 \right] dx \approx ES \left(\frac{w}{l}\right)^2 l \tag{10}$$

where w - the value of pipeline deflection.

Given equations form a system, the solution of which is determined by expression, obtained with the terms of approximation.

$$u(x) \approx \frac{F sh(\theta x)}{\theta ch(mL)} - \frac{\alpha_1 a_1}{\theta^2 - \alpha^2} \left(e^{\theta x} - e^{-\alpha x}\right) =$$

$$= \frac{1}{\theta L} \left(\frac{16}{\pi^2 \rho h} P \sum_{m=1}^{\infty} \sum_{n=1}^{\infty} \frac{\cos \alpha_m \xi \sin \varphi \sin \omega t}{mn(\omega_{mn}^2 - \omega^2)} \right) \frac{sh(\theta x)}{ch(\theta L)} - \frac{\alpha_1 a_1}{\theta^2 - \alpha^2} \left(e^{\theta x} - e^{-\alpha x}\right) \tag{11}$$

Interaction with the ground as an elastic base changes the frequency of a mode, by which instability of the horizontal pipeline position develops. The action of elastic ground across the pipeline can be modeled with spring substrate that resists cross motions (w). Consider the linearized expressions for the energy in the analysis of the stability of an elastic rod. To bending strain energy

$$\varepsilon_D = \frac{1}{2} EI \int_0^L \left(\frac{d^2 w}{dx^2}\right)^2 dx \tag{12}$$

and potential compressive load energy

$$U_D = -\frac{1}{2} P \int_0^L \left(\frac{dw}{dx}\right)^2 dx \tag{13}$$

the elastic base deformation energy is being added:

$$\varepsilon_k = \frac{1}{2} K \int_0^L w^2 dx \tag{14}$$

where K - stiffness of the base.

Looking at the initial stage of supercritical position, w can be determined by the expression:

$$w = Q \sin\left(\frac{\pi x}{L}\right) \tag{15}$$

where Q - mode amplitude.

As a result of transformations, it is possible to obtain an expression for ωi2 (i-th mode, by which instability develops):

$$\omega_i^2 = \frac{EI\left(\frac{i\pi}{L}\right)^4 + K - P\left(\frac{i\pi}{L}\right)^4}{m} \tag{16}$$

From this expression, it follows that the frequency of the mode, at which instability is realized, is larger due to the elastic interaction with the ground.

Interaction of pipeline with the ground affects the process of changing its geometric shape, i.e., a deviation from the design position. Researches [3, 8] show that at frequencies from 10 Hz to 100 Hz, maximum amplitudes of the vibrations, and the maximum reduction of tangential resistances to both the longitudinal and cross motions are observed. It was found that no matter how small the amplitude of the vibration is, anyway there is a noticeable decrease in the elastic soil impact.

3. Conclusion

Thus, the dynamic processes in the "soil-pipe" system have an impact on the process of extension of the linear part of the pipeline due to the growth of longitudinal compressive stresses. As approaching the system to the time of stability loss of the design position, there should appear certain changes. Thus, at the stage prior to the pipeline bending, low frequency vibrations (ωi) develop in it, which damp if pressure increases. Based on the above, it is possible to develop a method of forecasting the beginning of bending formation; more precisely one can define the range of frequencies at which instability can appear.

References

[1] Vagner V V 2008 *Changing the height position of the pipeline at a variable water cut of soil in the trench* The thesis for the degree of Candidate of engineering sciences (Tyumen) 119

[2] Voronin K S 2013 *Forecasting the development of damages on gas pipelines under the influence of dynamic load* Thesis for scientific degree of candidate of engineering sciences (Tyumen, TSOGU) 111

[3] Gorkovenko A I 2006 *Fundamentals of the theory of calculating the spatial position of an underground pipeline under the influence of seasonal processes* The thesis for the degree of Doctor of Sciences (Tyumen, TSOGU) 305

[4] Gumerov A G and Azmetov H A 2007 Strain calculation in the pipe system while eliminating deviations of their cross-sectional shape and diameter of the nominal values *Oil Industry* **1** 78-79

[5] Ibragimov A A, Podorozhnikov S Yu, Shabarov A B, Medvedev M V and Zemenkova M Yu 2014 Calculation model and algorithm for determining the residual life of the pipeline under periodic changes of strains and corrosion *Mining informational and analytical bulletin (scientific and technical journal)* (Moscow, Gornaya Book) **S4** 199-206

[6] Mikhalenko E S 2007 *Development of the method of calculating and forecasting critical parameters of main gas pipeline arcing* The thesis for the degree of Candidate of engineering sciences (Tyumen) 200

[7] Pulnikov S A 2007 *Interaction of vibro-loaded main gas pipelines with surrounding grounds* The thesis for the degree of Candidate of engineering sciences (Tyumen) 172

[8] Toropov S Yu, Dorofeev V S and Zemenkov Yu D 2013 On determining the longitudinal force in the pipeline at formation of the arch *Mining informational and analytical bulletin (scientific and technical journal)* (M , Gornaya book) **S3** 223-229

[9] Harinovsky V V 2000 *The reliability and resource of pipeline construction* (M, Nedra) 467

[10] N.N. Malushin, A.P. Kovalev, D.V. Valuev, E.A. Shats, I.F. Borovikov, Developing and studying the methods of hard-facing with heat-resisting high-hardness steels, J. IOP Conference Series: Materials Science and Engineering. 142 (2016) p. 1-6

[11] R.A. Gizatulin, D.V. Valuev, R.S. Dariev, V.A. Trifonov, I.F. Borovikov, Distribution of liquid flow rates in the process of bubbling with gas through gas-permeable inserts, J. IOP Conference Series: Materials Science and Engineering. 142 (2016) p. 1-6

[12] Zemenkova, M.Y.,Shipovalov, A.N.,Zemenkov, Y.D. 2016 Mathematic Modeling of Complex Hydraulic Machinery Systems When Evaluating Reliability Using Graph Theory IOP Conference Series: Materials Science and Engineering 127 (1) DOI: 10.1088/1757-899X/127/1/012056

[13] Zemenkova, M.Yu.,Zemenkov, Y.D.,Pimnev, A.L.,Kurushina, E.V. System of Controlling the Reliability of Hydraulic Machinery in Oil and Gas Facilities IOP Conference Series: Materials Science and Engineering 127 (1) DOI: 10.1088/1757-899X/127/1/012055

IOP Conf. Series: Materials Science and Engineering **154** (2016) 012020 doi:10.1088/1757-899X/154/1/012020

Forecasting and Evaluation of Gas Pipelines Geometric Forms Breach Hazard

K S Voronin

Industrial University of Tyumen, 38 Volodarskogo st., Tyumen, 625000, Russia

E-mail: voronin_tsogu@mail.ru

Abstract. Main gas pipelines during operation are under the influence of the permanent pressure drops which leads to their lengthening and as a result, to instability of their position in space. In dynamic systems that have feedback, phenomena, preceding emergencies, should be observed. The article discusses the forced vibrations of the gas pipeline cylindrical surface under the influence of dynamic loads caused by pressure surges, and the process of its geometric shape deformation. Frequency of vibrations, arising in the pipeline at the stage preceding its bending, is being determined. Identification of this frequency can be the basis for the development of a method of monitoring the technical condition of the gas pipeline, and forecasting possible emergency situations allows planning and carrying out in due time reconstruction works on sections of gas pipeline with a possible deviation from the design position.

1. Introduction

The main gas pipeline is a complex engineering construction operating in various natural and climatic conditions. Peculiarities of underground laying in conditions of high water-cut of ground cause formation of additional gas pipeline strength parameters in the construction technology. One of the parameters of reliable operation of the gas pipeline system in the northern regions is to provide occurrence depth of a pipeline [10].

As a result of upfloating and discharge of weighting agents, bending of the pipeline geometric shape takes place [9]. Clay and submerged soils, that a gas pipeline is filled with, have a significantly reduced density and, as a consequence, lower sqeezing ability. In such circumstances, the longitudinal motion of a gas pipeline in the ground lead to its elongation. Mechanisms for elongation of the pipeline associated with a sqeezing ability of the ground were considered in a number of studies [5, 6, 10]. Research [5] shows the calculation of longitudinal stability ballasted pipelines on longitudinal sections. However, in these studies, an impact of a variable load on the pipeline caused by pressure vibrations at gas transportation is not studied.

Works on construction of main gas pipelines in the swampland are usually carried out in winter, and putting the gas pipeline into operation is carried out in summer. This results in the a single elongation in the first spring-summer operating period. But in practice, the elongation is systematic.

2. Methods

In literature [6], the impact of alternating pressure on the pipe elongation process is mentioned. At the moment it can be considered generally accepted that the elongation process ultimately contributes to the emergence of geometric shape instability. Instabilities of this type are a consequence of the nonlinearity of dynamic processes. Researches [7, 8] analyzed accidents and catastrophic events, and it was shown that these events are described by a power law of distribution. Such a law of distribution of random events is typical for systems with feedback. In such circumstances, in the system phenomenon preceding emergency events, which are very important for the process of monitoring the work of the pipeline system, should be observed.

This paper researches the phenomena occurring in the pipeline before changing its geometric shape (bending). Let us study the bending vibrations of the cylindrical shell (see Fig. 1) with a radius r. Let us denote the movement of the point along the shell as *u*, along the tangent to the circle - υ, changes along the radius - *w*.

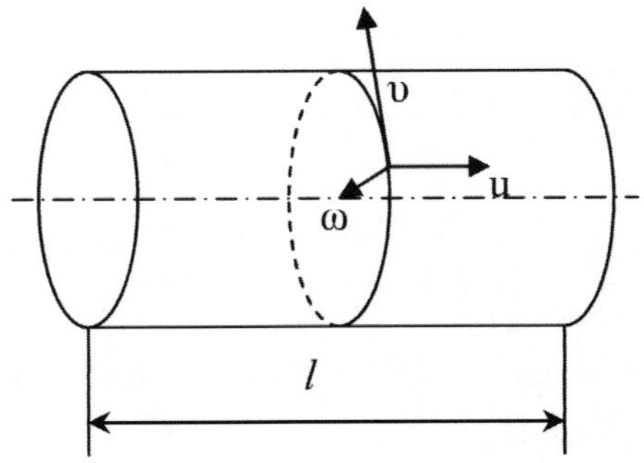

Figure 1. A model of the thin cylindrical shell. u - displacement of point on the surface along the shell, υ - displacement of point on the surface along the tangent to the circle, w - displacement of the point on the surface along the radius.

Original system of equations of forced vibrations of the cylindrical shell by the pressure P (t), neglecting the inertia forces along the shell and along the tangent to the circle $\left(\frac{\partial^2 u}{\partial t^2}=0; \frac{\partial^2 v}{\partial t^2}=0\right)$ can be written as follows [6]:

$$\begin{cases} \frac{\partial^2 u}{\partial \xi^2}+\frac{1-\mu}{2}\frac{\partial^2 u}{\partial \varphi^2}-\frac{1+\mu}{2}\frac{\partial^2 v}{\partial \xi \partial \varphi}-\mu\frac{\partial w}{\partial \xi}=0 \\ \frac{1+\mu}{2}\frac{\partial^2 u}{\partial \xi \partial \varphi}+\frac{\partial^2 v}{\partial \varphi^2}-\frac{1-\mu}{2}\frac{\partial^2 v}{\partial \xi^2}-\frac{\partial w}{\partial \varphi}=0 \\ \mu\frac{\partial u}{\partial \xi}+\frac{\partial v}{\partial \varphi}-w-k\nabla^4 w=\frac{\rho r^2}{B}\frac{\partial^2 w}{\partial t^2}+P(t) \end{cases} \quad (1)$$

where $\xi=\frac{x}{r}$, x - changing radius of the shell, φ - angle, $k=\frac{h^2}{12r^2}$, h - thickness of the shell, $B=\frac{Eh}{1-\mu^2}$, E - elastic modulus, μ - Poisson ratio, ρ - shell weight per unit of area.

Literature [9] studied methods of solution of this system. Assuming that $P(t)=P\sin\omega t$, one can find a solution to the system for changing the shell radius w as follows [2]:

$$w=\frac{16}{\pi^2\rho h}P\sum_{m=1}^{\infty}\sum_{n=1}^{\infty}\frac{1}{mn(\omega_{mn}^2-\omega^2)}\cos\alpha_m\xi\sin\varphi\sin\omega t \quad (2)$$

where $\alpha_m=\frac{m\pi r}{l}$, l - the shell length, m - number of half-waves in the direction of generator shell, n - number of waves along a circle,

$\omega_{mn}^2=\frac{Bk}{\omega^2\rho h}\left[(\alpha_m^2+n^2)^2+\frac{1-\mu}{k}\frac{\alpha_m^4}{(\alpha_m^2+n^2)^2}\right]$ - natural vibration frequency of a shell.

Elastic vibrations by changing the diameter, cause local bending of the pipe wall, whereby there emerges stress not only in radial direction but also along the pipe (longitudinal stresses -σ_x) [5].

$$\sigma_x \approx w\left(\frac{3}{1-\mu^2}\right)^{1/2} E \tag{3}$$

In their turn, the longitudinal stresses cause pipe elongation which can be estimated by the expression [2]:

$$u_x \approx \frac{\sigma_x}{E} \approx \left(\frac{3}{1-\mu^2}\right)^{1/2} \frac{16}{\pi^2 \rho h} P \sum_{n=1}^{\infty} \frac{1}{n(\omega_n^2 - \omega^2)} \cos \alpha_m \xi \sin \varphi \sin \omega t \tag{4}$$

Elongation results in the appearance of internal stresses force F, compressing the pipe:

$$F = \frac{ES}{2} \int_0^l \left[\left(\frac{dz}{dx}\right)^2 + \left(\frac{dy}{dx}\right)^2\right] dx \tag{5}$$

where S – cross-section area of a shell.

Let us assume that a small cross flexure of a pipe is about w. Then the derivatives $\frac{dz}{dx}$ and $\frac{dy}{dx}$ - about $\frac{w}{l}$, so the entire integral that determines the strength is:

$$F \sim ES \left(\frac{w}{l}\right)^2 l \tag{6}$$

Stable equilibrium position of the pipe will be maintained only as long as the compressive force F reaches its critical value. If a = b, rectilinear position of the pipe will be unstable. Then a small impact (e.g. damp soil buoyancy force) is enough to disrupt balance.

$$F_{cr} = \frac{\pi^2 EI}{l^2} \tag{7}$$

where I – axial moment of inertia of the pipe cross section.

At the stage preceding the the moment of bending of the pipeline, there should develop vibrations as a result of instability. Vibrations form must satisfy the geometric boundary condition of the problem. In a linear approximation, one can choose half of a sinusoid:

$$A(x,t) = A(t) \sin\left(\frac{\pi x}{l}\right) \tag{8}$$

where A(t) – amplitude of a pipe vibrations.

After differentiating (8) along the x and t, write down the expression for the potential and kinetic energy:

$$\varepsilon = \frac{1}{2} m \frac{\alpha}{2} \left(\frac{dA(t)}{dt}\right)^2 = \varepsilon_1 \frac{1}{2} \left(\frac{dA}{dt}\right)^2$$
$$V = \frac{1}{2}\left(EI \left(\frac{\pi}{l}\right)^4 \frac{l}{2} - F \left(\frac{\pi}{l}\right)^2 \frac{l}{2}\right) A_{(x,t)}^2 \tag{9}$$

From (9) we obtain an expression for vibration frequency [7]:

$$\omega^2 = \frac{V_1}{\varepsilon_1} = \frac{(\pi/l)^2}{m}\left(EI\left(\frac{\pi}{l}\right)^2 - F\right) \tag{10}$$

where $V_1 = \left(\frac{\pi}{l}\right)^2 \frac{l}{2}\left(EI\left(\frac{\pi}{l}\right)^2 - F\right)$, $\varepsilon_1 = m\left(\frac{l}{2}\right)$.

Given the (6) and (2) find the final equation:

$$\omega_0^2 \approx \frac{\pi^2}{2l}\left[EI\left(\frac{\pi}{l}\right)^2 - \frac{ES}{l}\left(\frac{16}{\pi^2 \rho h} P \sum_{m=1}^{\infty} \frac{\cos \alpha_m \xi \sin \varphi}{m\omega_{mn}^2} \sin \omega t\right)^2\right]$$

$$\omega_0^2 \approx \frac{\pi^4 EI}{2l^3}\left[1 - \frac{Sl}{\pi^2 I}\left(\frac{16}{\pi^2 \rho h} P \sum_{m=1}^{\infty} \frac{\cos \alpha_m \xi \sin \varphi}{m\omega_{mn}^2} \sin \omega t\right)^2\right] \tag{11}$$

From the expression (11) it follows that at the stage of the pipeline bending there occur low-frequency vibrations, which damp with increasing pressure. Allocation of frequency of these vibrations can become a basis for the development of the pipeline condition monitoring method at a stage preceding its bending. In this case, a breach of the geometrical shape is a consequence of the nonlinearity of pressure surges at an unstable gas flow.

Moment of bending of the pipeline as a result of stability loss corresponds to the condition $\omega_0^2 = 0$. Summand in equation (11) is actually the probability of realization of the following:

$$\rho = \frac{Sl}{\pi^2 I}\left(\frac{16}{\pi^2 \rho h} P \sum_{m=1}^{\infty} \frac{\cos \alpha_m \xi \sin \varphi}{m\omega_{mn}^2} \sin \omega t\right)^2 \tag{12}$$

The probability density is a function of distance and time of the pipeline operation:

$$\begin{cases} \xi(x,t) = \dfrac{d\rho(x,t)}{dx} = |A \cdot P_1 K| e^{-kx} + P_2 \\ P = P_0 + P_1 e^{-kx} + P_2 x \end{cases} \tag{13}$$

Figure 2 shows a comparison of the estimated probability density with the same value obtained after the processing of experimental data [8].

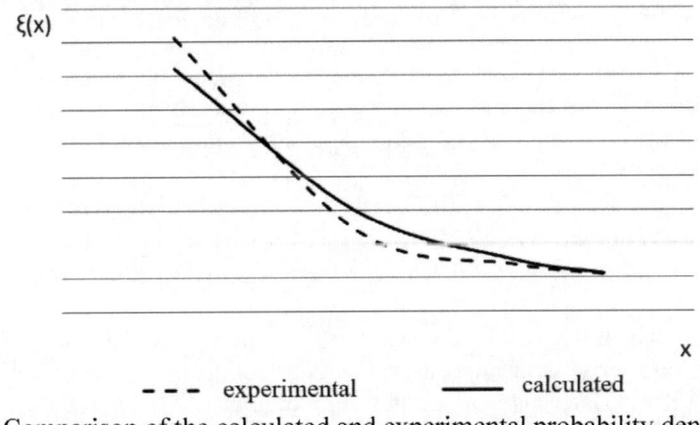

Figure 2. Comparison of the calculated and experimental probability density loss of stability.

In most cases, pipeline bending is associated with the degree of water cut, as its main cause. In paper [8], it is shown statistical analysis of the locations of 99 arch emissions, that possible to establish that hazardous in terms of loss of project positions are sections of the pipeline laid on the first 50 km from the exit of the pump station (76 emissions). Further, from 50 th to 80 th km, there was 15emissions, in the section from the 80th km to the next pump station there was 8 emissions in a fully water-cut trenches. From this it follows that in the area remote from pump station, where pressure surges influence is much lower and the number of pipeline bending is also reduced, despite the total water cut of these areas. Thus, the operation result is coherent with the experimental data that indicate the need to review bending mechanisms associated not only with the ground water cut.

3. Conclusion

As a result, the work carried out it was determined that the appearance of low-frequency vibrations is preceded by a change in the geometric shape of the gas pipeline and consequently a breach of its design position, which is of particular importance for pipelines laid in the soft soil. Forecasting possible emergency situations allows planning and carrying out in due time reconstruction works on sections of gas pipeline with a possible deviation from the design position. Diagnostics gas pipeline sections by identifying changes in the frequency of vibrations of the pipe wall can be the basis for the development of a method of monitoring the technical condition of pipeline.

References

[1] Bachernikov V F Paharukov Yu V 2005 *Vibration and destruction in immersion centrifugal pumps for oil extraction* (Tyumen, TSOGU) 325
[2] Voronin K S 2013 *Forecasting development of the damage on the main gas pipeline under the influence of dynamic load* Thesis for scientific degree of candidate of technical sciences. (Tyumen, TSOGU) 111
[3] Voronin K S and Zemenkov Yu D 2012 Dynamic indicators breach of geometrical shape of the gas pipeline *Proceedings of the higher educational institutions. Oil and gas* **3S** 70-2
[4] Voronin K S and Pavlov V P 2012 The impact of pressure vibrations in the gas pipeline on its deviation from the design position *Materials of the All-Russian scientific-practical conference "Problems of Operation of transport systems"* (Tyumen, TSOGU)
[5] Gorkovenko A I 2006 *Bases of the theory of calculating the spatial position of an underground pipeline under the influence of seasonal processes* Thesis for the degree of Doctor of Sciences (Tyumen, TSOGU) 305
[6] Dimov L A and Bogushevskaya E M 2010 *Pipelines in the conditions of swamps and water cut areas* (M, Gornaya book) 391
[7] Ibragimov A A, Podorozhnikov S Yu, Shabarov A B, Medvedev M V and Zemenkova M Yu 2014 Calculation model and algorithm for determining the residual life of pipeline in the conditions of the periodic variations of stress and corrosion *Mining informational and analytical bulletin (scientific and technical journal)* (Moscow, Gornaya Book) **S4** 199-206
[8] Kurganova I N 1990 The theoretical justification of the results of full-scale survey of gas pipelines in the non-project situation *The reliability of pipeline constructions* (M, VNIIGAZ) 147–55
[9] Oksogoev A A and Slepov B I 2003 *Applied physics. The vibrations of elements of constructions* (Tomsk, NTL) 300
[10] Harinovsky V V 2000 *The reliability and service life of pipeline constructions* (Moscow; Nedra) 467
[11] Gizatulin, R.A., Kozyrev, N.A., Valuev D.V., Valueva, A.V.Serikbol, A., Research on operational characteristics of imported rails under conditions of east siberian railway, J. IOP Conference Series: Materials Science and Engineering. 127 (2016) p. 1-7
[12] Antip'ev, Y.N., Nevolin, A.P., Zemenkov, Yu.D. 1981 Operation of intermediate pumping stations on transport of gas saturated oils. (1981) NEFT. KHOZ, 10, pp. 46 ISSN: 00282448

[13] Kurushina V and Zemenkov Y 2014 Innovative cyclical development of the Russian pipeline system WIT Transactions on Ecology and the Environment 190-2 881-8 DOI: 10.2495/EQ140822

IOP Conf. Series: Materials Science and Engineering **154** (2016) 012021 doi:10.1088/1757-899X/154/1/012021

Research and Design of Thermophysical Gas-Liquid Mixture Parameters in Product Pipelines

S M Dudin, Yu D Zemenkov, A V Maier and A B Shabarov

Industrial University of Tyumen, 38 Volodarskogo st., Tyumen, 625000, Russia

E-mail: srghome@mail.ru

Abstract. Operational problems are hard to overcome because of the temperature and pressure conditions of the hydrocarbon flow in the pipe, as well as the composition of the hydrocarbon system and the geometry of the pipeline. It is known that energy costs to pump a unit mass of RH in the form of gas 2-3 times exceed energy costs to pump a unit mass of RH in the form of liquid. As far as energy conservation during RH transportation is concerned, an important task is development and application of a method to calculate the gas-liquid hydrocarbons flow, and heat and mass transfer in process and trunk pipelines during their design and operation. The authors have developed a calculation method which is used to analyze the hydrodynamic state and composition of the hydrocarbon mixture in each ith section of the pipeline when temperature-pressure and hydraulic conditions change. The developed technique was tested on the hydrocarbon mixture of de-ethanized condensate and oil transported from northern oil and gas condensate fields via the main gas condensate line to the refinery.

1. Introduction

Gas condensate fields demonstrate a number of thermodynamic characteristics to be considered when they are developed, as well as when gas condensate is transported and processed. A complicated phase behavior of the gas condensate system, as well as the dependence of the extracted raw materials on the phase state of the deposit other conditions being equal, is a key aspect. Therefore, when designing gas condensate lines the crucial task is to select the most appropriate methods of calculating thermophysical properties and phase equilibrium of the transported gas condensate.

2. Methods, results and discussion

Methods of calculating thermophysical properties (TPP) and phase equilibria are reviewed in quite a lot of scientific publications and reference books [1, 2, 3, 4, 6, etc.]. However, when it comes to such complex systems as oil, gas condensate, natural gas and their by-products, using even the known techniques is a difficult task. As they are located in the near-critical region on the phase diagram, these complex systems are regarded as transition state systems that depending on temperature and composition can be in a single-phase liquid or gaseous state. This uncertainty complicates selection of methods (processes) for design and calculation of condensate lines.

Over the last decade the department of Transportation of Hydrocarbon Resources in Tyumen State Oil and Gas University (TSOGU) has been conducting research in the field of process monitoring in pipeline energy transportation systems. To date, a number of physical and mathematical models have been developed describing the process of transportation and storage of hydrocarbons in the oil and gas industry facilities. The latest development of the authors is a physical-mathematical model of a hydrocarbon mixture flow in the gas condensate line which allows you to perform an operational analysis of the hydrodynamic state of condensate in the pipeline. The main closing relations of the physical-mathematical model are the following.

1) The dependence of the gas condensate mixture density on pressure and temperature.

$$\rho_{C6+} = \frac{Nm_n}{\frac{1}{\rho sm_0} - \sum_{i=1}^{n-1} \frac{Nm_i}{M_i} \upsilon_i}, \tag{1}$$

where υi is the molar volume of other components calculated for both gas and liquid by a single Peng-Robinson equation of state.

2) The viscosity of the condensate mixture is calculated at temperature of 20°C and pressure of 1 atm by the formula:

$$\mu sm_0 = \frac{\sum_{i=1}^{N} z_i \sqrt{M_i} \mu ind_i}{\sum_{i=1}^{N} z_i \sqrt{M_i}} \tag{2}$$

Recalculation of viscosity on thermobaric conditions in different sections of the pipeline is carried out according to the formula:

$$\mu sm = 10^{\lg\left(\mu sm_0\right) + 1.4503\cdot 10^{-3}\cdot\left(\frac{P}{10} - 0.101325\right)\cdot\left(5.2054\cdot\mu sm_0^{0.278} + 0.0239\right) - 3}. \tag{3}$$

3) The isobaric heat capacity of the gas condensate mixture is calculated according to the additivity rule:

$$Cpsm = \sum_{i=1}^{N} z_i \frac{Cp_i}{M_i} \tag{4}$$

4) The soil heat transfer is found by:

$$\alpha_{\text{гр}} = \begin{cases} \dfrac{2\cdot\lambda_{\text{гр}}}{D\cdot\ln\left(\dfrac{2\cdot H}{D} + \sqrt{\left(\dfrac{2\cdot H}{D}\right)^2 - 1}\right)}, & if\ \dfrac{H}{D} > 3(\text{Forchheime r - Vlasov}) \\ \dfrac{2\cdot\lambda_{\text{гр}}}{D\cdot\ln\left(4\dfrac{H_{\text{пр}}}{D} + \dfrac{1}{Nu_{\text{гр}}}\right)}, & if\ \dfrac{H}{D} < 3(\text{Arson-Kutateladze}) \end{cases} \tag{5}$$

where H – the depth of the pipeline occurrence in the soil, m.

5) The composition and quantitative ratio of the equilibrium vapor and liquid phases is found using equations of phase concentrations of mixture components.

$$x_i = \frac{z_i}{V(K_i - 1) + 1}, \tag{6}$$

$$y_i = \frac{z_i K_i}{V(K_i - 1) + 1}. \tag{7}$$

6) The actual volumetric gas content is found by:

$$\alpha_g = \frac{1}{1 + \frac{\upsilon_L}{\upsilon_g}\frac{N_L}{N_V}} = \frac{1}{1 + \frac{\upsilon_L}{\upsilon_g}\left(\frac{1}{V} - 1\right)}. \tag{8}$$

7) The flow pattern of the gas-liquid mixture is determined by one of the known techniques, for example, by the methods shown in [7, 8]. For horizontal flows, various authors identified and described the following structural forms of the gas-liquid mixture flow (figure 1).

IOP Conf. Series: Materials Science and Engineering **154** (2016) 012021 doi:10.1088/1757-899X/154/1/012021

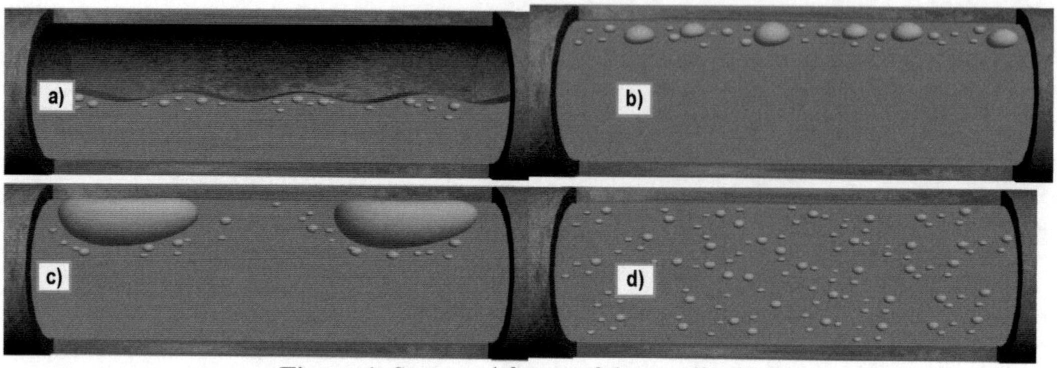

Figure 1. Structural forms of the gas-liquid flow:
a – separated; *b* – bubbly; *c* – slug; *d* – emulsion.

In commercial pipelines the most common structural forms of the flow are slug and emulsion. There are no clear interfaces between the individual forms of the gas-liquid mixture flow, as there are relatively broad transition zones both in speed and in the gas content.

A characteristic feature of the two-phase hydrocarbon flow patterns is pressure surge, for example in the case of pipeline transportation of gas-liquid mixtures. Pressure surge in a pipeline leads to disruption of normal operation of the pumping equipment, control and measuring devices, etc.

Operational problems in pipelines causes increase in friction loss, hydrostatic pressure drop and often a complete blockage of the pipeline cross-section (formation of local gas clusters, water slugs and crystalline hydrates).

Operational problems are hard to overcome because of the temperature and pressure conditions of the hydrocarbon flow in the pipe, as well as the composition of the hydrocarbon system and the geometry of the pipeline.

The developed physical and mathematical model of the hydrocarbon flow in the pipeline, taking into account a change in the composition of the liquid and gas phases, allowed us to conduct computational and parametric study of thermophysical parameters of oil and gas condensate mixture in various product pipeline sections.

To verify the adequacy of our technique for calculating operating parameters of the product pipeline, let us perform a comparison of the calculated values with the experimental values.

Let us consider the developed calculation technique as applied to the hydrocarbon mixture of de-ethanized condensate and oil transported from oil and gas condensate fields via the product pipeline.

Table 1. Performance characteristics of the product pipeline.

Composition, mass fraction %							
Methane	Ethane	Propane	i-Butane	n- Butane	i-Pentane	n- Pentane	Hexane +
The parameter values in the initial section of the pipeline							
0.09	0.67	12.98	7.02	11.63	5.9	6.33	55.38
The parameter values in the end section of the pipeline							
0.07	0.6	12.9	11.8	7.1	6.04	5.72	55.5

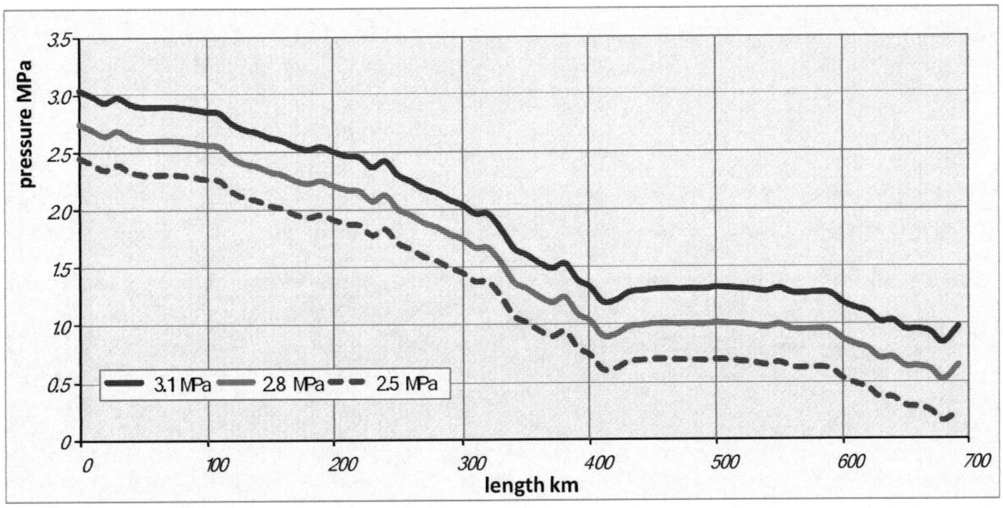

Figure 2. Changes in RH pressure along the condensate line route at various pressures in the inlet section.

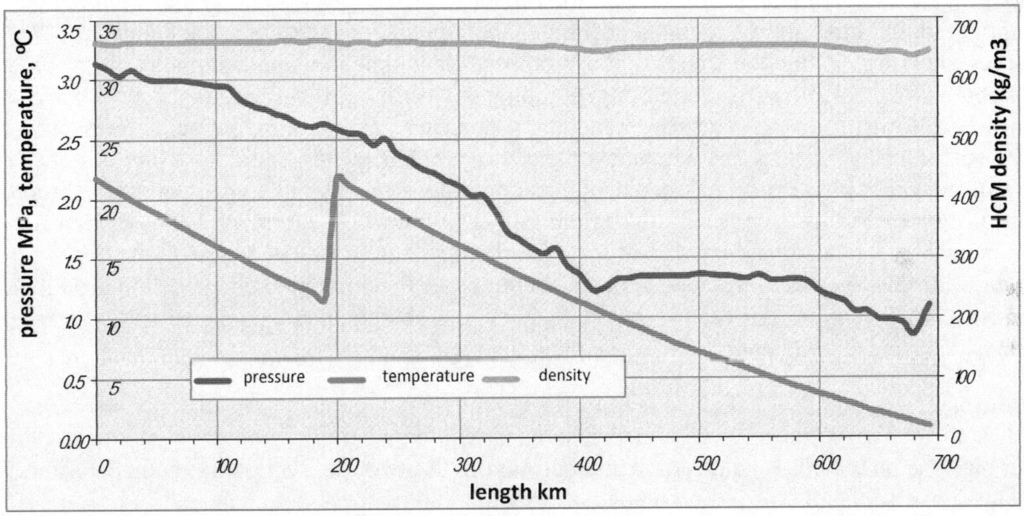

Figure 3. Hydrocarbon gas-liquid mixture parameters along the product line route.

Figure 3 shows the results of calculating hydrocarbon mixture parameters along the pipeline route. By comparing the calculated and actually observed differential pressure and temperature, we can conclude on the adequacy of the selected computational physical and mathematical model of the condensate line (relative deviation of the calculated differential pressure from actually observed is 4.6% and deviation of the estimated temperature change from the experimental was 1.5%).

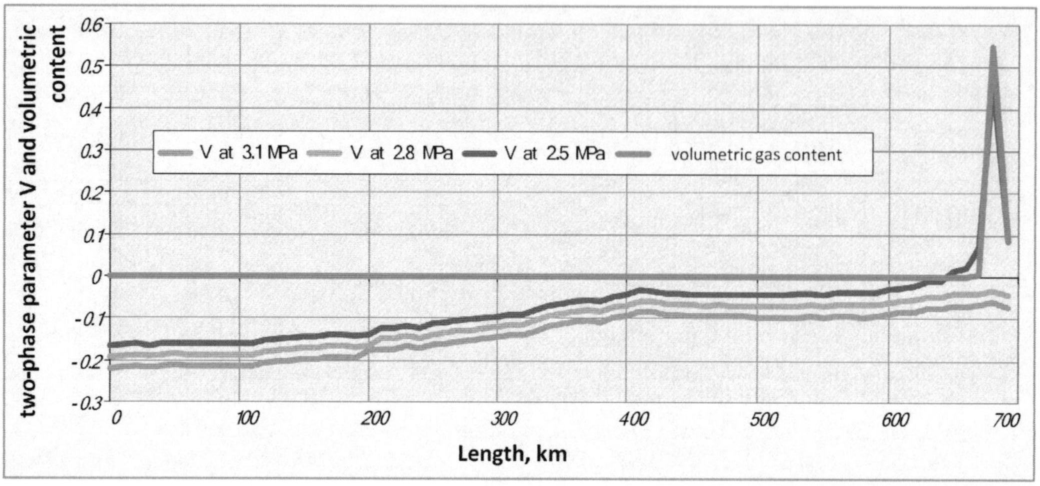

Figure 4. Changes in hydrocarbon mixture phase state parameters along the product line route at differential product pressure in the inlet section.

Figure 4 shows a computational and parametric study of RH phase state along the pipeline when the pump operating pressure in the initial section of the pipeline changes to 3.1, 2.8 and 2.5 MPa, with the curves showing changes in the two-phase parameter V and the true volumetric gas content of hydrocarbons when lowering the operating pumping pressure by 0.6 MPa at the beginning of the pipeline. It follows from these graphs that with a decrease in operating pumping pressure on the condensate line may arise areas where hydrocarbons enter the gas-liquid two-phase state (molar fraction of the vapor phase is $0<V<1$ and acquires a definite physical meaning, equal to the ratio of the number of moles of the mixture in the gas phase to that of the whole mixture). When the hydrocarbon mixture passes in the two-phase gas-liquid state the mixture moves in a two-phase state to the pipeline end section. In this case, the subsequent calculation of hydrodynamic parameters of the product line is carried out taking into account the two-phase flow of hydrocarbons. A two-phase flow pattern in the pipeline is determined by taking into account the known algorithm [8].

3. Summary

1. The influence of temperature and pressure conditions as the hydrocarbon mixture enters the product pipeline on the phase equilibrium and parameters in the outlet section has been determined.
2. The convergence of calculation results of pressure and temperature changes has been confirmed based on the developed model with known full-scale experimental data.
3. For the first time the possibility of using calculations and theoretical principles to determine the valid values of mass concentration of light hydrocarbons in the inlet section of a main condensate line on condition that a single-phase flow regime along the entire length of the pipeline is provided.
4. The developed calculation technique and algorithm allow us to perform a real-time analysis and prediction of the hydrodynamic condition of gas-liquid media in pipelines as the composition, temperature and pressure, and hydraulic conditions change. Therefore, the study results can be used by the control services of oil and gas companies during on-line monitoring of raw hydrocarbon pumping modes.

References

[1] Brusilivsky A I 2002 Phase transformations in the development of oil and gas fields (Graal, Moscow) 575p
[2] Guzhov A I, Titov V G, Medvedev V F and Vasilyev V A 1978 *Collection, transportation and storage of natural hydrocarbons* 223p

[3] Grigoryev B A, Gerasimov A A and Lanchakov G A 2007 *Thermophysical properties and phase equilibria of gas condensates and their fractions* (Publishing House MEI, Moscow) 344p

[4] 2002 *Diagnosis of damage and leakage in the pipeline transportation of multiphase hydrocarbons* Ed. Yu D Zemenkov (Vector Buk, Tyumen) 432p

[5] Dudin S M, Nekrasov V O and Zemenkov Yu D 2013 Physical and mathematical modeling of technological modes of transportation and storage of hydrocarbon fluids in pipeline systems. *Oil & Gas: Special issue of Mining informational and analytical bulletin* (Gornaya Kniga, Moscow) 53-62

[6] Reid R, Prausnitz J and Sherwood T 1982 *Properties of Gases and Liquids: A Reference Guide* Transl.ed. B I Sokolov (Chimia, Leningrad) 592p

[7] Guzhov A I, Titov V G, Medvedev V F et al 1978 *Collection, transportation and storage of natural hydrocarbon gases: Textbook* (Nedra, Moscow)

[8] Shabarov A B, Saranchin N V and Kutrunov V N 2005 Automated calculation of the flow of oil and gas mixture in the pipeline *Modernization of education in the context of globalization: Round Table "Education through science and innovation"* 126-29

[9] Valuev D.V. Malushin N.N.,Valueva, A.V. Dariev, R.S. Mamadaliev R.A., Traditional technology of chromium-tungsten steels facing, its disadvantages and suggestions for their eliminations, J. IOP Conference Series: Materials Science and Engineering. 127 (2016) p.1-7

[10] N.N. Malushin, A.P. Kovalev, D.V. Valuev, E.A. Shats, I.F. Borovikov, development and study of hard-facing materials on the base of heat-resisting high-hardness steels for plasma-jet hard-facing in shielding-doping nitrogen atmosphere, J. IOP Conference Series: Materials Science and Engineering. 142 (2016) p. 1-7

[11] Chekardovskiy, M.,Chekardovskiy, S.,Ilyukhin, K.,Gladenko, A. Upgraded Algorithm for Calculating the Turbo-Expander of Gas Distribution Stations 2016 MATEC Web of Conferences 73

Determining Fuel Losses in Storage Tanks Based on Factual Saturation Pressures

R E Levitin and R A Tryascin

Industrial University of Tyumen, 38 Volodarskogo st., Tyumen, 625000, Russia

E-mail: 89028130230@mail.ru

Abstract. At present, evaluation of fuel evaporative losses is based on a number of indirect parameters. Accuracy of such methods leaves much to be desired. The paper presents a method developed following the author's laboratory tests. An effective operation range of pressure vent valves in various tanks is provided, as well as low A92 gasoline losses for horizontal steel tanks at all operation temperatures.

1. Introduction

More than half of the operational losses are evaporative losses. They can be classified into two types: quantitative, caused by evaporation and qualitative, caused by changes in fluid composition. To assess qualitative losses a lab test of the substance composition must be carried out, but assessment of quantitative losses is done, as a rule, by calculation. To obtain factual values rather than calculated it is necessary to measure the volume of the displaced gas-vapor mixture and its vapor concentration. Such measurements must be conducted constantly over the entire life of the tank, which is connected to some technical complexity.

Thus, for instance, using air flowmeters is not always justified as the displaced mixture volume may be relatively small at small breathing and not fall within the flowmeter operating capacities. Determining vapor concentration in the displaced mixture requires constant sampling with subsequent lab analysis, which can be unprofitable and highly labor-intensive.

When determining oil and petroleum products emissions, Russian scientists F.F. Abuzova, P.A. Rybakov, N.I. Tikhonov and other researchers based their calculation methods on such parameter as saturated vapor pressure which was used to determine hydrocarbon tendency towards evaporation. [1,2]

Saturated vapor pressure (SVP) is a parameter that essentially depends on hydrocarbon composition, phase ratio and temperature and is described by the following dependences:

1. Simplified calculation of SVP within the temperature range from –30 to +100 °C according to P.A. Rybakov: $P_{SVt} = p38 \cdot 104.6\text{-}1430/T$, mm of mercury.
2. Dependence of petroleum products SVP on phase ratio, according to N.I. Tikhonov

$$p_{SVt} = p_{38}\left(\frac{14}{10+V_V/V_L}\right)^{0.31} 10^{4.0283-1252/T} \tag{1}$$

3. According to F.F. Abuzova,

$$P_{SVt} = 1.29(t/38)^{0.69}(V_L/V)^{0.19}p_{38}, \tag{2}$$

where T,t – the temperature of the liquid, P_{SVt} – the saturated vapor pressure, V_V – the gas phase volume, V_L – the liquid phase volume.

However, these dependences are significantly simplified. For example, P.A. Rybakov determines SVP depending on temperature but irrespective of phase ration. F.F. Abuzova and N.I. Tikhonov determine SVP depending on temperature and phase ratio, but in some cases the obtained dependences have a narrow temperature range. There is also SVP dependence obtained by the CSRL (Central science and research laboratory) which is directly proportional to temperature. However, the experimental data are outdated and require updating due to a significant change in hydrocarbon composition

2. Research object

To estimate the amount of losses we suggest determining the saturated vapor pressure (SVP). This value, with minimum indirect measurements, allows us to predict quantitative changes when storing fuel. Such multi-component substances as petroleum products have an extremely complex composition due to which their phase transformation happens in a more complicated way than that of chemically homogeneous substances. During vaporization, lighter cuts gradually evaporate and liquid phase becomes heavier. Thus, partial pressures of single hydrocarbons also change and total petroleum products saturated vapor pressure is always 10-30% higher than can be expected according to the law of additivity.

To determine motor spirits SVP based on the government standard an experimental unit was designed that allowed determination of pressure exerted by the gas-vapor mixture in a model tank at various phase ratios. The temperature of the tested motor spirits was changed within the operation range by means of a hydronic cryothermostat LOIP FT-311-80.

A series of lab test was conducted to determine motor spirits SVP depending on temperatures at various phase ratios within the operation range (figure 1). A full-scale experimental study involved placing gasoline A 92 in vertical cylindrical tanks 3000 m^3 in nominal volume.

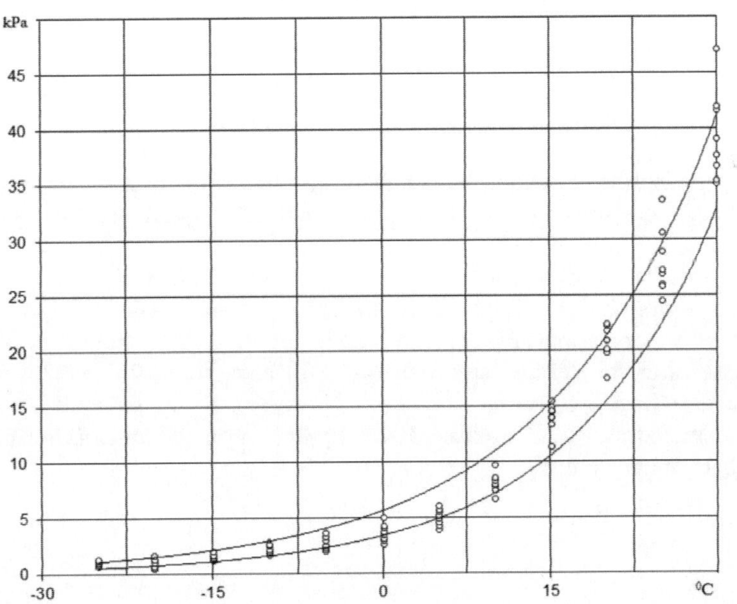

Figure 1. Dependence of gas-vapor mixture pressure (kPa) on temperature (^{0}C) when the tank is 20-95% filled.

For the purpose of precision, factual losses were measured in tanks with the same type pressure vent valves located at the same tank battery and within the same time and temperature interval.

3. Results and discussion

The study produced experimental data on changes in the gas space pressure and gas-vapor mixture volume.

The data were processed considering hydrocarbon concentration in the gas-vapor mixture and lost product masses were obtained (figure 2).

From the graphs we can see that gasoline has a high level of evaporation due to lack of means to reduce hydrocarbon losses in the test tanks. It should also be mentioned that the obtained test data strongly depend on temperature. Natural loss norms for petroleum products do not have such dependence.

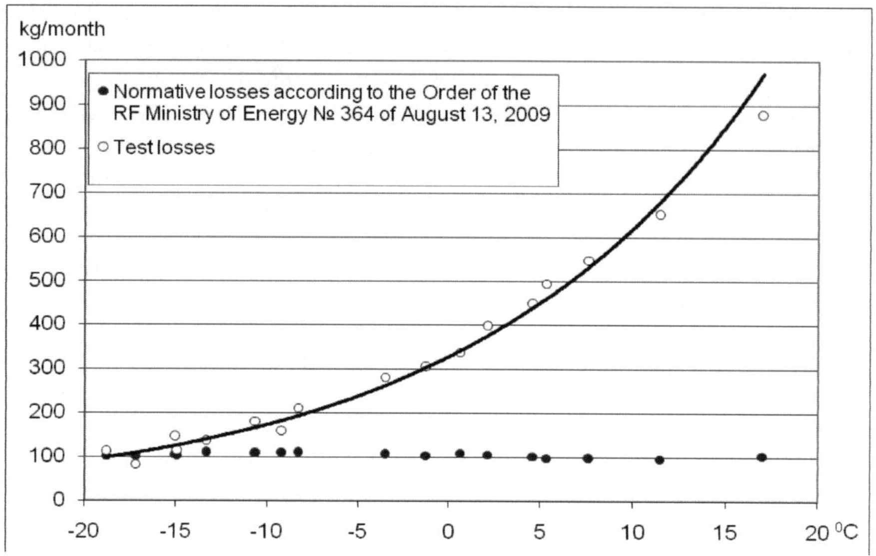

Figure 2. Gasoline vapor emissions in storage tanks.

Curves of factual gasoline losses are higher than the normative losses determined by the order № 364 "On establishing the norms of natural losses of petroleum products during storage" beginning from -15 ^{0}C.

Moreover, constructing a dimensional graph of hydrocarbon vapor pressure dependence on temperature and phase ratio for gasoline A92 (figure 3) we see that pressure vent valves are not always efficient. Their effective operation range (green area in the picture) is from 0 to –25 degrees Celsius depending on phase ratio for vertical steel tanks (VST) and from –25 to +25 degrees Celsius for horizontal steel tanks (HST).

Figure 3 also demonstrates that utilizing HSTs hydrocarbon vapor emissions can be eliminated without burying them underground.

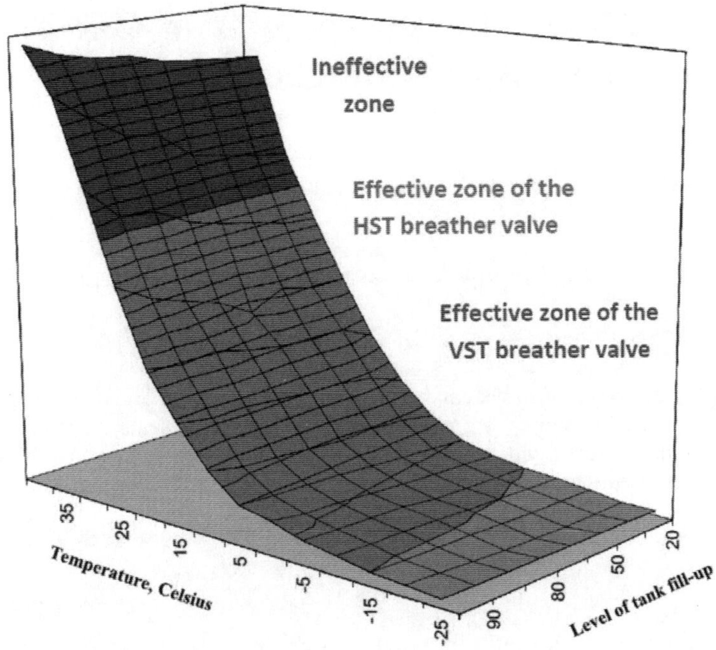

Figure 3. Effective zone of a tank breather valve (gasoline A92)

4. Summary

The obtained saturated vapor pressure dependences on temperature show that if there is even a slight change in average temperature for as little as 5 degrees Celsius, SVP can increase up to 100%. This results in a great error of the averaged norms.

The indirect dependence of natural petroleum product loss norms on temperature causes distortion of real hydrocarbon losses during storage.

References

[1] *Operation of Trunk and Process Oil and Gas Pipelines. Distribution and Accounting: Study Guide* ed Yu D Zemenkov 2014 (Tyumen: TSOGU) 370p.

[2] Patent 2500486 RU, MPK B05C7/04, B65D88/74. *A device to improve the performance properties of vertical steel tanks* V O Nekrasov, R E Levitin, I V Tyrylgin and Yu D Zemenkov (RU) -№ 2012125478/05, filed 10.12.13.

[3] API Manual of Petroleum Measurement Standards, Chapter 19 - Evaporative-loss Measurement, Section 1 - Evaporative Loss from Fixed-Roof Tanks Third Edition, March, 2002.

[4] API Manual of Petroleum Measurement Standards, Chapter 19 - Evaporative-loss Measurement, Section 2 - Evaporative Loss from Floating-roof Tanks, Second Edition, September, 2003.

[5] VDI 3479: *Emissionsminderung*. Raffinierieferne Mineralöltanklager, 2010.

IOP Conf. Series: Materials Science and Engineering **154** (2016) 012023 doi:10.1088/1757-899X/154/1/012023

A Frequency Model of Vibrational Processes in Gas-Turbine Drives of Compressor Stations of Main Gas Pipelines

M N Chekardovskiy, S M Chekardovskiy, A A Razboynikov and T G Ponomareva

Industrial University of Tyumen, 38 Volodarskogo st., Tyumen, 625000, Russia

E-mail: zemenkov@tsogu.ru

Abstract. At compressor stations, systematic measurements of noise and vibration of power equipment - gas compressor units - are carried out. The article presents basic equations for calculating natural and forced frequencies at which the main defects appear. According to the studied dependences, results of calculations are obtained on the following types of drives for gas-compressor units GTK-10-4, Avon-1534, DG-90.

1. Introduction

If vibration levels exceed the allowable, breakage of parts and components may occur, leading to failures of power equipment. The presence of excessive vibration and noise indicates the presence of various defects, which are one of the reasons for the accelerated wear and fatigue failure of parts. Measuring vibration and noise during operation can detect many defects at their early stages, prevent possible failures of equipment, as well as put it into operation and repair according to its actual state, instead of conducting planned preventive maintenance.

Methods of vibroacoustic diagnostics provide ample opportunities. They are based on the use of information contained in the oscillatory processes that accompany operation of machines and mechanisms. Vibroacoustic diagnostics is one of technical diagnostics sections. As a source of information, it uses not a series of static parameters characterizing the state of a mechanism, but a number of dynamic parameters causing the occurrence of vibration and acoustic waves. Vibration signal providing capacious enough information on the work of the unit and its components can be a good indicator of its status.

2. Research

Vibration and noise of the axial compressor (AC) of gas compressor units (GCU). The vortex noise of the compressor and the high-pressure turbine (HPT) arises from vortices occasionally shedding from the blades, the turbulent flow, and the turbulent jet. The spectrum of the vortex noise caused by periodic vortex shedding is determined from the equation [1, 2, 3]

$$f = 0.212\left(1-\frac{21.2}{Re}\right)\frac{D}{d}, \tag{1}$$

where Re=Vd/ν- the Reynolds number; d – the determining linear dimension; V – the linear flow velocity, ν - the viscosity of the medium.

The level of wideband noise can be determined according to the equation:

$$\alpha = 10\lg\left[k'(B\Delta T)^{1.5}\left(\frac{B\Delta T}{z^{0.27}\,Re^2}\right)10^3\right], \tag{2}$$

where α - the sound pressure level, B – the flow of steam through the HPT or air through the AC, ΔT – the temperature differential in the HPT or AC, z– the number of blades, k′ - the coefficient determined from the experimental data on the measured noise level and the geometric characteristics of the AC or HPT.

The level of noise of the discrete components [1, 2, 3]:

$$\alpha = 10\lg\left\{ k''B^2\eta\Delta T / \left[D_{р}^2(1-d_{в}^2)Re^2\,10^{-3}\right] \right\}, \tag{3}$$

where D_P – the impeller diameter, d_B – the hub-tip ratio, η - the efficiency of the first stage, k″ - the coefficient determined experimentally.

For multistage compressors, the acoustic power is determined by summing the power of each stage.

During operation, the AC rotates in a pulsating flow, whereby each blade transfers momentum to the working fluid that is a multiple of the number of blades. When summing these impulses and subsequently transmitting to the body a vibration is excited with frequencies that are multiples of the number of blades [1, 2, 3]:

$$f_{bl} = i \cdot z \cdot f_p \tag{4}$$

where i – the harmonic number, i=1,2…n, f_P – the rotor rotational frequency. The amplitude of the blading vibration depends on the steam or air density (the ratio of expansion or compression pressures) and some other factors; the smaller the density, the smaller the aerodynamic imbalance of the working fluid flow. The vibration frequency during a developed shedding is determined by the equation [1, 2]:

$$f_{cp} = (0.3 \div 0.6)\, f_p \tag{5}$$

Vibration of the rotor. The vibration state of the AC rotor can best be determined by the speed characteristics when accelerating or running down the rotor.

The oscillatory system rotor-body may be in a certain critical condition in the presence of a number of critical rotational frequencies of the rotor. These frequencies can be higher, lower or within the range of the operating frequency, which is highly undesirable. Reducing the vibration amplitude in this case is achieved by using different damping devices or by balancing.

The critical rotational frequencies of the rotor are calculated by the equation [4]:

$$f_{crit} = K\sqrt{\frac{EI}{L^2}} \tag{6}$$

where K – the coefficient obtained as a result of calculating the system of equations of small oscillations of the rotor, E – the modulus of elasticity, I – the reduced moment of inertia, L – the rotor length.

For the axial compressor blades, the range of static natural frequencies of the first three bending forms of vibrations in the plane of its least stiffness is determined by the equation of I.M. Meerovich:

$$f_{\text{bend}} = 0.0398\alpha_{i\text{bend}}\beta_{i\text{bend}}\frac{\delta_k}{l^2}\sqrt{\frac{0.8\psi^2 a_k^2\gamma_i}{\rho\delta_k^2(1+4\Delta_i\varphi_0^2)}} \tag{7}$$

where $\alpha_{i\text{bend}}$, $\beta_{i\text{bend}}$, ψ_i, γ_i, Δ_i – for each waveform there is a wedge, curvature and radial function, a_κ, δ_κ – the maximum curvature and thickness of the blade profile at the root section, φ_0 – the feather twist (the angle between the chords of the root and peripheral sections), l – the length of the blade feather.

The spectrum of the natural frequencies of torsional oscillations of the axial compressor rotor blades is determined by the equation [1, 2]:

$$f_{\text{cr}} = 0.343\alpha_{i\text{cr}}\beta_{i\text{cr}}\frac{\delta_k}{b_{cp}l}\sqrt{\frac{G}{\rho}} \tag{8}$$

where α_{icr}, β_{icr} – the coefficients determined by [1, 2], b_{cp} – the width of the blade chord in the middle part, G – the shear modulus.

The rotor unbalance, bending by the first form, the thermal instability of the rotor occur at a frequency [1, 2, 3]:

$$f_{o.r}=f_p \tag{9}$$

The uneven stiffness of the rotor cross-section, the resonant oscillations of the rotor blades due to loose connections, the misalignment of couplings, the rotor bending by the second form occur at a frequency [1, 2]:

$$f_{p.k}= 2f_p \tag{10}$$

Rotor elements touching the stator, separation of supports from the foundation [1, 2]:

$$f_{з.p}= (2.5\div 2.8)\ f_p. \tag{11}$$

Vibration and noise of the bearings. This noise is an indication of a malfunction, since it is not possible to determine it if the bearings are serviceable. The main defects of the bearings are the waviness on the bearing races, the increase in the radial clearance, oval balls and rings, the unroundness of rolling elements, increasing clearances in the sockets of separators, which we denote by (b). Then the forced frequency of faults is determined by the general equation:

$$f_n = b\cdot f_p \tag{12}$$

The value (b) is determined by the geometric dimensions of bearings. Specific equations are given in [1, 2, 3].

The frequency content of the vibration spectrum of the defective bearing depends on the type of defect. Working with defective bearings we connect the five main bearing frequencies:

1 – F_r- the rotor rotational frequency;
2 – F_{sep}- the separator rotational frequency;
3 – $F_{o.r}$- the ball (roller) frequency of passing the outer ring;
4 – $F_{i.r}$- the ball (roller) frequency of passing the inner ring;
5 – $F_{r.b}$- the frequency of natural rotation of the ball, roller.

The frequency of natural ball rotation is excited when the defect on the ball or roller meets the race track. The frequency may equal 2FTK, since its defect meets both rings. Excited frequencies seldom reach such high values as:

1. the ball is not always in the loading zone when the defect meeting the rings, and
2. the energy is lost when the signal passes through additional structural joints, when the defect meets the inner ring.

The frequency of natural rotation is also excited when the balls are pressed against the separator or the separator is broken. The separator rotational frequency is equal to the rotational speed of a set of balls or rollers and the separator around the axis of the shaft.

Noise of the reduction box gear wheels. This noise occurs when the gears collide in the short-term contact during operation. The short-term contact leads to the appearance of stress that is cyclical in nature and in turn causes stress wave propagation along the gear wheel and the emergence of noise. The level of noise of gear wheels depends on their residual imbalances, compression of air and oil, torsional oscillations of shafts, irregular shock loads, and uneven teeth surface.

The frequency of contact meshing noise is determined by the equation [1,2]

$$f_z = l \cdot z \cdot f_p \tag{13}$$

where z – the number of teeth.

The natural oscillation frequency is determined by the formula [1,2]:

$$f_c = [\lambda / (2\pi)]\sqrt{c/m}, \tag{14}$$

where λ - the factor considering the system flexibility; c - the teeth stiffness, m - the reduced gear wheel mass.

Vibration and noise of the combustion chamber. Due to the fact that the combustion chamber is located behind the air intake in the GTU and AC, its noise is the resultant noise caused not only by the combustion process, but also by the previous process. The combustion chamber noise, as well as the AC, is continuous in nature and has a wideband range. The most dangerous is the vibrating combustion that occurs as a result of strong vortices in the flow of gas mixture.

A stable combustion mode gives the maximum value of amplitudes in the low frequency zone of 40 ÷ 60 s^{-1} , and during the vibrating one the amplitude shifts to the zone of 200 ÷ 300 s^{-1} and above. Flame tubes of the combustion chamber of the investigated GCU are a cylindrical shell simply supported at the ends, so the approximate value of frequencies of natural oscillations is described by the equation [1, 2]

IOP Conf. Series: Materials Science and Engineering **154** (2016) 012023 doi:10.1088/1757-899X/154/1/012023

$$f_{ck} = \frac{\overline{\omega}}{R}\sqrt{\frac{E}{\rho(1-\mu^2)}} \tag{15}$$

where $\overline{\omega}$ - a dimensionless frequency of the normal mode of vibration supported at the ends of the cylindrical shell; R – the radius of shell curvature; E, μ,ρ- the modulus of elasticity, the Poisson's ratio and the density of the shell material.

Vibration and noise of the GCU turbine rotor. The first form of static natural bending vibrations for turbine blades is determined by the equation of A.E. Shneydman

$$f_{ct} = \frac{4}{l^2}\sqrt{\frac{E}{\rho \cdot x}} \tag{16}$$

where x - a function depending on the area and moments of inertia of the cross sections of the blades.

The relationship between the static and dynamic (f_d) natural oscillation frequency of the blades [1, 2]:

$$f_p = \sqrt{f_{cm}^2 + \beta f_p^2} \tag{17}$$

where β - the dynamic frequency coefficient determined from the literature data [1, 2].

Table 1 shows the results of the calculation of natural and forced oscillation frequencies of main units of the gas generator Avon-1534 at f_p=125 Hz.

Combination frequencies. In addition to the abovementioned fundamental frequencies excited on the body of a product by rotors, blades, bearings, a gearbox and drive units, the occurrence of combination frequencies is possible, the general formula of which is given by:

$$f_k = K_i \cdot n_i + K_j \cdot n_j \tag{18}$$

where n_i and n_j- fundamental frequencies, K_i and K_j- integer-valued coefficients.

A combination frequency corresponds to a certain physical model of excitation, different for each individual case.

Table 1. An example of calculation of the frequency components of vibration processes in the main units of the gas generator Avon-1534.

Axial compressor									
Stages	f_{bl}, Hz	f_{cp}, Hz	f_{crit}, Hz	f_{bend}, Hz	f_{cr}, Hz	$f_{o.r}$, Hz	$f_{p.k}$, Hz	F_n, Hz	$f_{z.p}$, Hz
00	2750	37.5÷75	Rotor of the gas generator f_{crit1}=509 f_{crit2}=2088	196	865	125	250	>250÷55000	312÷360
0	3625			198	984				
1	4625			201	1187				
2	6625			216	1267				
3	7875			219	1290				
4	9125			270	1536				
5	11125			298	1864				
6	12125			323	1898				

7	12125			327	1945				
8	12125			334	1978				
9	12625			339	1989				
10	12625			370	2162				
11	12625			390	2230				
12	14125			420	2290				
13	14125			450	2440				
14	14125			500	2500				
15	14125			560	2560				

Combustion chamber	f_{ck}, Hz
	741÷4072

High pressure turbine			
Stages	f_{bl}, Hz	f_{ct}, Hz	f_d, Hz
1	10375	1027-921	1067
2	8875	423-379	492
3	7375	246-220	285

3. Conclusion

With the help of the vibrating diagnosing method, a direct control of the dynamic force action is carried out, making it possible to detect and prevent the development of a defect at an early stage. So, damage to individual elements of the gas turbine unit causes an instantaneous change in the level and frequency characteristics of the vibration spectrum. The system of diagnostics by the frequency model will increase the service life of individual components, reduce the number of unscheduled repairs and downtime, and improve the reliability and efficiency of gas compressor units in general.

References

[1] Antonova E O, Ivanov I A, Stepanov O A and Chekardovskiy M N 1998 *Monitoring of power units at compressor stations* (St. Petersburg; Nedra) 216

[2] Krylov G V, Chekardovskiy M N, Yakovlev E I and Bloshko N M 1990 *Technical diagnostics of gas transmission mains* (Kiev; Naukovadumka) 303

[3] Rusov V A 1996 *Spectral vibrodiagnostics* (Perm; Vibro-Centr) 176

[4] Chekardovskiy S M, Razboynikov A A and Chekardovskiy M N 2014 *Diagnostics and elimination of equipment vibration in oil and gas facilities* (Tyumen; TyumGNGU) 108

[5] Zemenkova, M., Shalay, V., Zemenkov, Y., Kurushina, E. 2016 Improving the Efficiency of Administrative Decision-Making when Monitoring Reliability and Safety of Oil and Gas Equipment MATEC Web of Conferences 73 Volume 73, 11 August 2016, Article number 07001 Code 123284; DOI: 10.1051/matecconf/20167307001

[6] Shalay, V.,Zemenkova, M.,Zemenkov, Y.,Toropov, S. 2016 Modeling Parameters of Reliability of Technological Processes of Hydrocarbon Pipeline Transportation MATEC Web of Conferences 73 Volume 73, 11 August 2016, Article number 01029 DOI: 10.1051/matecconf/20167301029

[7] Valuev D.V. Malushin N.N.,Valueva, A.V. Dariev, R.S. Mamadaliev R.A., Traditional technology of chromium-tungsten steels facing, its disadvantages and suggestions for their eliminations, J. IOP Conference Series: Materials Science and Engineering. 127 (2016) p. 1-7

[8] Zemenkova, M.Y.,Shipovalov, A.N.,Zemenkov, Y.D. 2016 Mathematic Modeling of Complex Hydraulic Machinery Systems When Evaluating Reliability Using Graph Theory IOP Conference Series: Materials Science and Engineering 127 (1) DOI: 10.1088/1757-899X/127/1/012056

IOP Conf. Series: Materials Science and Engineering **154** (2016) 012024 doi:10.1088/1757-899X/154/1/012024

Modern Prototyping as a Basis for Industrial Design

I A Chekardovskaya and L N Bakanovskaya

Industrial University of Tyumen, 38 Volodarskogo st., Tyumen, 625000, Russia

E-mail: zemenkov@tsogu.ru

Abstract: The paper presents methodological bases for designing object-oriented processes as applied to industrial design via the targeted activity and system approach to make production efficient for the petroleum industry. Possibilities of introducing scientific advancements into practice have been overviewed.

1. Introduction

Currently, design methods [1] rely on the research system–sociological analysis [2], logic modeling [3], system approach [4]. The desire to solve problems (eg., stylistic solution of shape in deeper structural links between a man and a work suit) remains decisive for designers.

2. Research

In order to take into account and link in a single system multiple targets, converging on the subject of a work suit for the petroleum industry as a design object, you need to build targeted programs. This may be a "tree" of goals, a matrix ша goals, a list of requirements to the subject in a certain sequence and interrelation, structural and functional model of goals. Prototypes play a crucial rolein traditional design. This concept, as a rule, is the basis of the design solution and largely defines it and allows efficient management of production using modern methods of information processing.

Analysis of prototypes from the position of industrial production suggests that in traditional design they allow you to develop standard products meeting the requirements of the social functioning of the object and processes for its production. However, naming the design process a stage of clothing manufacture we encounter prototypes that have no analogues in standard design concerning manufacturing garments for the petroleum industry, and as a result, no possibility to manufacture single products based on them.

Production specialist skills and knowledge are aimed at manufacturing specific products, samples of which are based on the continuity of culture and traditions. The division of labor required for sewing complex industry-oriented products (coats, suits of mass production, etc.) has led to the fact that the work of individual masters came to regard specific sites of the garment (welt pocket in the frame) or certain operations (for example, connecting collar to a neck hole). The product as a whole fell outside the scope of competence of one master. The appearance of design as a stage of the production process is connected with the need to organize the division of labor of many masters to achieve the ultimate goal - the manufacture of the garment of mass production. Being thus functionally linked to the organization of work and responsible for the final result of the product manufacture, design must be provided with appropriate professional tools. These tools were in the form of drawings, diagrams, standards, calculation methods, knowledge, etc.Instead of "samples" there were complex systems of symbolic tools, knowledge and principles that are transmitted both in writing and orally (with the help of the costume language). These tools, combined by the type of the object and traditional forms of design, organization and sewing activities, are the basis of prototypes of "specific" objects of design.

The specificity of the professional design activities is the need to reproduce a plurality of similar objects. Consequently, their samples and prototypes were created so that the expert could make necessary refinements and adjustments, in accordance with the cultural values, conditions and requirements (also considering the specifications of the petroleum industry).

A prototype includes the most important characteristics of the object necessary for the organization of work. It captures not a single object, but some very specific "type" of an object (jacket, pants, etc.). Expansion of tasks of professionals (designers, fashion designers), complexity of social life and development of material culture led to an increase in the number of "model" prototypes.

Social dynamics of the present day, intensive and, as a rule, independent and uneven development of various areas of production, science and technology creates a situation of activities disagreement and leads to the formulation of tasks for which there are no prototypes in modern design (especially if talk about clothes for the oil and gas industry).

Schematically, the role and place of design [5] in the new environment can be represented in methodological models of the social reproduction cycle (Figures 1, 2). They contrast two different mechanisms of social reproduction, conventionally called "society with culture" and "society with design".

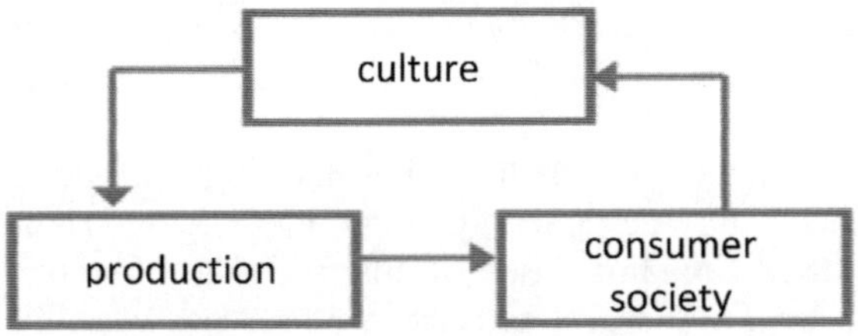

Figure 1. Interaction cycle of "society" and "culture" (culture, production, consumer society).

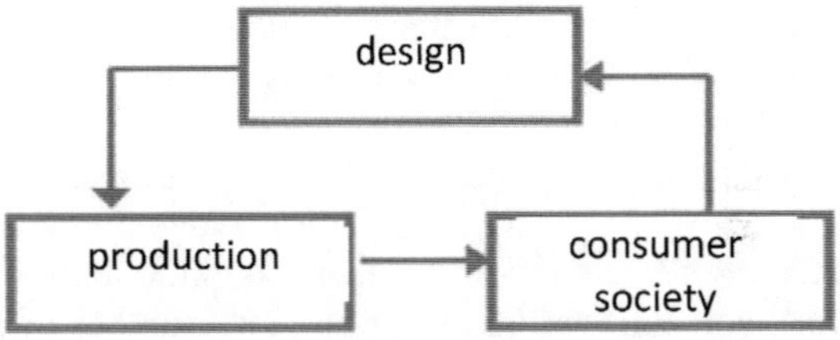

Figure 2. Interaction cycle of "society" and "design" (design, production, consumer society).

Reproduction of "society with culture" is provided by the fact that "culture" stores standards, samples of objects of production and consumption. In the mechanism of "society with design" production patterns are developed in design. If reproduction with "culture" implies social statics, reproduction with "design", by contrast, focuses on the dynamic type of society.

The presented methodology models define the semantic opposition of "design" and "culture". A combination of these mechanisms of social reproduction in developing social systems is schematically presented in the configuration diagram (figure 3), in which "design" and "culture" interact.

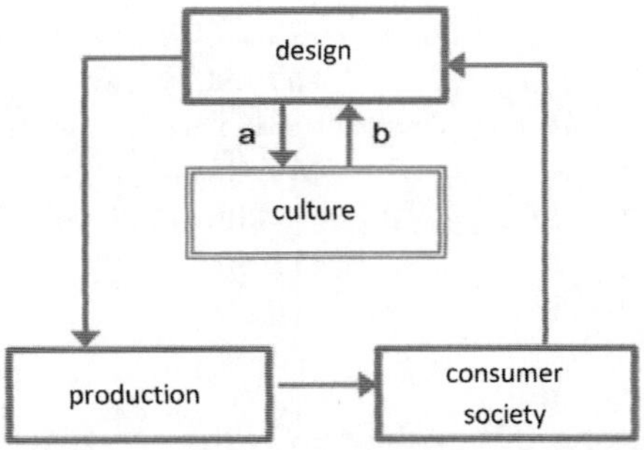

Figure 3. Interaction cycle of "design" and "culture" (design, culture, production, consumer society).

Analysis of the interaction cycle of design and culture suggests that specific prototypes of design objects is a special form of presenting cultural norms in the design.

Thus, the functions of the suit as a subject are manifested in different contexts of social and cultural human activity: transformation of the environment (instrumental function), maintaining the environment in a state that ensures the normal course of life processes (adaptive function), social integration of people (integrative function). Each of these functions is a special theme in the design of costume objects and creation of a holistic design image.

In [6, 7] the authors presented a methodology of targeted orientation of clothing design on the example of a classification of men's business suit design. Customer satisfaction is a result of the company's activity to identify the target of its products depending on the price group that indicates the status of the owner of the suit and the company they work for. Products of low-price group comply with fashion trends and modern requirements, but they have a simple, economical construction with minimum wet-heat treatment and the minimum number of finishes, which is insignificant for industrial clothing.

Average-price group costumes have a typical simple design (for example, one slit on the back) for industrial production of standardized parts and standard manufacturing. These products are a compromise between the high and low price groups: a more complex design compared to the low price level, but manual types of work are replaced by the use of special machines.

Higher-price group products are characterized by a complex design (that is worked by hand), and a large number of seams, which results in a perfect fit on the body (company's dress-code policy). Such suits both outside and inside do not tolerate any compromise; details of the jacket lining repeat the design of details of the top fabric. Original finish is used (hand finishing stitches and perhaps with contrasting threads).

Using the targeted orientation of design on a particular consumer proves the logic of using prototypes in industrial production and makes it necessary for enterprise to plan product release. There are currently a large number of economic and mathematical methods used by enterprises to plan their work and find optimal solutions to a number of problems: the largest amount of output per year, the highest profit or the lowest expenditures, the lowest cost, etc. As solving these problems is carried out under limited conditions, it takes form of optimizing the use of the available resources, i.e. their most advantageous use under these limiting conditions.

Table 1 shows an example of calculating the cost of production, taking into account the sum of all financial expenses for production and sales of OOO «Fashion studio «Silhouette», Tyumen.

Table 1. Initial data and calculation results

№	Notion of expenditure items	Expenditures		Sensitivity, thous.rub/pc	Expected expenditures, thous.rub.
		in the base period, thous.rub	in the reporting period, thous.rub		
1	Mz	1026.50	1452.85	1.37	525.01
2	Za.c	1175.50	1296.80	0.39	1032.83
3	Po.т.	2050.00	1850.00	-0.64	2285.25
4	Oc.c.	59.45	53.65	-0.01	66.27
5	Oм.c.	65.60	59.20	-0.02	73.13
6	Op.f.	410.00	370.00	-0.13	457.05
7	Ostr.vz.	20.50	18.50	-0.01	22.85
8	Oaм.	10.00	10.00	0.00	10.00
9	Pp.	5.00	20.00	0.05	-12.64
10	Пkr.	50.00	70.00	0.06	26.48
11	Zrem.	0.00	30.00	0.10	-35.29
12	Zdr.	15.00	50.00	0.11	-26.17
13	$\sum_{i=1}^{n} Z_i$	4887.55	5281.00		4424.76
14	n, pcs			311.00	
15	V_{opt}, pcs				366.00
16	A, %			7	9

Analysis of Table 1 shows that the cost in the reporting period compared to the base one fell by 7% with the release of 311 units of products.

IOP Conf. Series: Materials Science and Engineering **154** (2016) 012024 doi:10.1088/1757-899X/154/1/012024

The calculated sensitivity of each expenditure to the released 311 pieces of products made it possible to calculate the maximum number of products (366 pcs) in order to reduce the cost to a minimum value of the expected 4424.76 rubles, which reduces cost by 9% compared to the base option. Such results can be achieved due to better organization of production and effective use of internal resources.

3. Conclusion

In conclusion, it should be noted that the design that is based on prototypes is more properly termed "design by prototypes". Prototype as the form and way of existence of design should be distinguished from various specific prototypes of design objects. A particular prototype as the object meaning and essence can be regarded as knowledge, models, tools, techniques, etc., at the same time understanding it as the structure of the project activity. This work aims at reducing terminological confusion and creating an effective system approach to designing.

References

[1] Chekardovskaya I A, Bakanovskaya L N and Toropova I A 2015 An effective system of design production processes *Nauchnoe obozrenie* **5** 272-8

[2] Ivanov V A, Volynec I G and Chekardovskaya I A 2009 Methods of determining the performance of organization of production using graphic-analytical method *Higher Educational Institutions News. Oil and Gas* **1** 48-52

[3] Ahmadulin R K and Turenko S K 2005 Building a quality indicator based on formalization of expert ideas *Estestvennie i tekhnicheskie nauki* **5** 135-7

[4] Ahmadulin R K 2016 Experience in applying the principles of object-oriented programming in the development of applications in geology and geophysics *Automation, telemechanics and communication in the oil industry* **1** 38-42

[5] Chekardovskaya I A and Chekardovskiy M N 2015 Golden square. Methods of determining the economic performance of oil-gas transportation production *Business journal Neftegaz.RU* **11-12** 72-7

[6] Mokeeva N S, Zaev V A and Sennikova E V 2008 Ways to reduce the duration of the production cycle in a flexible cutting production *University News. North-Caucasian Region. Technical Sciences Series* **5** 104-6

[7] Mokeeva N S and Bakanovskaya L N 2009 Creation automatic worker of the place of the technologist for designing the technological process *Programmnye produkty i sistemy* **4** 52-54.

IOP Conf. Series: Materials Science and Engineering **154** (2016) 012025 doi:10.1088/1757-899X/154/1/012025

Calculation of the Thermal Parameters and Change in The Share of Emissions of Harmful Substances During Combustion of Water-Fuel Emulsion

S Y Toropov, V I Berg, V A Petryakov and N A Mostovaya

Industrial University of Tyumen, 38 Volodarskogo st., Tyumen, 625000, Russia

E-mail: allbert-@mail.ru

Abstract. The article investigates the possibility to improve efficiency and ecological compatibility of fuel oil with use water-fuel emulsion. The calculation of the thermal parameters and change in the share of emissions of harmful substances during combustion of water-fuel emulsion and optimal water content in water-fuel emulsion.

1. Introduction

Recently in the world is observed the essential growth of consumption of energy resources, such as oil, natural gas, fuel oil is Fuel oil is the most widespread type of the liquid fuel intended for combustion in heating boiler room to generate heat.

So, Oil Transporting Joint Stock Company «Transneft» uses fuel oil in volume of not less than 10 000 tons per year for heating of production and administrative buildings and for heating of viscous and high-viscous oils Furnace fuel oil has similar structure with crude oil and it is characterized by considerable viscosity, density, content of high-molecular substances and petroleum resins. Currently fuel oil is considered ineffective and ecologically dirty type of fuel [1].

One of the promising ways to improve efficiency and ecological compatibility of fuel oil is the use of water-fuel emulsion, which is a compound of fuel oil and water at the molecular level, where the water becomes the catalyst for improving and accelerating the combustion process [2].

The aim of research is the calculation of the thermal parameters and change in the share of emissions of harmful substances during combustion of water-fuel emulsion.

2. Materials and methods

In the course of research were studied physical- chemical properties of furnace fuel oil of the M-100, including conditional viscosity which was studied with the help of the viscometer at 80 and 100 °C. Theoretical dependence was constructed using of standard values of viscosity at 80 and 100 °C has taken from GOST 10585-99. On the basis of the obtained results was constructed experimental dependence of conditional viscosity from temperature , it is shown at Figure 1.

In the process of calculating the thermal parameters were investigated the dependence of the maximal temperature of combustion water-fuel emulsions from the percentage of water in it [3].

To determine the amount of heat in the combustion process, coefficient of efficiency and change in the share of emissions of harmful substances (emissions of oxide of nitrogen) was made a comparison of experimental data from water-fuel emulsions and furnace fuel oil M-100 by combustion, the results are presented at Figures 2-4. Experiment was made in the closed furnace chamber with using the thermometer and a gas analyzer.

To determine thermal parameters such as heat capacity, the number of formed heat and coefficient of efficiency was created the mathematical model of the combustion process and energy conservation equation in a closed combustion chamber [4].

We write the continuity equation:

$$\frac{\partial\rho}{\partial t} + \mathbf{div}(\rho\vec{u}) = \mathbf{0}\ , \tag{1}$$

where t - independent variable, ρ - mass density, $\vec{u}$ – rate of movement

the conservation impulse equation:

$$\rho\left(\frac{\partial u}{\partial t} + u\nabla u\right) + \nabla p = 0, \tag{2}$$

energy conservation equation:

$$\rho T\left(\frac{\partial S}{\partial t} + u\nabla S\right) = \frac{\rho x}{\tau}q - Q(T), \tag{3}$$

where τ - burning time, S - entropy, q - heat of combustion

After transformation of this system, we write the energy conservation equation:

$$C_v \rho T u \frac{d\,\mathbf{ln}\frac{\rho}{\rho^{\gamma}}}{dl} = \frac{\rho x}{\tau}q - Q(T), \tag{4}$$

3. Results and Discussion

Figure 1 shows the theoretical and experimental dependences of relative viscosity from temperature.

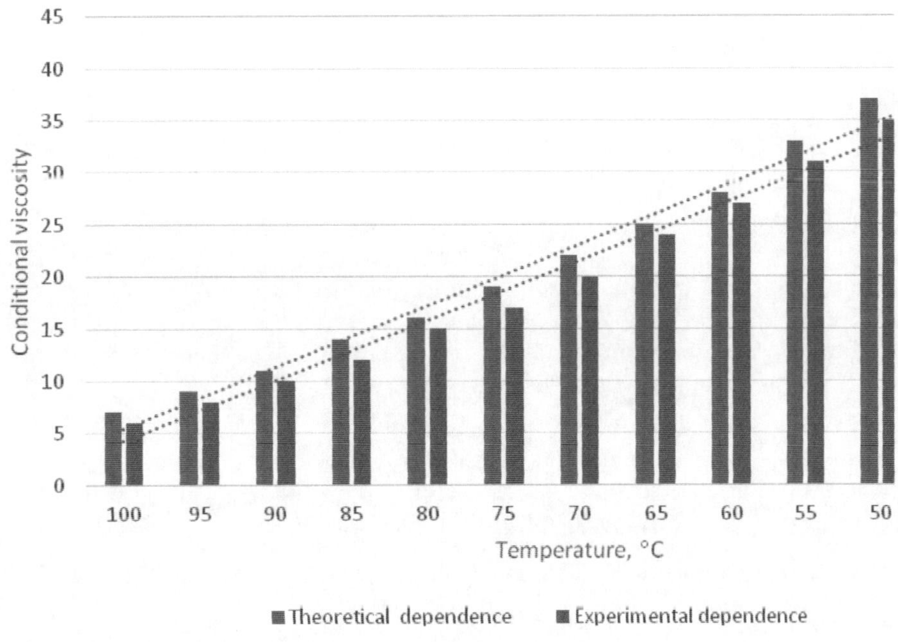

Figure 1. The dependences of the relative viscosity of fuel oil M 100 from the temperature.

The figure shows that the theoretical and experimental dependences differ. It can be a reasonably high percentage of paraffin and petroleum resin in the fuel oil, which negatively influence the combustion process.

IOP Conf. Series: Materials Science and Engineering **154** (2016) 012025 doi:10.1088/1757-899X/154/1/012025

Figure 2 shows the graphical dependence of the amount of heat formed by the combustion of water-fuel emulsion and oil fuel from the excess air.

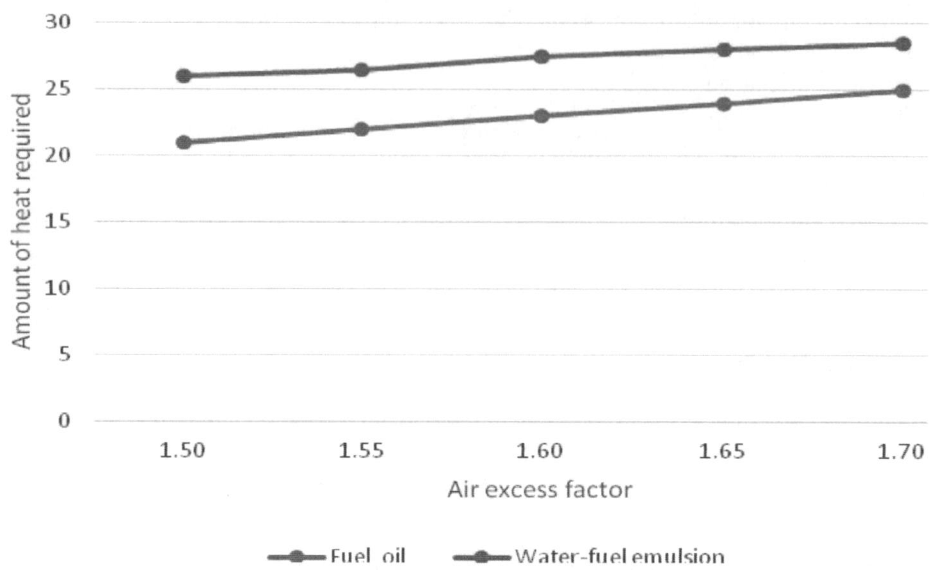

Figure 2. Dependences from heat loss q from the coefficient of excess air.

Figure 3 shows the dependences of the coefficient of efficiency of water-fuel emulsion and fuel oil from the excess air.

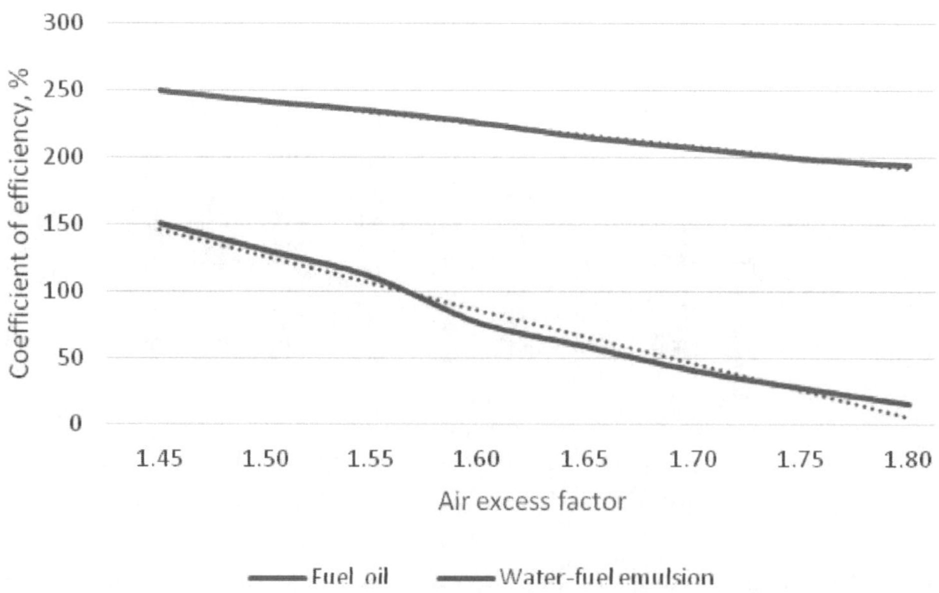

Figure 3. The dependence the coefficient of efficiency of boiler from the coefficient of excess air.

Figure 4 shows the dependence of the emissions of nitrogen oxides water-fuel emulsion and fuel oil from the coefficient of excess air.

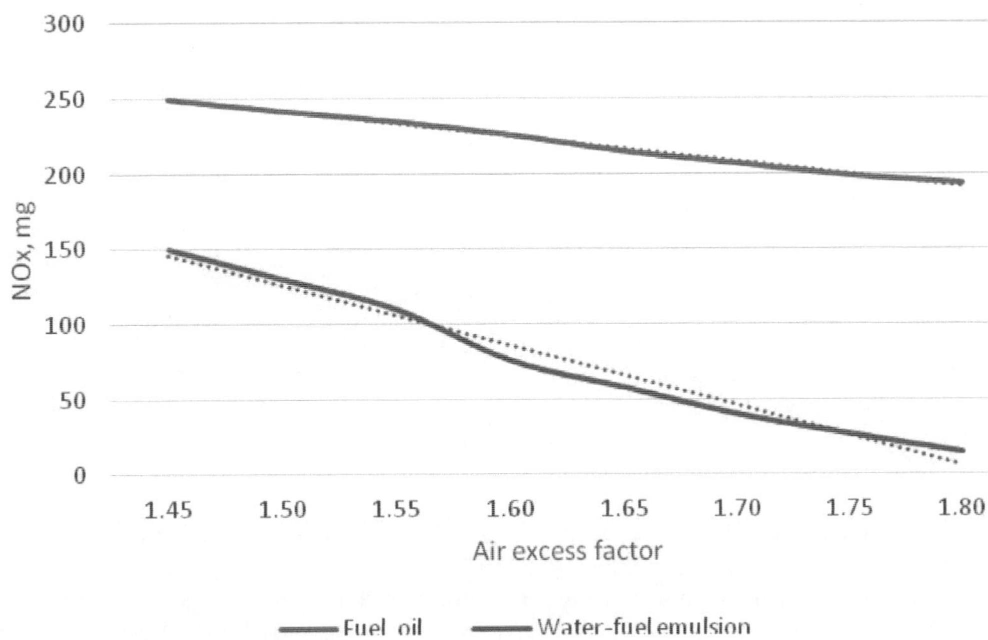

Figure 4. The dependence of the emissions of nitrogen oxides from the coefficient of excess air.

In the research, process was also applied additive MgO. In the contact zone of water and material of the boiler presents vanadium, which stimulates the destruction of the protective layer of Fe_2O_3 at the metal, leads to corrosion of the boiler elements [5,6].

The magnesium oxide binds sulfur oxide SO_3, is formed of magnesium sulfite, which is destroyed.

$$3MgO^{2760\,°C} + V_2O_5^{675\,°C} = Mg_3V_2O_8^{1190\,°C} \quad (1)$$

$$SO_3 + H_2O = H_2SO_4 \quad (2)$$

$$SO_3 + MgO = MgSO_4 \quad (3)$$

As a result of decrease in the concentration of sulfuric acid in the flue gases lowers the dew point. Owing to interaction of additive with sulfur the amount of harmful emissions is reduced significantly

4. Conclusion

Consider the use of water-fuel emulsions based on fuel oil for combustion in boiler rooms, heating buildings and industrial objects. As a result of researches, it is revealed that the water-fuel emulsion has higher coefficient of efficiency and allocates bigger amount of energy in the combustion process. As a result of the analysis of data of the thermometer the dependence of a heat of water share in a water fuel emulsion, it is presented in figure 5.

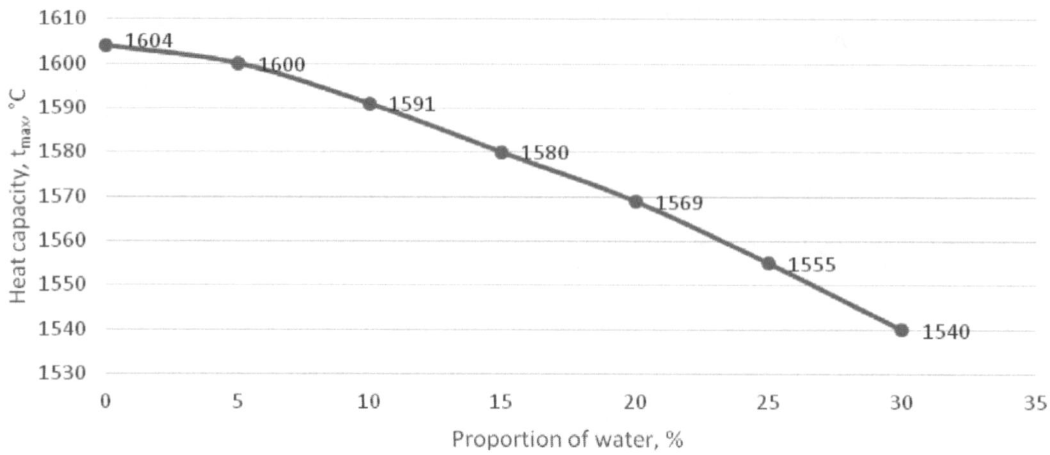

Figure 5. The theoretical dependence of heat from the water content.

The figure demonstrates that with increasing proportion of water the heating process decreases, that negatively influences heating process. It is revealed that in order to avoid more reduction of heating process optimal is the addition of 15% water. Thus, the article is shown of the calculation of the thermal parameters of the combustion of water-fuel emulsion, which it is a promising kind of fuel.

References

[1] Lebedeva E A, Kocheva M A, Luchinina A E, Sharov A V and Khokhlova E N 2010 Energy-saving technologies of consumption and production of heat *Privolzhsky Scientific Journal* **3** 82-88

[2] Geller C V 2010 Preparation of water-fuel emulsions by wave dispersion *J. News of heat supply* **4** 21-23

[3] Vengerov A A and Brand A E 2014 Oil cavitation treatment to prevent formation of paraffin deposits *J. Natur. and intellect. resource. of Siberia.* **15** 36-37

[4] Melnik A Yu, Minikaeva S N, Pavlov S B and Kharlampidi Kh E 2012 Effect of water on diesel fuel specifications *Bulletin of Kazan Technological University* **24** 213-215

[5] Brand A E, Vershinina S V, Vengerov A A and Mostovaya N A 2015 Applying the technology of hydrodynamic cavitation treatment of high-viscosity oils to increase the efficiency of transportation *IOP Conf. Ser.: Mater. Sci. Eng.* **93 012005**

[6] Isakov A Ya 2004 Simulation of microexplosion of droplets of water-fuel emulsion *J. Izvestiya vuzov - Technical Science* **4** 94-97

[7] Mamadaliev, R.A., Kuskov, V.N., Popova A.A., Valuev D.V., Alloying elements transition into the weld metal when using an inventor power source, J. IOP Conference Series: Materials Science and Engineering. 127 (2016) p. 1-7

[8] Valuev, D.V., Malushin, N.N., Valueva A.V., Dariev, R.S. Mamadaliev R.A. Manufacturing component parts of mining equipment with application of hardening technologies, J. IOP Conference Series: Materials Science and Engineering. 127 (2016) p. 1-7

[9] Antip'ev, Y.N., Nevolin, A.P., Zemenkov, Yu.D. 1981 Operation of intermediate pumping stations on transport of gas saturated oils. (1981) NEFT. KHOZ, 10, pp. 46 ISSN: 00282448

[10] Kurushina V and Zemenkov Y 2014 Innovative cyclical development of the Russian pipeline system WIT Transactions on Ecology and the Environment 190-2 881-8 DOI: 10.2495/EQ140822

IOP Conf. Series: Materials Science and Engineering **154** (2016) 012026 doi:10.1088/1757-899X/154/1/012026

Application of Technology of Hydrodynamic Cavitation Processing High-Viscosity Oils for the Purpose of Improving the Rheological Characteristics of Oils

Y D Zemenkov, M Y Zemenkova, A A Vengerov and A E Brand

Industrial University of Tyumen, 38 Volodarskogo st., Tyumen, 625000, Russia

E-mail: Allbert-@mail.ru

Abstract. There is investigated the technology of hydrodynamic cavitational processing viscous and high-viscosity oils and the possibility of its application in the pipeline transport system for the purpose of increasing of rheological properties of the transported oils, including dynamic viscosity shear stress in the article. It is considered the possibility of application of the combined hydrodynamic cavitational processing with addition of depressor additive for identification of effect of a synergism. It is developed the laboratory bench and they are presented results of modeling and laboratory researches. It is developed the hardware and technological scheme of application of the developed equipment at industrial objects of pipeline transport.

1. Introduction

Viscous oils (VO) and high-viscous oils (HVO) concern to hardly recoverable types of hydrocarbons and differ from conventional oils by hyper dynamic viscosity in natural conditions and by composition [1]. In the total structure of reserve stocks of the Russian Federation the share of VO and HVO on the basis of the Ministry of Energy of the Russian Federation grounds to 2014 has reached 30% of Russia's total. The most part of stocks is concentrated in the Volga-Ural and West Siberian petroliferous basins and the Volga and Urals regions contain about 70.8 per cent of the Russian reserves [2].

The feasibility analysis of the transport system VO and HVO has demonstrated its low efficiency characterized by a low transport speed, by the necessity of pre-treatment and continued support of high temperature oil, and a short operational resource of the system in conditions of low ambient temperatures. Widespread methods of pre-processing to achieve big speeds of transportation and improvement of rheological properties of oils are:

- thermal heating
- addition of additives and thinners
- application of electromagnetic radiation

Thermal heating is the most widespread and effective method, but at the same time and the most expensive method of processing, as by heating of oil is used natural gas as fuel. Thus for the heating of 1 million tons of VO on 1 °C will require about 241 thousand rubles and 100 thousand m^3 of natural gas on the basis of 2016. In the conditions of the low prices of oil and the economic instability, it is very actual to research and to develop new power - and resource-saving methods of increase of efficiency of transport [3].

The authors have investigated a possibility of application of hydrodynamic cavitational processing of VO and HVO for the purpose of a break of complex hydrocarbonic chains, decreasing of the dynamic viscosity and tension of shift. According to foreign scientists, such as Hemmit F, Daley J., from the American society of mechanical engineers, hydrodynamic cavitation processing has economic prospects and high efficiency.

2. Research

The cavitation phenomenon occurs at the moment of approaching the values of pressure of liquid to value of pressure of saturated steam of liquid in the process of which there are copious bubbles (cavities) filled with the dissolved gas. The collapse of cavities is characterized by sharp jumps of temperature and pressure, and the distribution of energy wave contributes to the destruction of nearby carbon chains and molecular compounds. Number of bubbles can vary from 10^4 to 10^6 m^3 [4].

Figures 1 and 2 shows simulation of the hydrodynamic cavitation process due to pipeline section change by means of ANSYS/FLUENT software. The obtained data and dependences have been taken into consideration when conducting laboratory tests.

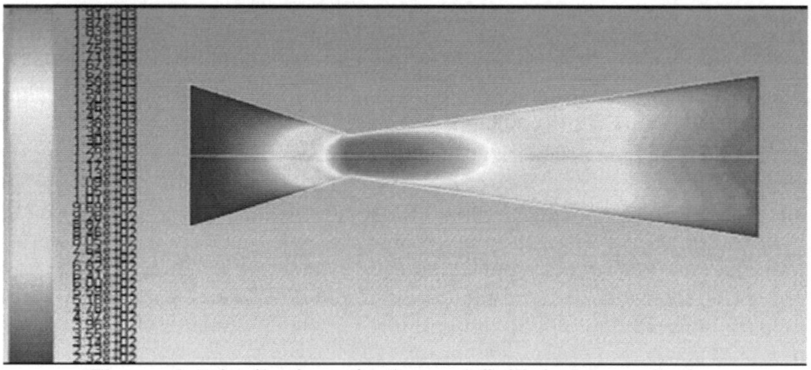

Figure 1. Distribution of a thermal field during cavitation
Zone 1 – a preheat zone, zone 2 – a zone of cold boiling,
zone 3 – flow and thermal field stabilization.

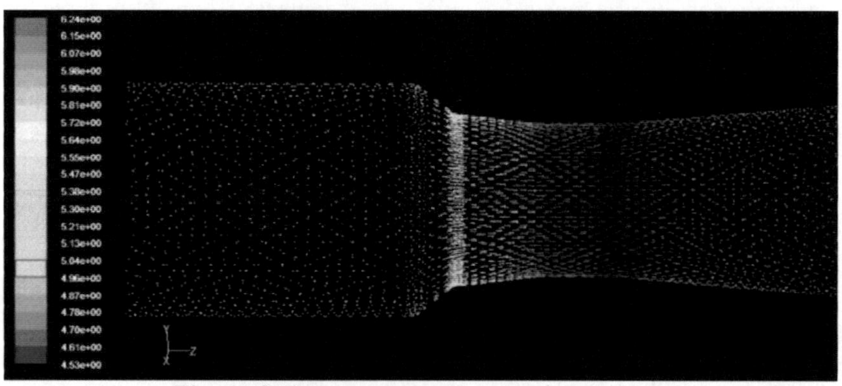

Figure 2. Pressure wave propagation vectors

The simulation has showed that in the process of passing a liquid through a narrowing phase occurs the local temperature increase. The cavitational equipment created by the simulation results is presented at Figure 3.

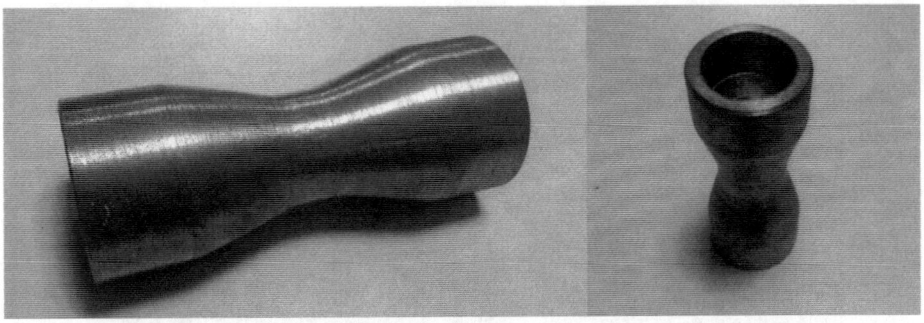

Figure 3. Cavitation equipment.

The energy of the collapsing cavitation bubble, which was spherical, determined by the formula (1):

$$E=\frac{4}{3}\cdot\pi\cdot P_{kr}\cdot R_{\max}^{3} \quad (1)$$

where R_{max} – the radius of a cavitational cavity before a collapse which is calculated by the formula (2):

$$\frac{dR_{\max}}{dt}=\sqrt{\frac{2(p_n-p_{kr})}{3\rho}}, \quad (2)$$

We will write down the total energy emitted at a collapse of cavitational cavities through number of cavitational cavities (3):

$$N=K\cdot V, \quad (3)$$

K – concentration of cavitational cavities (1/m^3)
V – the volume of fluid without cavitation (m^3)

Considering formulas 1-3, we will write down the formula of the energy released due to collapse of cavities at the developed cavitation (4):

$$E=K\cdot V\cdot\pi\cdot\frac{85.33\cdot p_n}{g\cdot\upsilon}\left(\frac{29\cdot\rho\cdot g}{p_n}\right)^{1.9}\cdot\int_0^t\left[\frac{2(p_n-\frac{64\cdot p_n}{g\cdot\upsilon_{kav}}(\frac{29\cdot\rho\cdot g}{p_n})^{1.5})}{3\rho}\right], \quad (4)$$

where $t=0.915\cdot R_{\max}\sqrt{\frac{\rho}{p_{rk}}}$ - rayleigh times of a full collapse, sec.

It is demonstrated the laboratory setup at the Figure 4. The testing process the following equipment involved these: centrifugal pump by head 50 m and flow rates up to 70 p/a, cylindrical tubes with a diameter of 3 cm^2, temperature sensors and pressure, the cavitational equipment [5, 6]. During the tests was processed about 200 litres of high viscosity of the Usinsk oil ρ = 945 kg/m^3 and V = 150 mPa•s.

Figure 4. Laboratory setup.

Figure 5 shows that complex processing of oil has high efficiency,

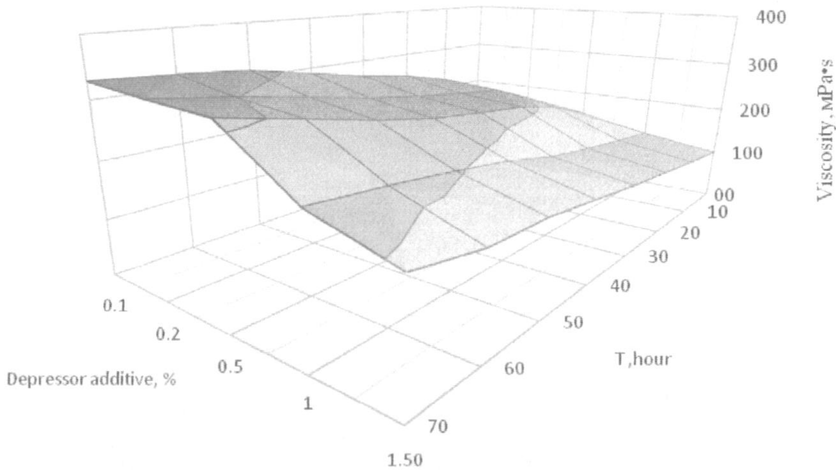

Figure 5. The dependence of the viscosity from depressor additive and time.

3. Conclusion

Using the findings by cavitational processing of oil with the use of additives, by the authors was developed a hardware-technological scheme on installation of the cavitational equipment at Figure 6.

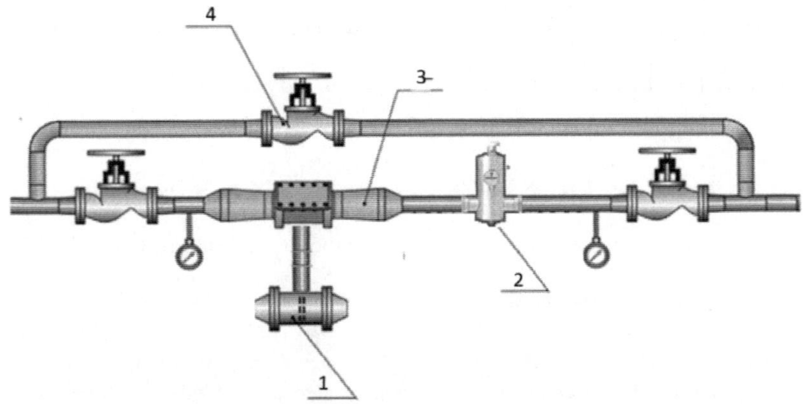

Figure 6. Cavitational equipment hardware-technological scheme
Zone 1 – a preheat zone, zone 2 – a zone of cold boiling, zone 3 – flow and thermal field stabilization.

The author's analysis and calculations show that the method of cavitation of cavitational processing improve the rheological properties of VO and HVO is effective and promising [7]. The feasibility of application of complex processing of oil in pipeline transport system consists in the following:

• increase of reference temperature of heating of high-viscosity oil;
• reduction of viscosity of oil
• reduction of the spent energy of traveling heaters and fuel consumption;
• improvement of rheological properties of oil at the expense of a rupture of carbonaceous communications.

The applied technology can be used on trade or technological pipeline systems with diameters up to 300 mm and viscosity of oils transporting 150-200 mPa·s.

References

[1] Shiryaev A M and Zholobov V V 2015 *J. Science and Technology* **24** 60-68
[2] Ilyin E G and Ivantsova S G 2014 *J. Crude oil pipeline* **6** 67-70
[3] Vyboyschik M A and Ioffe AV 2015 *J. JSC Rosneft* **36** 45-52
[4] Vengerov A A and Brand A E 2014 Oil cavitation treatment to prevent formation of paraffin deposits *J. Natur. and intellect. resource. of Siberia*. **15** 36-37
[5] Tarasenko M A and Brand A E 2014 Reducing oil viscosity by a method of hydrodynamic *J. Natur. and intellect. resource. of Siberia*. **15** 44-45
[6] Brand A E, Vershinina S V, Vengerov A A and Mostovaya N A 2015 Applying the technology of hydrodynamic cavitation treatment of high-viscosity oils to increase the efficiency of transportation *IOP Conf. Ser.: Mater. Sci. Eng*. **93 012005**
[7] Vershinina S V, Brand A E and Mostovaya N A 2015 Application of hydrodynamic cavitation treatmentof high-viscosity oils for the purpose of increase of efficiency of transportation *J. Higher Educ. Instit. News. Oil and Gas* **4** 97-101
[8] Valuev D.V. Malushin N.N.,Valueva, A.V. Dariev, R.S. Mamadaliev R.A., Traditional technology of chromium-tungsten steels facing, its disadvantages and suggestions for their eliminations, *J. IOP Conference Series: Materials Science and Engineering*. 127 (2016) p. 1-7

IOP Conf. Series: Materials Science and Engineering **154** (2016) 012027 doi:10.1088/1757-899X/154/1/012027

Methods for Determining the Thermodynamic Parameters of Gas Compressor Units of Main Gas Pipelines

M N Chekardovskiy, S M Chekardovskiy and K N Ilyukhin

Industrial University of Tyumen, 38 Volodarskogo st., Tyumen, 625000, Russia

E-mail: zemenkov@tsogu.ru

Abstract. Methods for evaluating the state of gas compressor units of main gas pipelines by the values of the thermodynamic parameters measured during operation, that is indirectly, should be developed. In this regard the authors have conducted a comparative analysis of various techniques. Based on the results conclusions have been made on how to develop the thermodynamic method of diagnosis.

1. Introduction

During operation of gas compressor units of main gas pipelines (further on – GCU of MGP) the equipment is gradually deteriorating because of natural wear-out. Therefore, the development of efficient methods for determining the technological and diagnostic GCU parameters during operation, identification of defects at an early stage of their occurrence is a very urgent problem, which is complicated by a great variety of automatic data collection and management systems for nominally controlled parameters. Given the complexity of the processes occurring in GCU it is necessary to develop specific diagnostic methods for further application in complex systems to optimize the maintenance and repair of units that will lead to the improvement of quality, reliability and efficiency of GCU of MGP. One of the most promising methods for determining the technological and diagnostic parameters of GCU is the thermodynamic method. This is due to the current lack of sensors that allow one to directly determine the technical condition of the GCU elements (erosion of centrifugal blower impellers, radial clearance of turbines, combustion chambers burnouts and others). Such sensors may never appear due to their complexity, uniqueness (depending on the design of a unit), and high cost. It is necessary to develop methods for assessment of the GCU technical condition by the values of process parameters measured directly during operation, that is, indirectly. For this purpose, a comparative analysis of different techniques for determining the main thermodynamic parameters (TDP) was conducted. Based on the results we made our conclusions on the direction of the thermodynamic method development.

2. Research

In the system approach to developing diagnostic techniques based on the thermodynamic method, in most cases, a GCU as a diagnostic object is divided into elements according to their functions and processes:

1. gas turbine unit (GTU) compressor;
2. GTUcombustion chamber (CC);
3. GTUturbine;
4. gas blower or compressor;
5. GTU recuperator;
6. exhaust gas heat exchanger of GTU.

The mathematical apparatus of the thermodynamic method makes it possible to explore the parameters of GCU both element by element, and as a whole as a single object, ie to solve problems of integral and differential diagnosis.

IOP Conf. Series: Materials Science and Engineering **154** (2016) 012027 doi:10.1088/1757-899X/154/1/012027

Calculating the integral TDP by various techniques, such as the effective power (N_e) and the effective efficiency of the unit (η_e) is performed for six units GTK-10-4 with a blower 370-18-1. Techniques were used based on the TDP calculations of the following authors: Zarickiy S. P., Lopatin A. S., Porshakov B. P., Ryabchenko A. S., Stepanov O. A., Chekardovskiy M. N., Shabarov A. B. Calculation of parameters was carried out in order to compare and identify their strengths and weaknesses in the practice of diagnosing gas compressor units. Figure 1 provides the design scheme. The design scheme shows basic structural elements of the GCU and the parameters measured in units using both regular equipment, and specially installed instrumentation required to obtain sufficient initial information used in computational methods. A more detailed description of the scheme is given below.

From the suction chamber air is taken with the parameters $T_a$and P_a, which are recorded by regular devices. Before the first step of the axial compressor (AC) there is a slight change in the intake air parameters to the level of T_0, P_0. In the AC the air is compressed and acquires the output parameters T_C, P_C, which are recorded by regular devices. Compressed air is supplied to the heat exchanger (recuperator) and heated to a temperature T_v, pressure is P_v. The compressed and heated air enters the combustion chamber where it is mixed with the fuel gas, which is supplied through a separate fuel line. The fuel gas parameters T_d, P_d, ΔP are recorded by additional devices. The combustion products after the CC enter the HP turbine with the parameters T_Z and P_Z. T_Z is calculated, P_Z is recorded by regular devices. The temperature of the combustion products ($T`_S$) between the HP turbine and the LP turbine is calculated. The combustion products at the LP turbine outlet have the parameters T_S and P_S recorded by regular devices. In addition, a flow meter for combustion products is installed, which records ΔP in the gas duct before the recuperator. Via the blower, regular devices record the gas parameters T_1, P_1 and T_2, P_2, as well as the rotor speed of the blower $n = n_{ТНД}$ and the AC rotor ($n_{OK} = n_{ТВД}$). The composition of the transported gas components according to the CC data is given in table 1.

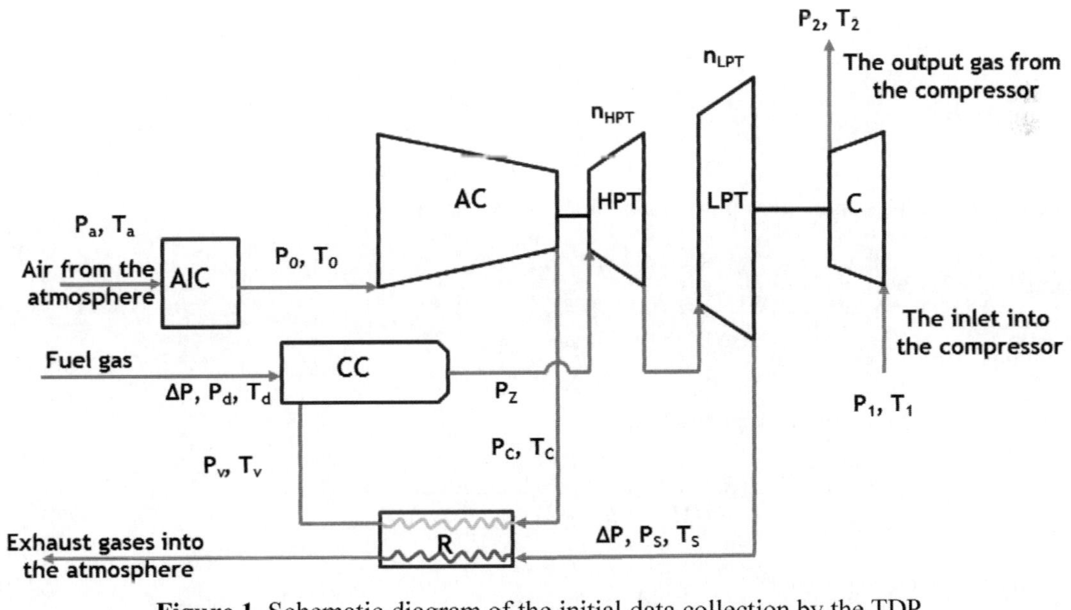

Figure 1. Schematic diagram of the initial data collection by the TDP

In the studies were used data on the composition and characteristics of the natural gas components of the Urengoiskoye field and the initial data by the TDP - GTK-10-4 at the Bogandinskaya CS.

Techniques for TDP calculation:

1. Calculation of the TDP based on the technique by Chekardovskiy M. N. and others. (Technique 1)

Sequence of calculating Ne and ηe [4]:

Using the initial data from tables 1 and 2 let us calculate the following characteristics of the gas by the equations [2]:

$$\mu_m = \frac{1}{100} \cdot \sum r_i \cdot \mu_i \quad (1.1)$$

Elemental fuel weight composition in percentages:

$$C^P = \frac{12.01}{\mu_m} \cdot \left(r_{C_1} + 2r_{C_2} + 3r_{C_3} + 4r_{C_4} + r_{CO_2}\right) \quad (1.2)$$

$$H^P = \frac{1.008}{\mu_m}\left(4r_{C_1} + 6r_{C_2} + 8r_{C_3} + 2r_{H_2S}\right) \quad (1.3)$$

$$S^P = \frac{32.06}{\mu_m} r_{H_2S} \quad (1.4)$$

$$N^P = \frac{28.016}{\mu_m} r_{N_2} \quad (1.5)$$

$$O^P = \frac{32}{\mu_m} r_{CO_2} \quad (1.6)$$

where C^P, H^P, S^P, N^P, O^P – weight concentration of, respectively, carbon, hydrogen, sulfur, nitrogen, oxygen.

$$E = \frac{12.01 \cdot H^P + 0.126 \cdot O^P}{4.032 \cdot C^P + 0.3755 S^P} \quad (1.7)$$

$$L_0 = 0.1151 \cdot (1 + E) \cdot C^P \quad (1.8)$$

$$Q_H^P = \frac{\sum r_i \cdot Q_i}{100 \cdot \mu_m} \quad (1.9)$$

$$R = \frac{8.314}{\mu_m} \quad (1.10)$$

The density of the combustion products behind the LP turbine:

$$P_S = P_a + P_{exc} \quad (1.11)$$

left gas duct

$$\rho_L = \rho_{15} \cdot \frac{P_S \cdot 288}{98.1 \cdot T_S} \quad (1.12)$$

right gas duct

$$\rho_R = \rho_{15} \cdot \frac{P_S \cdot 288}{98.1 \cdot T_S} \quad (1.13)$$

The consumption of combustion products:

$$M_{CP_L} = F_L \cdot \sqrt{2 \cdot \Delta P_L \cdot 10^6 \cdot \rho_L} \quad (1.14)$$

$$M_{CP_R} = F_R \cdot \sqrt{2 \cdot \Delta P_R \cdot 10^6 \cdot \rho_R} \quad (1.15)$$

$$M_{CP} = M_{CP_R} + M_{CP_L} \quad (1.16)$$

$$M_{AIR} = M_{CP} - B \quad (1.17)$$

$$\alpha = \frac{M_{AIR}}{B \cdot L_0} \quad (1.18)$$

$$C_{P_{AIR}} = 0.9379 + 0.000198 \cdot \frac{T_S + T_C}{2} \quad (1.19)$$

Let us determine the temperature of the combustion products before the HP turbine:

$$T_Z = \frac{0.98 \cdot Q_H^P + C_{P_{AIR}} \cdot (\alpha \cdot L_0 \cdot T_v + T_0)}{C_{P_{AIR}} \cdot (\alpha \cdot L_0 + 1)} \tag{1.20}$$

The capacity of the axial compressor:

$$N_{AC} = M_{AIR} \cdot C_{P_{AIR}} \cdot (T_C - T_0) \tag{1.21}$$

The capacity of the high pressure turbine on condition that the HP turbine and the AC are on the same shaft [1]:

$$N_{HPT} = 1.015 \cdot N_{AC} \tag{1.22}$$

The temperature of the combustion products behind the HP turbine:

$$C_{P_{CP}} = 0.9796 + 0.000283 \cdot \frac{T_S + T_C}{2} \tag{1.23}$$

$$T`_S = T_Z - \frac{N_{HPT}}{M_{CP} \cdot C_{P_{CP}}} \tag{1.24}$$

The effective capacity of the GCU:

$$N_e = M_{CP} \cdot C_{P_{CP}} \cdot (T`_S - T_S) \tag{1.25}$$

The effective efficiency of the GCU:

$$\eta_e = \frac{N_e}{B \cdot Q_P^H} \tag{1.26}$$

2. Calculation of the TDP based on the technique by Shabarov A. B. (Technique 2)

Sequence of calculating N_e and η_e [5]:

Using the data in Table 3.2 let us determine the absolute pressure of the combustion products behind the LP turbine:

$$P_T = \frac{P_{S_R} + P_{S_L}}{2} \tag{2.1}$$

The expansion ratio of the combustion products in the turbine:

$$\pi_T = {P_G}/{P_T} \tag{2.2}$$

To calculate the temperature of the combustion products before the HP turbine let us take k = 1.33 [1]:

$$T_G = \frac{T_T}{1 - \left(1 - \pi_T^{\frac{1-k}{k}}\right) \cdot \eta_L} \tag{2.3}$$

By (1.19) and (1.23) let us find the heat capacity of the air and combustion products C_{Pair} and C_{PG} correspondingly, at temperatures T_P and T_G, then determine the relative fuel consumption and air consumption:

$$g_f = \frac{C_{P_G} \cdot T_G - C_{P_B} \cdot T_P}{Q_H^P \cdot G_f - \left(C_{P_G} \cdot T_G - C_{P_G} \cdot T_O\right)} \tag{2.4}$$

$$G_{AIR} = {G_f}/{g_f} \tag{2.5}$$

To determine the specific work of the compressor and the turbines by (1.19) and (1.23) us find the heat capacity of the air and combustion products C_{Pair} and C_{PG} correspondingly, within the predetermined temperature range: T_C, T_a (for the air in the compressor) and T_G, T_S (for the combustion products in the turbines):

$$L_{AC} = C_{Pair} \cdot (T_{AC} - T_a) \tag{2.6}$$

$$L_T = C_{P_G} \cdot (T_G - T_T) \tag{2.7}$$

The effective capacity of the GCU:

$$N_e = L_T \cdot (G_f + G_{AIR}) \cdot \eta_M - L_{AC} \cdot G_{AIR} \tag{2.8}$$

To calculate the effective efficiency of the GCU let us determine the specific power and specific fuel consumption:

$$N_{e\,sp} = N_e / G_{AIR} \tag{2.9}$$

$$C_e = \frac{3600 \cdot g_m}{Ne_{sp}} \tag{2.10}$$

Note: Q_H^P is taken from the results of the calculation technique 1:

$$\eta_e = \frac{3600}{C_e \cdot Q_H^P} \tag{2.11}$$

3. Calculation of the TDP based on the technique by Porshakov B. P. (Technique 3).

Sequence of calculating N_e and η_e [6]:

Using the equations from technique 1 and the data in tables 1 and 2 let us calculate the change in enthalpy during the compression of gas in the blower:

$$C = 360 - \frac{(66 \cdot r_{CH_4})}{100} \tag{3.1}$$

$$\Delta C_P = \frac{6 \cdot C \cdot P_1 \cdot 10^3}{\left(\frac{T_1 + T_2}{2}\right)^3} \tag{3.2}$$

$$C_{P_0} = 4.187 \cdot \left(0.334 + \frac{0.66 \cdot r_{CH_4}}{100}\right) \cdot \left(0.519 + 0.000583 \cdot \left(\frac{(T_1 + T_2) - 546}{2}\right)\right) \tag{3.3}$$

$$\Delta h = (C_{P_0} + \Delta C_P) \cdot (T_2 - T_1) \tag{3.4}$$

Let us find the effective capacity of the GCU, the nameplate capacity and the technical condition factor of the GCU:

$$N_e = \sum_{j=1}^{n} b_j \left(\frac{\Delta h}{60}\right)^{j-1} \cdot \rho_C \cdot \left(\frac{n}{n_{nom}}\right)^3 + N_M \tag{3.5}$$

$$N_{e\,pr} = N_{e\,nom} \cdot \left[1 - 4.2 \cdot \left(1 - \frac{T_Z \cdot T_{anom}}{T_{Z_{nom}} \cdot T_a}\right) \cdot \frac{T_Z \cdot T_{a_{nom}}}{T_{Z_{nom}} \cdot T_a}\right] \cdot \frac{P_a}{P_{a_{nom}}} \cdot \sqrt{\frac{T_a}{T_{a_{nom}}}} \tag{3.6}$$

$$k_{\text{GCU}} = N_e / N_{e\,pr} \tag{3.7}$$

The effective efficiency of the GCU determined by the heat received from the combustion of fuel, taking into account the technical condition factor:

$$B \cdot Q_H^P = \frac{0.25 \cdot [1 + (1 - k_{ITY}) \cdot C`]}{\eta_{e_{HOM}}} \left(N_{e\,HOM} \cdot \frac{P_a}{P_{a_{HOM}}} \cdot \sqrt{\frac{T_a}{T_{a_{HOM}}}} + 3 \cdot N_e\right) \tag{3.8}$$

$$\eta_e = \frac{N_e}{B \cdot Q_H^P} \tag{3.9}$$

4. Calculation of the TDP based on the techniques by Zarickiy S. P. (Technique 4A, 4B)

Sequence of calculating N_e and η_e (variants A and B) [3]:

Variant A.

The expansion ratio of the combustion products in the turbines, necessary for determining the power parameter B, is determined by the following formula similar to (2.2):

$$\varepsilon_T = {P_Z}/{P_S} \tag{4.1}$$

By [3] let us determine B depending on ε_T and calculate the effective capacity:

$$N_e = B \cdot P_S \cdot 101971 \cdot \sqrt{T_S} \tag{4.2}$$

The effective efficiency is determined by (1.26), using the missing data from technique 1 (B, Q_H^P):

$$\eta_e = {N_e}/{B \cdot Q_H^P} \tag{4.3}$$

Variant B.

Let us determine the factor K by the following formula, where P_{HOM} and T_{HOM} – the pressure and temperature under normal conditions:

$$K = 1.0164 \cdot \left(\frac{P_a}{P_{HOM}}\right)^{1.5} \cdot \left(\frac{T_a}{T_{HOM}}\right)^{0.5} \tag{4.4}$$

$$N_e = A \cdot K \cdot (101936 \cdot \Delta P_K)^{1.5} \tag{4.5}$$

$$\eta_e = {N_e}/{B \cdot Q_H^P} \tag{4.6}$$

5. Calculation of the TDP based on the technique by Shapoval Yu. A. (Technique 5)

Sequence of calculating N_e and η_e [2]:

The effective capacity and the effective efficiency are calculated as follows:

$$N_e = N_{e\,HOM} \cdot \left(\frac{n}{n_{HOM}}\right)^3 \tag{5.1}$$

Note: missing data are taken from technique 1 (B, Q_H^P).

$$\eta_e = {N_e}/{B \cdot Q_H^P} \tag{5.2}$$

Comparison of the techniques for calculating the TDP for the purpose of diagnosing the technical condition of the GCU is given below.

The results of calculations by techniques 1, 2, 3, 4, 5 are given in table 1[7].

Table 1. Results of calculations.

Unit №		1	2	3	4	5	6
Parameter	Unit of measurement	Values					
Technique 1							
N_e	kW	7895	7882	7263	6304	8585	9763
η_e		0.287	0.289	0.268	0.227	0.266	0.285
Technique2							
N_e	kW	6471	5856	6184	6724	8521	9282
η_e		0.234	0.214	0.228	0.241	0.263	0.271

		Technique3					
N_e	kW	7279	7235	6545	6317	7126	9104
η_e		0.236	0.237	0.230	0.240	0.244	0.256
		Technique4A					
N_e	kW	7510	6916	7560	7280	9473	9859
η_e		0.272	0.253	0.279	0.261	0.293	0.287
		Technique4B					
N_e	kW	9077	8335	8430	8471	8885	9164
η_e		0.329	0.305	0.311	0.304	0.275	0.267
		Technique5					
N_e	kW	7189	6990	6232	6795	7290	1000 0
η_e		0.260	0.255	0.230	0.244	0.225	0.292

To compare the techniques for calculating the TDP for the purpose of diagnosing the GCU we calculated deviations in the values of the effective capacity (table 2) and efficiency (table 3) of the units.

Table 2. Deviations in the values of the effective capacity of the GCU (kW) from the average value.

Techniques	Unit					
	1	2	3	4	5	6
1	324	679	226	-677	271	234
2	-1099	-1346	-851	-258	207	-246
3	-291	32	-490	-664	-1187	-424
4A	-60	-286	524	298	1160	330
4B	1506	1132	1394	1489	571	-364
5	-381	-212	-804	-187	-1023	471
Mean value	7570	7202	7036	6982	8313	9529

Table 3. Deviations in the values of the effective efficiency of the GCU (%) from the average value.

Techniques	Unit					
	1	2	3	4	5	6
1	1.64	2.95	1.02	-2.65	0.43	0.84
2	-3.52	-4.46	-2.96	-1.14	0.24	-0.56
3	-3.38	-2.20	-2.73	-1.32	-1.66	-2.03
4A	0.24	-0.58	2.12	0.85	3.19	1.12
4B	5.93	4.61	5.33	5.14	1,36	-0.90
5	-0.91	-0.31	-2.78	-0.88	-3.57	1.53
Mean value	27.0	25.9	25.8	25.3	26.2	27.7

3. Conclusion

From tables 2 and 3 we can see that for on-line determination of only N_e and η_e we should use technique 5, as it does not require any material expenses for carrying out the experimental work. Other techniques require additional installation and commissioning works to obtain the necessary initial data for individual nodes (AC, HP turbine, LP turbine, recuperator, centrifugal blower). The technique developed by the authors, if material expenses for installation and commissioning are available,

IOP Conf. Series: Materials Science and Engineering **154** (2016) 012027 doi:10.1088/1757-899X/154/1/012027

provides data on individual elements of the GCU for monitoring and diagnosing their technical condition. Therefore, this development solves the problem of determining the technological and diagnostic parameters of the GCU during operation, identification of defects at an early stage of their occurrence.

References

[1] Ahmedzyanov A M, Dubravskiy N G and Tunakov A P 1983 *Diagnosing the state of the ABE for thermogasdynamic parameters* (Moscow; Nauka) 321

[2] Shapoval Yu. A. 2000 *Development of technical performance evaluation methods of functioning of* compressor *stations of gas mains equipment* Author's abstract of Cand.Sc. Eng. Disseration (Tyumen; TSOGU) 20

[3] Zaritskiy S P 1987 *Diagnostics of gas pumping units with gas turbine drive* (Moscow; Nedra) 197

[4] 1999 *Development of an automated calculation program for thermogasdynamic parameters of GTU on* the *basis of initial information* Report (Tyumen; TSOGU)

[5] Shabarov A B 1983 *The methodology of calculation of gas turbine engines* (Moscow; Bauman MVTU) 31

[6] Krylov G V, Matveev A V, Stepanov O A and Yakovlev E I 1985 *Exploitation of West-Siberian gas pipelines* (Leningrad; Nedra) 288

[7] Chekardovskiy S M 2001 *Development of thermogasdynamic methods for monitoring and diagnostics of gas supply system equipment* Author's abstract of Cand.Sc. Eng. Disseration (Tyumen; TyumGASA) 22

[8] Antip'ev, Y.N., Nevolin, A.P., Zemenkov, Yu.D. 1981 Operation of intermediate pumping stations on transport of gas saturated oils. (1981) NEFT. KHOZ, 10, pp. 46 ISSN: 00282448

[9] Zemenkov, Yu.D., Shalay, V.V., Zemenkova, M.Yu. (2015) Expert Systems of Multivariable Predictive Control of Oil and Gas Facilities Reliability. Procedia Engineering, Volume 113, pp.312-315 DOI: 10.1016/j.proeng.2015.07.271

[10] Mamadaliev, R.A., Kuskov, V.N., Popova A.A., Valuev D.V., Alloying elements transition into the weld metal when using an inventor power source, J. IOP Conference Series: Materials Science and Engineering. 127 (2016) p. 1-7

[11] Valuev, D.V., Malushin, N.N., Valueva A.V., Dariev, R.S. Mamadaliev R.A. Manufacturing component parts of mining equipment with application of hardening technologies, J. IOP Conference Series: Materials Science and Engineering. 127 (2016) p. 1-7

Innovative Methods of Anthropogenic Landscape Reconstruction in the Urbanized Oil and Gas Region Environment

L N Shabatura, N V Bauer, N I Speranskaya and O E Iatsevich

Industrial University of Tyumen , 38 Volodarskogo Str., Tyumen, 625000, Russia

E-mail: maru-safronova@rambler.ru

Abstract: The article deals with the problems of the urbanized environment appearing as a result of intensive region developing. The state neglect towards people affects the population life quality of the of oil and gas extraction areas as well as problems resolving, and it provokes enormous losses for manufacturing and the whole region. The environment influences the person behaviour, one's perception and space understanding. The city environment is considered as the human existence space influencing on it directly, so it is necessary to renovate it. The authentic region developing cannot be reduced to "pure" economics (for not to be deserted), but needs to be fully mastered. To renovate the destroyed landscapes, it is necessary to use the landscape design methods making a cultural landscape. They help to increase the natural components of the city environment and to make it more harmonious, more harmless, more comfortable for residents.

1. Introduction

The urban environment is the cultural relations carrier among people and between people and the environment as well, and it is the population lifestyle reflection of each historical era, the social ideology, helping the city residents to integrate culturally. At the same time, the urban environment imprints the people's community activities: any change in the citizens cultural life corresponds with the transformations in the urban environment and these changes are compared with other projects born in the economic, managerial and social spheres of the city life. You must understand that the urban environment can contribute to harmony, comfort, satisfaction, person-environment feelings, and vice versa, may cause an alienation feeling rising social anomie and the destruction of the traditional social cultural code.

Constructing human habitat (creative, comfortable, attractive, safe, harmonious, connecting the benefits of city living and the normal wish of citizens to nature, traditions preservation and developing, introducing new technologies) is one of the main challenges facing modern culture.

2. The methodological basics of the urban landscapes renovation

The development of the Western Siberia oil and gas region caused a real demographic explosion. The Tyumen region population during three decades from 1961 to 1990 has increased by almost 3 times. The leading factor of urbanization was allocating the manufactory, dedicated to the deposits, and people were considered to be mobilized and distributed the required amount resource for its constructing and functioning.[6, p. 56].

A distinctive feature of the Tyumen Region settling process became the settlements forming in the immediate closeness to deposits, that is, in the hard- to - reach places which were not always suitable for permanent living and performing labor functions in a stationary mode. The principle "the first is a plant - then a city" was often used in this region at that time. New town-planning decisions were difficult to apply: the territory development process and the settling preparation concepts were taking at the same time, it rose miscalculations inevitably. Therefore, throughout the 1970-ies there was a correction of the General plans of the Middle Ob towns. The Tyumen General plan is still being corrected.

The migration and demographic processes analysis in the Western Siberia oil and gas region shows that it was mostly possible to solve the task of attracting labor resources to the Tyumen North due to external migration. The problem of retaining people in the North appeared to be more difficult. The low survival rate was determined by the underdeveloped social infrastructure (housing and communal services, transport system, cultural-welfare, healthcare let a settlement be a city) was in a poor condition.

Departments often regarded cities as an expensive application to the oil and gas fields, investing only a tenth of the total cost of large deposits development in the urban construction [5, p. 115]. Therefore, settlements formed around oil and gas fields, hasn't been cities, even towns for a long time. The state neglect towards people affects the population life quality of the oil and gas extraction areas as well as problems resolving. During the second half of 1970-ies the crisis in the regional leading industries development confirmed the inviolability of the law: "If you lose a man, you lose everything". It eventually turned into huge losses for both the territory and ultimately, the industry itself.

The creating experience of the Western Siberia oil and gas complex convincingly demonstrated once more that the real region developing cannot be reduced to "pure" economics(for not to be deserted), but needs to be fully mastered.

As cities grow many problems revealed and exacerbated: air, soil and waters pollution, noise, effluent discharge, waste disposal, transport, pipelines, energy supply, etc. As a result of excessive technogenic and recreational loads many natural landscapes were perished

The simplest and most effective way of recovering contemporary urban ecology is developing the optimally shaped system of green planting.

Nowadays taking into account the historic urban environment destruction, the actuality of the modern landscapes transferring into the value category increases; it is associated with moving from external deformations to internal ontogeny accompanied by a new spiritual space fulfillment.

Designers' orientation to a person requires a philosophical comprehension of urban environment landscape objects advantages, defined by the man's attitude to nature psychology and relationship, understanding culture. The landscape object symbolic significance associated with the semantic essence of a place, its spirituality.The forming urban environment method should include searching new signs in the traditional and transformed urban landscape objects context.

The optimum city environment should assist to improve the population's physical and spiritual health. D.L.Armand pointed out that the cultural landscape must be not only productive and healthy but beautiful as well: «The aspiration for beauty is an ineradicable instinct of human spirit» [1, p. 196]. The cultural landscape is considered to bethe favorable for the population nature-cultural formation. V.I.Vernadsky's concept about a noosphere as a result of a person's creative, intellectual and constructive activity on transforming and purposeful developing of a biosphereisa methodological basis ofthe term «a cultural landscape» interpretation. E.J.Kolbovskyappeals to noosphere concept of a cultural landscape andinterprets a cultural landscape as «the nature model which is generated and works in the cultural environment as «the target shadow (culture) on the nature»[12, p. 24].

V.L.Kagansky has formulated the more capacious definition of the cultural landscape comprehended as space in which «pragmatic is inseparable from semantic». He develops the approach interpreted as landscape hermeneutics in the works. «The vital environment of the big enough people'sgroup is a cultural landscape if this space is integral and differentiatedat the same time, and the group has mastered this space utilitarian, semantically and symbolically» [8, 244]. The landscapehermeneutics is an «active landscape reading», focused on conceptual development of its variety and critical interpretingthe landscape space as a kind of society "works".

The domestic scienceexamines themorphologicalstructureanddynamicsofacultural landscape mostly. However the cultural landscape (separate special space places) hierarchy practically is not developed. But traditional for Russia understanding of a cultural landscape emphasizes in it «splicing corporal and symbolical components provides sense, integrity and value» [6, p.40]. The determinative factor of the city environment formation should be a system of spiritual, moral, aesthetic, intellectual values on which the orientation of landscape-formative processes depends.

The culture, relaxing or practically transforming a natural landscape, allocates with its images, the values, new properties and creates its materialized and virtual landscapes with their relief, rhythm, sense. The symbolism in a cultural landscape arises as impact spiritual and material, created by a man who brings new sense in the developed image of an environment.

Measures on creation, restoration of cultural landscapes should assist to develop harmonious relations between the nature and a human society andto transform modern anthropogenic landscapes into qualitatively new spaces – economically effective, ecologically safe, possessing high aesthetic potential and «place spirit».

To form a stable city environment the most effective way of its improvement is considered to be creatingthe optimum system of green plantings. Thus it is necessary not only to increase the green plantings areas– parks, parkways, gardens, squares, protective green zones and forest parks, but also to form the city society spirituality. InXIX century in the work «Ethics of a life» English philosopher Thomas Carleyl called up people «Make light shine, and we shall see the green blossoming world instead of a hell» [10, p.308]. The author means not only direct understanding«the green blossoming world», but also more deeper spiritual sense. The city environment forming is necessary for a life, the person's spiritual development.

The new millennium person should be an adequate member of a society, that is onemust live in harmony both with oneself and with surrounding people and an environment.

Russian philosopher A.S.Ahiezer characterizes a urbanization of the country «not only and not so much as increasing sizes and numbers of cities, but first of all, forming and more widely spreading the city life way and the city culture» [3, p.44]. Ignoring this fact has led to technocratic skews in the stable environment formation in Russian cities, that arisesemergence of serious ecological and social problems.

The increasing number of experts on the city environment formation comes to conclusion, that a basis of well-being is the citizens' individual culture. It is marked, that nanotechnologies appear in a cultural, university city, and cannot arise in city type settlements. Japan inhabitants paid attention to that fact that quality of any industry production directly depends on aesthetic education of those whoperform this production. Therefore, having adopted much from experience of 20th years of the last century (Russian Vkhutemas, German Bauhaus), Japanese have made it general property.

Culture increasing in the city environment improvement by landscape design methods has got a purpose to form people's sensation of harmony, visual comfort and environment spaces recognition. Emotional orientation of the city environment objects generated by landscape design means is connected with the person's perception process of one's environment, thus, the transformed nature plays the same role as the kept nature. That's whythe great urgency gets searching new ways of visual expressiveness in the open city spaces.

Realizing that today's economic opportunities are rather limited, the construction of the traditional type parks and gardens is becoming more difficult. This situation forces to seek solutions in a non-traditional direction, without resorting to the creation of large-scale facilities, and distributing funds for the landscaping of local city fragments.

3. Landscape design methods of the urban landscapes renovation

The realization of the new landscape design possibilities provides fora person a chance to live in the individual (by its image) space, possessing ecological stability, long time keeping harmonious communications of natural and artificial components of a cultural landscape. The innovative landscape design methodsare expediently focused on increasing the natural components in the architectural objects structure. It is reachedthanks to usingthe various ways [13], which are aimed at revealing plastic object characteristics and strengthening its figurativeness:

- the natural addition of architectural forms method may be applied to soften the contrast between a vertical building-walls surface and a horizontal plane basis due to creating the transitive composite elements from various vegetation forms;
- the method of application in the form of imposing a vegetation surface expands the vertical gardening treatment due to using various versions of the twisted breeds plants accommodation on building skeletons and facades;
- the method of the entrance directions accentuation assumes concentrating achievement means of the art expressiveness with the help of the landscape design along the basic ways of the buildings approach;
- the morphological similarity method consists in using the similar pattern forming elements of buildings and artificial geometric landscape;
- the morphological similarity method consists in using the similar pattern forming elements of buildings and artificial geometric landscape;
- the object weight visual absorption by vegetation method allows to remove partially an acuteness of a scale discord of a construction and its environment;
- the contour contact of a building and a water mirror method can shade the silhouette expressiveness the façade relief effectively enough, bringing in the architecture object perception unexpected foreshortenings.

Achievingthe composite communication with an environment depends on introducing aspects of landscape design methods in decorating boundary spaces.

The idea of the natural components consecutive revival in the architectural objects structure finds the increasing urgency as it is directed on the problem - solution of the ecology environment maintenance. The actual factor of the similar approach was defining the terms meaning movement to the new quality of architectural process, in particular, such as "green" style, "green" architecture, "green" walls, ecoarchitecture,ecodesign in the last decades mainly in the foreign practice and theory.

A person'sconformity to high culture conducts to one'saccord with an external world. The concept «To live in accord with the nature» stated in an ancient philosophy, remains true in the widest sense. It has been told by A.Schweitzer: «I am a life, wishing to live among a life» - that means «ethics of life awe» [14, p.117].

The urbanization process, increasingthe urban population assists washing out the traditional culture and forming the new culture, new valuable installations.

The developed situation in a biosphere ecological condition in many aspects is defined by culture marginality of modern mankind. First of all culture tasks should be solved which success provides the environmental problems decision. Creating new valuable ideals, comprehending the person's new status in the world, formingthe new ideology and new culture, education, new technologies and new interoperability with the world around is a main aim forthe present and the nearest generations, - otherwise, the future generations cannot «live among a life» any more, that is among the natural environment.

The society existence is impossible without the nature transformation. But, changing the nature mankind oughtn't to weaken its generating force. Only creative nature transformingshould be thesingle way of disclosing the person's intrinsic forces, realization one's nature and culture.

The certain spatial objectimage formed by landscape design means, possessing alongside with aesthetic appeal, the natural components composition originality (tree-bush compositions, the relief, water sources characteristic lines) has the aesthetic importancefor a person. The non-standard components accommodation, appropriating the certain plot, offer the definiteschemes environmental person'sbehaviour, motivate one's movement, action, supervision, leaving visual impressionsin memories, fixing place sensation. The place category plays a special role in social- cultural disclosing people feelings, their behaviour in the city environment.

It is necessary for a person to see around oneself greens, water, light, the bright paints interesting, aesthetic forms, as a man isnatureharmonious. It also is a life quality when it would be a great desire

IOP Conf. Series: Materials Science and Engineering **154** (2016) 012028 doi:10.1088/1757-899X/154/1/012028

to go to a court yard, a street, a city to admire them, to communicate with friends, to see beauty, to live with it and in it [4, p.159].

The demands of unique natural objects risein the process of the people's culture level growth. The decision listed social- cultural problems is possible under the condition of increasingthe culture forming of the stable city environment, using the best world tendencies of landscape design as new design culture

The social- cultural design potential is contained in its possibility to introduce advanced technologies achievements in thesocial- cultural environment keeping cultural society values. Consequently, the design represents the unity of material and spiritual culture of a society where innovations are the results of the person's design work.

The necessity of the qualitative city environment transformation by means and methods of landscape design is also connected with the questions suspension of the person's emotional perception of a spatial environment. The functionalism predominance in the landscape organization of the city environment has led to displacement of accents aside rational problems to the detriment of emotional ones.

In our opinion the modern cityquality is defined by the humanistic approach to the urban environmentorganization, its orientation to the population various categories demands with different possibilities (including limited). The reference points of modern landscape design are as follows: the respect for a historical and cultural heritage, the attention for residents, a humane combination of old and new in architecture, an active introduction of natural components with their detailed study, skills to work in any building context.

The social –cultural potential saved up during centuries and concentrated in such public phenomenon as city isthe society progress source, its major "capital". The term «the city culture» defines the artificial existence and self-realization environment created by people. This is the physical objects complex, normative and valuable formations, technologies including not only material components, but also, as N.P. Antsiferov characterized, «the city soul» [2, p.29]. In the modern science practice «the city soul» is characterized by such culture elements as social psychology of city community, norms and values, the citizens' life way and mentality, the social communications and social design, a security and social well-being.

The person expresses the aesthetic attitude to the world by design means. The design is a special activity. It defines the person's culture. E.Kassirer, a German thinker, emphasizes, that the most important person's characteristic is one's activity. The work, the activities system also defines area of "humanity" [11]. We consider activity as culture, and design activity is one of the major factors of culture which can be considered as a method as a process and as a result in the stable city environmentformation.

Today architecture, fashion, art, new technologies - everything, that changes ideology,influenceon landscape design as the type of joint work of the person and the nature. The landscape design art develops synchronously with the processes occurring in a society: new materials, technologies, possibilities appear in our life. The novelties reflecting present ideology and philosophy should supply organically already developed landscape design directions and improve, not destroy culture. The city environment formation becomes at the same time both philosophy, and the industry. The city should not be grey, sad and boring. The environment creation for a person – creative, comfortable, attractive, safe, connecting advantagesof city life and citizens' natural aspiration to the nature, keeping traditions and introduction, development of new technologies is a problem which faces modern culture of the stable urban environmentformation

City environment harmoniously organized during design work is perceived by a person as beautiful. Only the true beauty creates the emotional environment multiplying people's capacity to create beauty, to develop good, taste, aesthetic feeling, culture

Now city residents are surrounded not with landscape art (design)works as it was meant in XIX century, butwith detail-spatial integrity where imposing of environment impressions is more valuable, than art originality of composed components. Therefore nowadays not a volumetricdesignedmasterpiece of one "professional", but creating conditions for coauthorship of artists and any creative people who can form art environment aspects by theirown participationbecomes the objective of the city environment cultural landscape renovation As a result the unexpected solution to the problem, so desirable by artists of all generationsarises, the new form of arts synthesis appears too. Synthesis of surrounding components attitude at a new level means also new knowledge of the world emotional development ways; awakening the creative attitude to the validity in a person; the new technologies formation created «the second nature » and its ecological unity with a natural landscape basis. So the historical mission of landscape design opening philosophical sense of variations in design environmentmethods, rendering direct influence on the person culture clears up.

The designer realizes mission on keeping and augmenting cultural images, bears the responsibility before the consumer for design results. He/she specifies assessment criteria and offers hierarchy and the sum of values. The design provides the cultural value continuity of the detail-spatial environment in time, provides the way for a person to feel indissoluble communication of the modern phenomenon with elements of the last world. Russian philosopher O.I.Genisaretsky speaks about design as project culture: «the design becomes a conductor of human culture achievements» [7, p.40].Agreeing with his opinion, we consider that the landscape design purpose is filling the city environment open objects by social-cultural sense, by forming their image, the using way comprehensible fora consumer. Such design projecting brings a social, emotional, aesthetic measure and culture in the stable city environment formation. The social - cultural design potential concludes its possibilities, keeping the cultural society values and introducingthe advanced technologies achievements in thesocial - cultural environment [7].

4. Conclusion

Giving to landscape objects the certain aesthetic and functional properties, the designer "projects" a person who will live in this environment; formsone's culture, thinking, understanding, the attitude to surrounding. So the important social – cultural landscape design function is building a better urban environment, not only in time but in spirit, contributing to the creation of a favorable emotional climate, improving quality of life, aesthetics, harmonizingpeople's everyday lives.

The landscape design, being philosophically meaningful action, assists projecting the city environment as system, social - cultural, dynamically developing object of the intellectual attention. The city environment imageformed by innovative landscape design methods should become more sensitive to the person's perceptions and to one's attitudes with the world around.

References

[1] Armand D L 1975*Science about a landscape* (Moscow; Idea) 286
[2] Antsiferov N P 1991 *Incomprehensible city* (Moscow;Canon)227
[3] Ahiezer A S, KoganL B and Yanitskiy O N 1969 The urbanization, a society and scientific and technical revolution*Questions of philosophy***2** 44-52
[4] BauerN V and Shabatur L N 2014 Culture and tradition in landscape design city environment*Values and sense* 155-61
[5] GavrilovaN Yu and Karpov V P 2003 The experience of social development in oil and gas extraction Western Siberia region (1960-80-s)*Domestic history***2**111-8
[6] Ganapolskiy M G and Litenkova S P 2005 The structure of Tyumen region settling: особенности genesis peculiarities and developing perspectives *Reports of the Russian Academy of Science. Geographical Series* **3** 56
[7] Genisaretskiy O I 1987 Design culture and conceptualism*Social - cultural problems of the life style and the subject environment* (Moscow; VNITE)39-53
[8] Kaganskiy V L 2008 Cultural landscape*Humanitarian geography: the Scientific and cultural-educational almanac*Ed.5 (Moscow; College) 243-6

IOP Conf. Series: Materials Science and Engineering **154** (2016) 012028 doi:10.1088/1757-899X/154/1/012028

[9] Kantor K M 1986 *Truth about design. Design in a context of the pre-perestroika thirtieth culture (1955-1985.)* (Moscow;ANIR) 286
[10] Carleyl T 2013 *Life ethics. The author's collection*(Moscow;Art Prezent) 672
[11] Kassirer E 1998 *Experience about a person. Introduction in the human culture philosophy*(Moscow; Gardarika) 784
[12] Kolbovskiy E J 2006 *Landscape science*(Moscow;Academy)480
[13] Nefedov A F 2002 *The landscape design and environment stability* (Saint Petersburg) 295
[14] SchweitzerA 1992 *Culture and ethics / Life Awe*(Moscow; Progress) 572

IOP Conf. Series: Materials Science and Engineering **154** (2016) 012029 doi:10.1088/1757-899X/154/1/012029

The Formation of the Urban Environment in the West Siberian Oil and Gas Region: Problems and Challenges

N V Bauer, N I Speranskaya, L N Shabatura and O E Iatsevich

IndustrialUniversity of Tyumen, 38 Volodarskogo st., Tyumen, 625000, Russia

E-mail: maru-safronova@rambler.ru

Abstract: The article deals with the problems faced by the person in relation to the culture of creating a sustainable urban environment. The urban environment is seen as space of human existence, influencing its spiritual and physical health. All the experts involved in the urban environment, think about the culture of its formation, unfortunately, only from a departmental point of view. Differently organized human environment inevitably affects behavior, emotional state of a man, his perception and understanding of space. The cultural heritage of society fits in the space of an artificial environment created by a man, and determines the behavior of people.For the solution of problems of formation of the urban environment, it is necessary to apply a set of interrelated technical, social, environmental, aesthetic and other measures. In this capacity, landscape design as a set of methods has to build a sustainable urban environment to ensure harmonious living.Current trends, methods, and techniques of landscape design are to create sustainable environment for a person – creative, spiritual, comfortable, attractive, safe, harmonious.

1.Introduction

Social settings of the urban life are the most importantmodern research objects. Analyzing the city life, we essentially study the society's life in general, because the large part of the population is citizens or preparing to become citizens. Being a symbol of culture, the city is a mechanism of its conversion to real life-, a socio-psychological community- and of a human personforms.

The urban environment has a direct impact on social well-being, social relations, forms of behavior, the townspeople's culture. And at the same time, the urban environment is a reflection of the lifestyle of the population in each historical era, the social world, contributing to the cultural integration of the city residents.

At the same time, the urban life brings its own contradictions. These include the problem of the organization of the urban spaces, which runs the whole life of the modern citizen. Modern cities produce many acute social problems related to inequality, the rising cost of living, crime, worsening traffic problems, increasing resource consumption, pollution, nature destruction in the city, etc. Different populations show differential requirements for the urban environment organization, wanting to form it according to our preferences, tastes and interests. In this situation it is necessary to understand that the urban environment can contribute to feelings of harmony, comfort, satisfaction, person-environment, and vice versa, may cause an alienation feeling that leads to social anomie and the destruction of the traditional cultural code of the society.

In these conditions the special importance has the determination of the optimal ratio of the primary nature in the cultural landscape — a robust strategy and a systematic organization in interactions of the society with the natural environment.

The city attractiveness is determined by the absolute advantages of the urban life, which increases the pace of socio-cultural development of the modern society, but the urban life brings its own contradictions in it. These include the problem of the urban spaces organization, which runs the whole life of the modern citizen.

2. Methodological bases of the formation of the urban environment

The formation of the Tyumen region as the object of oil and gas civilization occurred as the result of the industrial development. The concept of "development" can be found in studies of different subject orientation. It permeates many aspects of culture. There is a reason to consider the development as the essential trait of Russian culture. Along with the usual combinations of "new industrial development", "development of new territories", found their full right the following expressions: "development of new materials and technologies", "space", etc. Today researchers even say about the development of the world by man that is the concept of "development" to which is given a special philosophical meaning.

Intensive industrial development of the West Siberian oil and gas province has transformed the landscape of the sparsely populated areas in the Tyumen region, led to the formation of cities. For a short historical period, there was created a unique civilizational complex for the extraction and transport of hydrocarbons. Distinctive features of this complex are: the predominance of the oil and gas extractive industry, with its reference to a particular place, a utilitarian attitude to the natural landscape, a special practice of the design and construction of oil and gas production enterprises and residential complexes. Thus, methods of work organization apply in other spheres of life. Even the place of residence is, as a rule, the departmental settlement with production projected at it by the hierarchy. About it in his works says a well-known Tyumen scientist Ganopol'skii M. G.: "This unity of man and technology, both rooted in both the natural and social environment in oil and gas industry, allows us to consider the Tyumen region as the object of oil and gas civilization" [5, P. 207].

The urban settlement was formed in the Tyumen region for three- four decades in the new industrial development in the sparsely populated, and in some places and uninhabited space. The development of oil and gas fields in Western Siberia caused a real demographic explosion. The population of the Tyumen region increased from 1965 to 1990 2.5 times, and in the centers of oil and gas production in Khanty-Mansi (KhMAD) and Yamalo-Nenets (YNAO) region — its growth amounted to, respectively, 9.2 and 7.6 times. The process of urbanization was rapid: during 25 years have grown 16 new cities in the 1990's and innoughties — six [6]. The design was consistent with the matrix of geological exploration of the territory. People were considered as a quantitative resource for the construction and further operation of production facilities.

Hydrocarbons arteries are an integral part of the industrial landscape of the Tyumen region, destroying its natural components. At the same time, a place of life and work is specialized; it is closely intertwined with daily household and a high degree of responsibility for the results of labor. From social well-being of these places depends a lot of things, including energy and environmental security not only of our region, but of the whole country.

The problem of retaining people in the North was difficult. The low survival rate is a result of underdeveloped social infrastructure. That eventually turned into huge losses for the territory, people and ultimately for production itself.

The experience of creating WSOGK once again convincingly demonstrated that the real development of the region cannot be reduced to "pure" economics, especially in its sectoral understanding. The North could be easily depopulated, and it was necessary to develop it comprehensively, not turning into a technological Playground" [7, C. 207].

Settlement of the Tyumen North demonstrates the imperfection of its current structure. It is primarily the fragmentation and different levels of sophistication of those fragments. Unfortunately, when building the cities of the North of Tyumen was not fully exploited neither domestic nor foreign experience.

Handicap provisions can be seen in the capital of WSOGK. Tyumen is a city which was not looking into the future like newly built young cities of the Tyumen North, it does not live its glorious past. It is made up of different pieces, fragmented in time and space. The space of the city is not connected, needs to become the whole unity.

The other thing is more important. About the consciousness of the townspeople says the Tyumen philosopher M. G. Ganopolsky. The cultural landscape of Tyumen Michael Grigor'evich describes as the satin of different manners: "Its special temper had recently not only every district, but almost every route connecting it with the urban transport. Cultural life of the city in its everyday life is not able to withstand this stream of morals..." [4, p. 204] environment inevitably affects the behavior, psycho-emotional state, perception and understanding of space and its culture. Residents, coming into contact with the city, fill and endow the environment with their senses, emotions, soul, mind, attitude, extracting from it what they find interesting and valuable. This environment, marked by the presence of human values and transformed by them, is called the human environment, in relation to the city – the urban environment. The urban environment has a direct impact on social well-being, social relations, shaping behavior, and the culture of townspeople. And at the same time, the urban environment is a reflection of the population lifestyle in each historical era, the social world contributing to the cultural integration of the population.

Modern urban environment should be a harmonious, socio-cultural space for better organization of work and rest of the population. The transformation of the urban environment in general is connected not only with questions of production economics but also affects the ideological sphere, since we are talking about a new man, the bringing up of certain aesthetic tastes for beauty and harmonization of the surrounding space. The urban environment as a system, social space which shapes the worldview of the person, aims to improve his culture.

Rapid urban growth in the twentieth century, the increasing complexity of functional and spatial structure of the urban centers brought with them problems of human orientation in the urban space. High-rise development of centers as a whole sets the dilemma of psychological and physiological comfort/discomfort of the person in the subject-spatial environment of the modern city. The wide spread of industrial methods of construction based on the model design during the second half of the twentieth century has become one of the causes of the destruction of the unique architectural ensembles of the cities and natural landscapes in appearance of dullness and monotony of the urban environment, reducing its aesthetic qualities. Particularly acute raised the issues of preserving the historically valuable environment of the city and its artistic expression. In the process of urbanization continued to increase the isolation of man from the natural landscape. Particular difficulties were compounded by the fact that along with the process of urbanization has been actively implemented technicalization, increasingly penetrating deep into the human subject environment, thereby violating the ecological balance of anthropogenic and natural components, leading to a shortage of the latter. And today on the background of this man-made world more and more we can talk about the environment of the spiritual. Incurred to the XIX century landscape architecture, as a kind of panacea for "diseases of urbanization" today is already unable to cope with the task of maintaining the ecological equilibrium of the human environment, the affected "man-made metastasis" in the post-industrial society. There is an urgent need to find new means of solving this problem. Urban environment, as a structural component of the reproduction of society and as they are generated by relations of the subject and its relations, can only coexist in this tandem. The degree of this interaction, including its transformation, depends on the measure of its meaningfulness, the culture of its formation. Consequently, the environment appears as an objective reality, which includes the set of methods and forms.

The urban environment is a synthesis of architecture and landscape design that should be considered as a new artistic reality of our being which needs to be culturally and architecturally designed. This required a new kind of creative activity – landscape design, which is a specific activity involved in the design of open objects environment. The meaning of landscape design of the urban environment - drawing up of the emotional- technological scenario that determines the efficiency and artistic focus of everything permanently happening. In this regard the professionals ' attention should be drawn to those elements of the urban environment which must satisfy aesthetic, functional and cultural needs, to promote the creation of favorable conditions for their life.

Seriously not doing a conversion of the environment using modern techniques and means of landscape design, a person runs the risk to "preserve" him in the past. Cultural history says that ongoing developments in international practice offers a wide range of modern technologies for landscape arrangement and makes a new meaning in the interpretation of the urban environment.

3.Ways of solutions of the problems of forming a harmonious urban environment

The urban environment is determined by a set of cultural parameters of human life. The formation of the urban environment occurs as a result of changing human culture, i.e. together with changes in the needs and preferences of a person associated with quality, level and way of living. Versatile culture of human and society as a whole, defines today the possibilities of urban development and reasonable transformation of the urban environment methods of landscape design.

The urban planners, designers have a massive and responsible task - the city entirely woven out of the problems and contradictions that needs to be created harmonious and comfortable, with a full-fledged urban environment. Any change in the city requires a reliable scientific basis. New connections that are created by a man must be closed to ensure the permanence of the basic parameters of the planet Earth system that affect its environmental stability. Therefore, at the regional level, each municipality should use the principles of ensuring the competitiveness and sustainability improvements.

In these conditions the special importance should be paid to the determination of the optimal ratio oftheprimary nature with the cultural landscape. A sound strategy and a systematic organization in interactions of society with the natural environment - a new stage of environmental management. In modern conditions a special value of all forms of activityshould have in the aesthetic reconstruction of the natural environment. It is primarily the culture of the clearance space, increasing areas for parks, the development of the art, of creating gardens and parks, recreation and small groups. All of the above listed activities and processes create individual artifacts are only a small portion of those activities on the culture of the environment, which are called landscape design. Landscape design is humanistically directed activity related to the spirit of the times through the medium of subject – semantic, emotional. Landscape design of the surrounding space, as the object of intellectual attention, the object with content, with its own meaning. Today the principles and technology of landscape design are formed and perfected. Professor G. Knabe says that "the design organizes the spatial environment that gives to it a cultural meaning" [8]. In order not to be "conquerors" of Northern spaces and not be in "the graveyard of his victories" it is necessary to understand the current trends in the field of landscape design for the radical transformation of the urban environment in the interests of the individual: the transition to a holistic interpretation of the urban space; efficient use of the landscape components of the urban open spaces; consideration of human needs to ensure the necessary level of comfort; the achievement of functional orderliness, visual expression of landscape objects urban environment; improvement of environment of urban spaces; creative transformation of nature; the formation of the urban space recognition; improvement of personal culture, the formation of a new worldview, a new co-evolutionary interaction with the environment [2].

The main task of a contemporary man, as the main task of the landscape design is to make the Earth a beautiful home. You must remember the F.M. Dostoevsky'swords: "Beauty will save the world!" But this will not happen by itself. A focused, organized work is needed throughout the country, on all landscape objects. It is clear that highly qualified and socially responsible professionals in the field of landscape design are necessary. Residents should relate to the urban environment not as to something ready and given, but as to a space on which it is possible to influence in accordance with nature. No system procedures by itself, without the creative activities of citizens, cultural relations, cultural behavior, cannot make a qualitative change in the structure of the urban environment.

It is essential to change the environment into an object that, by increasing in it the artistic-figurative contents, must promote the humanization, humanitarization, ecologization of culture at all levels – from the artistic image of a single recreational facility to figurative contents of the ecological landscape system of the urban environment.

Socio-cultural potential of a landscape is composed in its possibilities for the conservation of cultural values of open spaces of the urban environment and the introduction of technical innovations in the sphere of human culture. Therefore, the landscape design appears as a specific kind of activity developing its own cultural principles and techniques of project activity for building a sustainable urban environment [3].

The authors propose a matrix model of the urban environment formation using landscape design methods

Factors of sustainable urban environment formation	Social and cultural problems of the urban environment	Modern trends in the development of the sustainable urban environment	Principles of the sustainable urban environment formation	Landscape design methods for building a sustainable urban environment
1	2	3	4	5
1	2	3	4	5
Socio-cultural: -socio-demographical, aesthetic, -ethical; -traditional innovative; -historical; -humanistic; -emotional-psychological; -natural-climatic; -environmental and town-planning; -technical-economic; -normative-legislative	-increasing environment pollution; -aesthetic inexpressiveness; -functional disorder; -violation of architectural harmony with the person's needs; -elimination of functional yard spaces; -the erosion and disappearance of the spiritual urban space fulfillment	-the achievement of the functional orderliness of visual expression, improvement of the urban environment landscape features; -creative transformation of nature; -formation of awareness of the urban space; -individual improvement of culture, the formation of a new worldview, a new co-evolutionary interaction with the environment	-ecological stability; -aesthetic improvement, diversity; multi-level vertical structure; spatial integration; -complexity (the unity of the natural and built environment); -humanization of the urban environment; harmonization; environmental object's integrity; bionic principles	Visual expansion of the space; planned change of a landscape; imitation of the natural landscape; integrating of the architecture with the environment; renovation of historic landscapes; creation of a "spirit of place"; the diversity of urban nature

Giving to the landscape objects certain aesthetic and functional properties, the designer "projects" the person who will live in this environment; creates a culture of cognition, thinking, understanding, attitude to the environment and further the translation of the formed values.The urban environment formed by the methods of landscape design undoubtedly becomes more humane, sensitive to a person, his relationships with people, nature and the world around him. Value responsiveness, sensitivity, humanity, willingness with the data of the cultural tradition and natural environment – this is what is valued now in the concepts of any kind [10,p. 63].

There is no coincidence in the philosopher's words by Jean Baudrillard in his work "System of things" [1], where he considers design as an approach, as a principle," which "can apply to everything, anything... in the end, on far-reaching social processes. This is the formation of broadly understood environment: open valued behavioral, structural... And the main object of formation is the man himself". Design, including landscape, is directly related to the structural, functional, process relations, which transform the system into a coherent unity which is able to satisfy the material, social and spiritual needs of the person. Meeting these needs is a powerful motivational basis. "The motivational bases, in turn, provide the work efficiency (economic, social, environmental)" [9,p. 57].

4. Conclusion

Landscape design, being philosophically significant effect, it contributes to the design of the urban environment as a system, socio-cultural, holistic, dynamically evolving object of intellectual attention and values including the main value - the person.

IOP Conf. Series: Materials Science and Engineering **154** (2016) 012029 doi:10.1088/1757-899X/154/1/012029

This implies important socio-cultural functions of landscape design – improving the quality of life, cultural production and consumption, harmonization of the relations of production, of the aestheticization of existence, creation of favorable psychological climate for people. Man vivificates when he is surrounded by harmony. Not only a man influences the urban environment, but the environment is able to shape the meanings, feelings, moods, desires, goals and values. This is dedicated to a new direction in science - philosophy of design, which considers design as a way of thought, action, as a way of ordering the world in a holistic, harmonious structure and the solution of universal problems of essence and existence of a man. But in general, the nature and content of the design philosophy is advanced and spiritually ascending reproduction, preservation and comprehensive development of society and its culture on a planetary scale.

References

[1] Baudrillard J 1999 *The System of things* (Moscow; Rudomino)224

[2] Bauer N V 2015 *The Culture of creating a sustainable urban environment using landscape design methods*(Tyumen;TSOGU) 151

[3] Bauer N V and Shabatura L N 2014 Innovative directions of the development of the urban landscape design *Bulletin of the Chelyabinsk Academy of culture and art***1 (37)** 63-7

[4] Ganopol'skii M G and Litenkova S P 2005 The structure of the settlement in the Tyumen region: the features of the genesis and prospects of development *Izvestiya RAN Geographical series***3**56-62

[5] Ganopol'skii M G and Fedorov R Y 2011 On the way to oil and gas producing civilization *Bulletin of archaeology, anthropology and Ethnography***1 (14)**203-11

[6] Zemenkov Y D, Prokhorov A D and Vasil'ev G G 2007 *History and prospects of development of the oil and gas industry*, fuel power complex (Saint Petersburg; Nedra) 224

[7] Karpov V P 2011 *The Tyumen industrial "explosion": the history of a megaproject (*Tyumen) 207-8

[8] Knabe G S 1986 *Ancient Rome - history and everyday life: Essays* (Moscow; Iskusstvo) 206

[9] Kolomoets G P 2003 *Design in modern socio-cultural space*: dis. ...candidate. Culturology: *24.00.01*(Krasnodar)160

[10] Shadrina S Z and Shabatura L N 2002 *Economic culture in the structure of the humanitarian education* (Ekaterinburg)

Tomography in Geology: 3D Modeling and Analysis of Structural Features of Rocks Using Computed Micro-Tomography

A A Ponomarev, R A Mamadaliev and T V Semenova

Industrial University of Tyumen, 38 Volodarskogo st., Tyumen, 625000, Russia

E-mail: ponomarev94@mail.ru

Abstract. The article presents a brief overview of the current state of computed tomography in the sphere of oil and gas production in Russia and in the world. Operation of computed micro-tomograph Skyscan 1172 is also provided, as well as personal examples of its application in solving geological problems.

1. Introduction

Computed micro-tomography, or micro-CT, is recreation of X-ray binary models in three-dimensional images. Figure 1 illustrates how it works. A sample is placed inside the scanner on a table which rotates during scanning. An X-ray source generates beams that travel through the sample leaving shadow projections on the detector (camera). Following each rotation a separate projection is captured (a separate two-dimensional image corresponding to the X-ray intensity passing through the sample in question is called shadow projection).

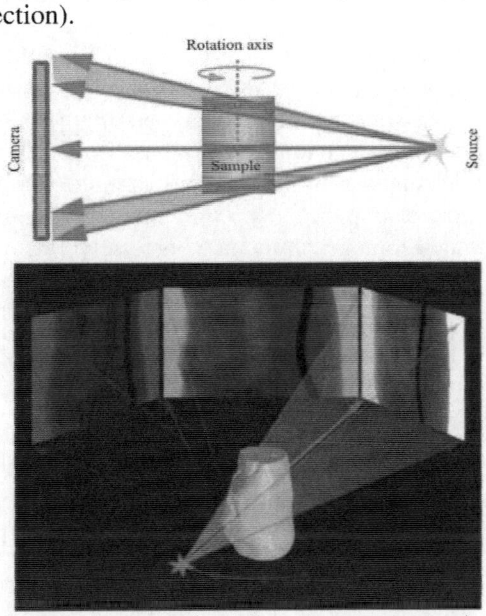

Figure 1. Operation of a micro-tomograph.

Brightness (varying grayscale) on an X-ray shadow projection represents X-ray radiation attenuation due to dispersion and absorption of the signal passing through the sample. [1,3]

Attenuation depends on the density and thickness of the object in question, as well as on the effective atomic number the object is composed of.

When X-ray radiation passes through some material the level of absorption can be related to four types of interaction: photoelectric absorption, Compton scattering, electron-positron pair creation and coherent Rayleigh scattering. In case of studying geological objects, when radiation sources with up to 100 keV are used, photoelectric absorption is the primary process determining X-ray radiation attenuation.

Upon completion of scanning and during reconstruction, projections are converted into slices of single-voxel thickness, from which a 3D model is then constructed and analyzed. For SkyScan 1172 the maximum size of the sample in length/width should not exceed 64 mm, maximum height – 70 mm.

In Western Siberia, the time when wells were flowing is gradually coming to an end. The new era of depleted deposits and hard-to-recover reserves dictates the necessity to introduce innovative high-precision computer technologies into a complex of standard petrophysical core analyses. [3]

2. Current state of computed tomography application in studying petophysical rock properties

Developed Western countries use a comprehensive approach to apply this technique. In Russia computed micro-tomography has been developing for over five years now; the first micro-tomograph was introduced by Schlumberger. At present, a majority of well-developed oil and gas companies utilize tomographs to evaluate large-size pore space features in full-size core samples (cavern porosity, large jointing). However, the resolving power of a tomograph is not sufficient to obtain precise porosity data that is why computed micro- and nano-tomography is resorted to. Main advantages are:

- determining porosity and pore space morphology of varying genesis to predict permeability and porosity properties of core samples;
- studying core samples which lithological and petrophysical properties are impossible to study accurately, samples with defects that make it impossible to carry out standard laboratory tests (chipping, fractures), cuttings, and samples from sidewall core sampler;
- studying reservoirs categorized as hard-to-recover which can't be studied by standard laboratory tests (lack of equipment, considerable variability of sample properties during testing, extreme PVT parameters) or test results do not provide an accurate study of the test object (exclude processes at a pore or micro-pore level), as well as studying core samples basic analyses of which are significantly time-consuming. [2]

Over the last couple of years complex tomography has become a mainstream technique in core analysis. International companies apply it to analyze and simulate various kinds of petrophysical and lithologic rock properties. For example, such companies as Shlumberger, iRockTechnologies, INGRAIN, FEI can use computed tomography data to precisely calculate porosity, permeability and wettability, and to simulate multi-phase filtration displacement processes and enhanced oil recovery techniques. This proves that for Russian oil and gas sector it is necessary to develop its own software and new methods of introducing the reviewed technique into petroleum field geology.

3. Personal experience in applying micro-tomography in geology

TSOGU research geochemical laboratory is equipped with a computed micro-tomograph Skyscan 1172. Its standard software package is unable to directly determine permeability and model filtration, EOR techniques and rock mechanics. However, even without this it is possible to obtain a fair bit of information on rock structure or determine porosity in samples with difficult sample preparation.

Some of the analysis examples are given below.

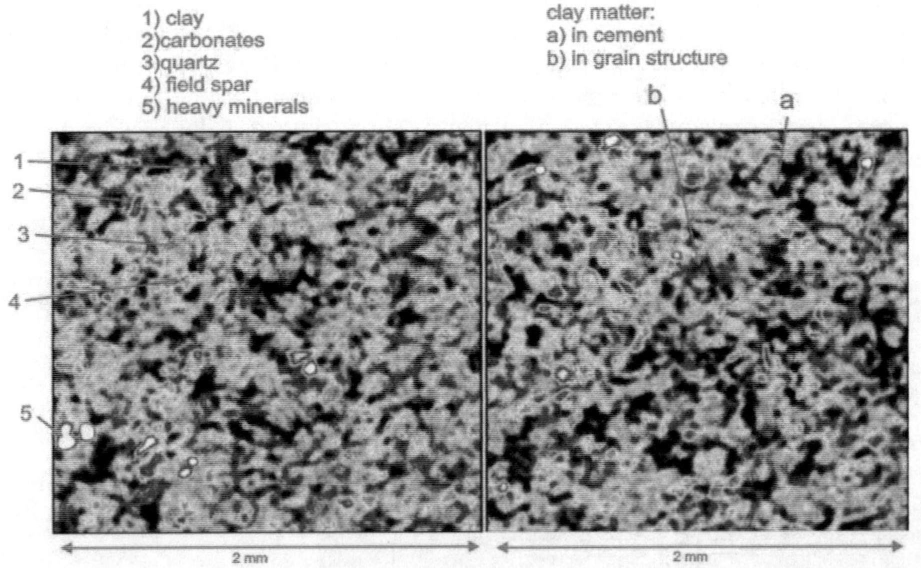

Figure 2. A sample determination of rock mineral composition, with cement segregation.

The percentage of all these minerals, including clay in grain structure and in rock cement, relative to the entire sample volume, can be calculated using the software.

In geology, to calibrate tools used for well logging it is necessary to know the shale volume. Calibration and re-interpretation of the obtained data for the entire formation are performed according to the amount of clay matter in core samples. A standard technique of determining shale volume is not precise enough, as it determines clay matter in cement. It is due to the actual experiment procedure of grain size analysis. Therefore, data on clay matter in grains are lost. Computed microtomography allows for both correct determination of shale volume and its segregation into two categories: in cement, in grain structure. [3]

Aside from calculations, computed tomography allows for construction of the sample 3D models. Modeling, as well as calculations, is performed based on the selected range of thicknesses, i.e. all we can see with a naked eye on tomograms (figure 3) can be simulated in 3D. Let us look at the example of the simulated pore structure of a microcement-based rock sample (figure 3).

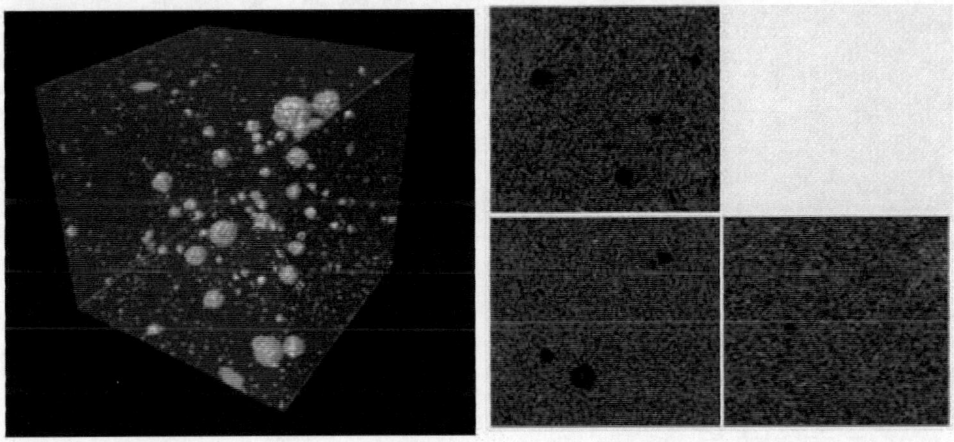

Figure 3. A sample pore space 3D model and its development in 3 cross-sectional planes.

As seen from the 3D model, the sample is mostly comprised of pores isolated from each other, which is a success for the cement rock developers as permeability will be almost equal to zero.

The device is unique in that it provides non-destructive control of the sample. To calibrate well logging tools precise calculations of rock properties must be available. As a rule, they are obtained from core analysis; after that correlation dependence is built and interpreted for the entire formation.

Core analysis is done in stages; in this particular case, salt content (halite (NaCl) and sylvite (KCl)) in terrigenous rock had to be determined by two methods: standard (leaching) and micro-tomography. But prior to conducting analysis, rocks were saturated with formation water to determine electrical resistivity necessary for logging tools calibration. Subsequently, micro-tomography was performed (figures 4-5).

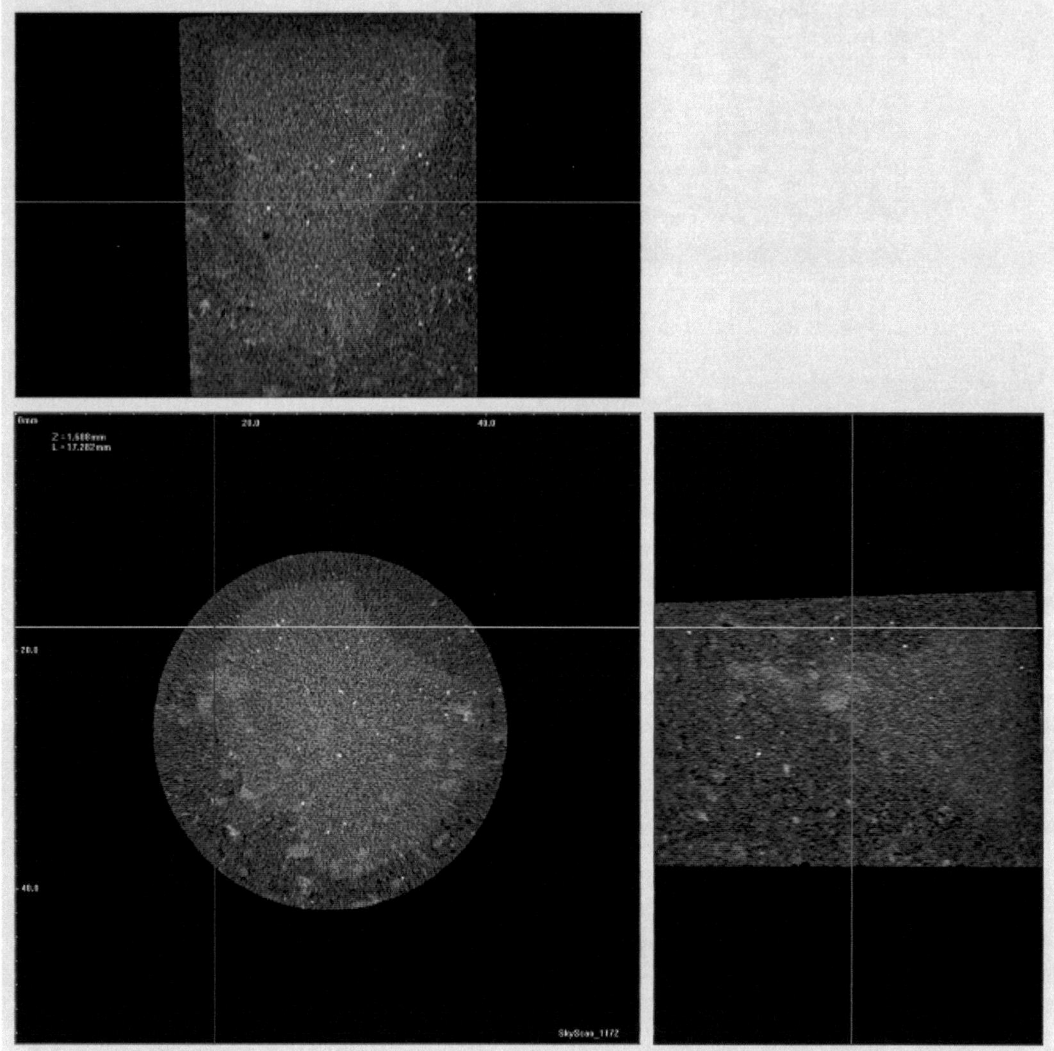

Figure 4. Development in 3 cross-sectional planes, sample 1 (presence of industrial salinization).

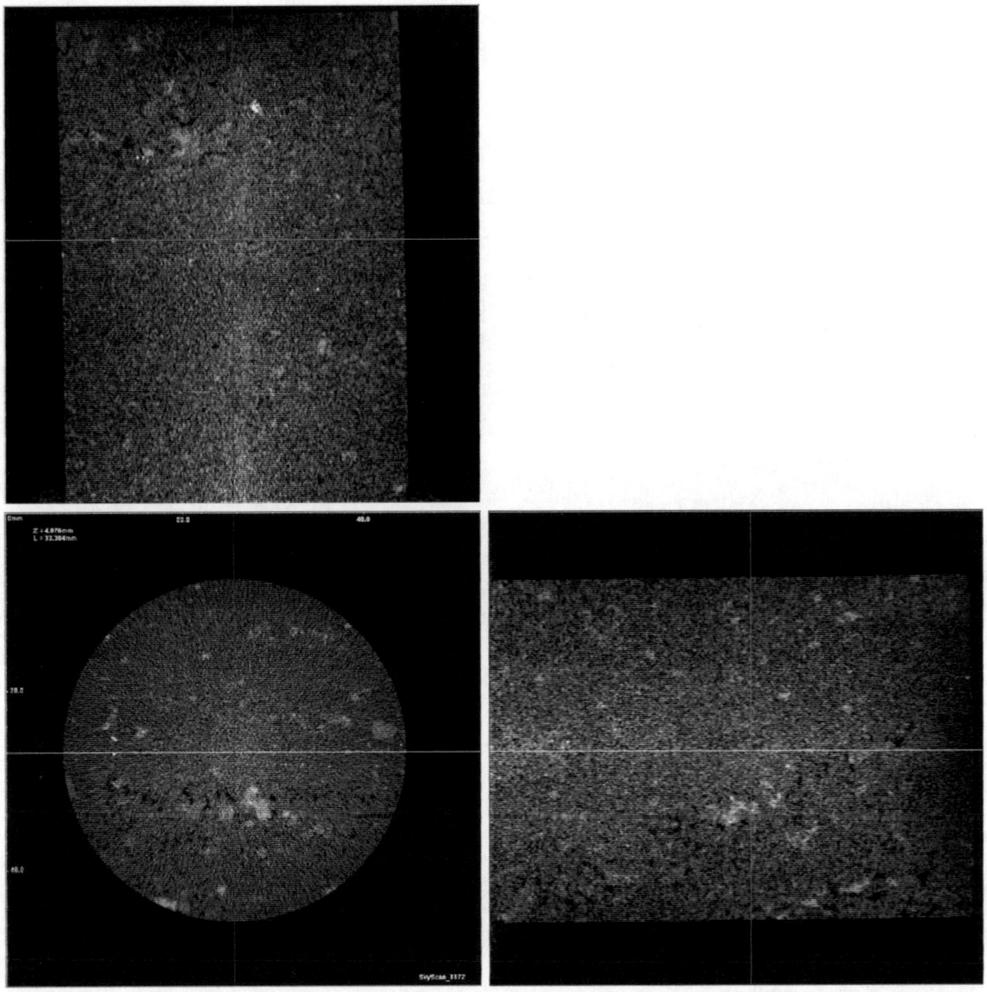

Figure 5. Development in 3 cross-sectional planes, sample 2 (absence of industrial salinization).

As the analysis discovered, during formation water saturation one of the samples experienced industry-induced deformation of the rock structure, namely salt precipitation on rock walls (figure 5). This phenomenon causes distortion of all physical and mechanical rock properties during further analysis. Only computed microtomography is capable of identifying such industry-induced irregularities.

4. Conclusion

The paper covered key aspects of tomography and micro-tomography development. The operating principle of a computed micro-tomograph Skyscan 1172 was overviewed, as well as some results of its work. There are no ready to use GOST techniques for the micro-tomograph, wherefore the accuracy of the obtained results depends almost completely on the interpreter's experience.

Based on this work we can make a conclusion that oil and gas companies should realize the need to advance research in this area in order to gradually improve the capacity and preciseness of analysis on computed tomographs, micro-tomographs and nano-tomographs.

Applying this method makes it possible, in some cases, to simplify the process of obtaining petrophysical properties of rocks (such as cuttings, loose soil, cavernous and fractured reservoirs).

Micro-tomography is also a helpful tool of comparative analysis, allowing researchers to study rocks before and after some tests and make conclusions based on its results.

References

[1] Korost D V *Heterogeneity of terrigenous reservoir structure and types of pore space structures (on the example of the upper Tyumen formation of the Urnenskoye oil field in West Siberia)* Dissertation

[2] Vakhrusheva I A, Gilmanov Ya I and Patrakov D P 2015 Digital core – new area of petrophysical core analysis. Current state in Russia and in the world. *Proceedings of the Vth research and applied science conference Supercomputer technologies in the oil and gas industry. Mahematical methods, software and hardware.* 112-16

[3] Ponomarev A A and Zavatskiy M D 2015 Methods of applying computed micro-tomography in geology *Proc. of Institutions of Higher Education. Oil and Gas* **3** 29-33

IOP Conf. Series: Materials Science and Engineering **154** (2016) 012031 doi:10.1088/1757-899X/154/1/012031

Dynamics of Pipeline Pulling Process By Horizontal Directional Drilling

V S Toropov, O.S. Tamer, S Yu Toropov and N A Nikiforov

Industrial University of Tyumen, 38 Volodarskogo St., Tyumen, 625000, Russia

E-mail: 264301@mail.ru

Abstract. For using in horizontal directional drilling, the authors developed a model of a cyclical pipeline pulling process, enabling to perform calculations of resistance forces of pipeline movement considering the well profile geometry, mass characteristics of drill pipes, pulled pipeline and mud, as well as the velocity parameters of pullback. Analytical dependences were derived to determine absolute values and time changes in pipeline pullback force considering the well geometry, physical properties of interacting elements of the pullback process. In addition, analytical dependences were obtained for certain power consumptions of a drilling rig during pipeline pulling considering the pullback process model.

1. Introduction

In-service pipeline pullback is the most difficult [1] step in horizontal directional drilling (HDD) for construction especially if we speak about a large diameter steel pipeline. A combination of factors determining the features of pullback is reviewed in [3]. Basic approaches to the technological process control are stated in [5]. This article develops the ideas of decreasing the power consumption and other resources used in trenchless construction by pullback control in real time mode.

It is necessary to determine the degree of main parameters influence on the extent of power consumption required for the execution phase by the rig to develop the model of pipeline pullback process. The main controlled parameters are pullback velocity, starting acceleration, execution phase time, pullback process forces [8].

Obviously, instantaneous velocity, acceleration and force of pullback can change in the execution phase process and, eventually, determine the necessary power and time for the execution phase. So, controlling the pipeline pullback process will involve adjusting such parameters as pullback velocity and starting acceleration.

2. Materials and methods

The scheme of pipeline pullback into the borehole is illustrated in figure 1. The total length of the borehole is denoted L_{bh}. The pipeline pullback length – L_p. The length of the pipeline below ground – L. The level difference – ΔH. The pullback force – $\mathbf{F}_p$. The moving coordinate is denoted dx.

Considering the process recurrence, it is advisable to examine the pipeline movement in one cycle, because cycles will repeat and their number will equal the number of the drill pipes used for borehole drilling[4]. The distance that pipeline overcomes during one cycle and that equals the length of one drill pipe is designated l_{dr}.

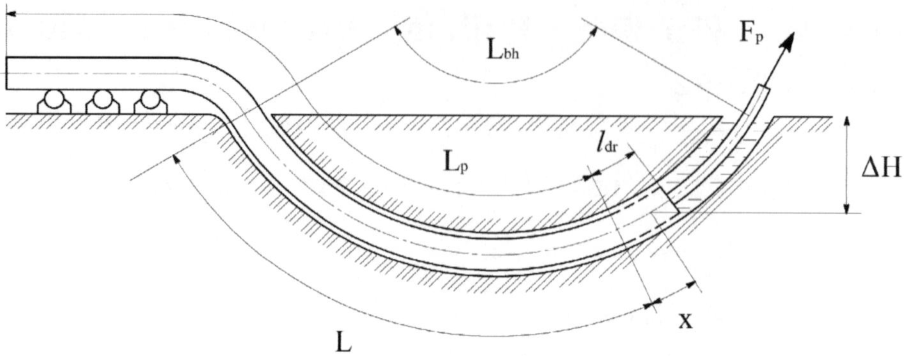

Figure 1. The scheme of pipeline pullback into borehole.

In general, every cycle will consist of three periods. Starting the pipeline and drill string and their acceleration to the maximum velocity will be at the first period, the pipeline will move at a steady pace at the second period and braking to a standstill will be at the third. The dependence of the pipeline movement velocity on the distance per cycle would generally look like the one described in illustration (figure 2).

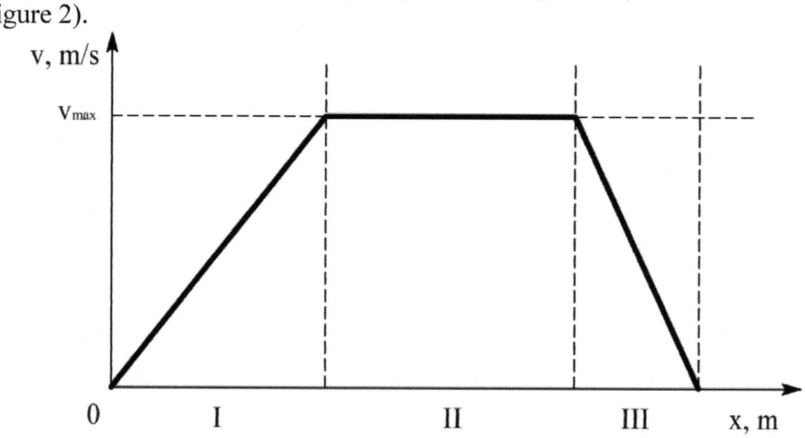

Figure 2. The dependence of the pipeline movement velocity on the distance per cycle at its different periods in general: I - acceleration period; II - steady movement period; III - breaking period; $\mathbf{v_{max}}$ – maximum cycle speed.

According to the D'Alembert principle, we can state that at every period it will be right for a moving drill string and pipeline in the borehole:

$$\sum F_{act} + \sum F_{res} + \sum F_{mom} = 0, \quad (1)$$

where$\sum$ $\mathbf{F_{act}}$ – the sum of active forces;
$\sum$ $\mathbf{F_{res}}$ – the sum of resistance forces of pipeline movement;
$\sum$ $\mathbf{F_{mom}}$ – the sum of inertial forces.
Then, for the first period:

$$\sum F_{act} = F_{rig},$$
$$\sum F_{res} = F_{res.I}\left(F_{fhr}, F_{fcr}, F_{fsr}, F_{ssr}\right), \quad (2)$$
$$\sum F_{mom} = \left(m_p + m_{ds} + m_{fc}\right)\frac{d^2x}{dt_I^2},$$

where $\mathbf{F_{rig}}$ – the pulling force of the drilling rig;
$\mathbf{F_{res.I}}$ - the movement resistance force at the first period;

$\mathbf{F_{fhr}}$ - the hydraulic resistance force of the drilling fluid movement in the borehole;
$\mathbf{F_{fcr}}$ - the drilling fluid column resistance force;
$\mathbf{F_{fsr}}$ – the movement resistance force of the free pipeline segment, located on the day surface;
$\mathbf{F_{ssr}}$ - the resistance force of the drilling fluid shear stress surmounting;
m_p - the mass of the pipeline;
m_{ds} - the mass of the drill string;
m_{fc} - the mass of the drilling fluid column;
$\frac{d^2x}{dt_I^2}$ - the acceleration at the first period.

For the second period:

$$\begin{aligned} &\sum F_{act} = F_{rig}, \\ &\sum F_{res} = F_{res.II}\left(F_{fhr}, F_{fcr}, F_{fsr}, F_{ssr}\right), \\ &\sum F_{mom} = 0. \end{aligned} \tag{3}$$

So, for the second period:

$$F_{rig} = -F_{res.II}\left(F_{fhr}, F_{fcr}, F_{fsr}, F_{ssr}\right). \tag{4}$$

For the third period:

$$\begin{aligned} &\sum F_{act} = 0, \\ &\sum F_{res} = F_{res.III}\left(F_{fhr}, F_{fcr}, F_{fsr}, F_{ssr}\right), \\ &\sum F_{mom} = \left(m_p + m_{ds} + m_{fc}\right)\frac{d^2x}{dt_{III}^2}. \end{aligned} \tag{5}$$

Then, for the third period:

$$F_{res.III}\left(F_{fhr}, F_{fcr}, F_{fsr}, F_{ssr}\right) = -\left(m_p + m_{ds} + m_{fc}\right)\frac{d^2x}{dt_{III}^2}. \tag{6}$$

During the pipeline pullback process, some part of it stays on the day surface out of well right to the last moment. This part of pipeline is usually located on rollers. We call it the free segment of pipeline. As the pipeline moves, the drilling rig has to overcome the rolling friction force of this free segment. The free segment movement resistance force equals:

$$F_{fsr} = \frac{\left(L_p - L - x\right)}{L_p} \cdot m_p \cdot g \cdot k_{rf}, \tag{7}$$

where L_p - the length of the pipeline;
k_{rf} – the rolling friction coefficient of the pipeline;
L - the length of the pipeline section located in the well at the beginning of the cycle;
g – the free fall acceleration;
x – the current coordinate along the well profile.

This component of the movement resistance force is proportional to the mass of the pipeline part located out of the well and the length of which decreases in the course of pipeline headway, therefore, $\boldsymbol{F_{fsr}}$ decreases during pullback.

Thereby, taking into account (7), conclusions contained in [3] and that we analyze the mud shear stress overcoming the resistance force only at the beginning of every cycle, the sum of resistance forces of the pipeline movement in the well on the border of the second and third periods equals:

$$\begin{aligned} \sum F_{resII-III} = {} & \lambda \cdot \rho_f \left(L_{bh} - L - x\right) \cdot \frac{\pi\left(D_{bh}^2 - D_{dr}^2\right)}{8D_{bh}} \cdot \left(\frac{dx}{dt}\right)^2 + \frac{\left(L_p - L - x\right)}{L_p} \cdot m_p \cdot g \cdot k_{rf} + \\ & + \frac{\pi}{4}\left(D_{bh}^2 - D_{dr}^2\right) \cdot \left(\sqrt{R_c^2 - x^2} + H_{\max} - R_c\right) \cdot \rho_f \cdot g, \end{aligned} \tag{8}$$

where λ - the coefficient of hydraulic resistance;
ρ_f - the density of drilling fluid;
D_{bh} – the borehole diameter;
D_{dr} - the drilling rod diameter;
L_{bh} - the borehole length;
R_c - the bending radius of the well profile;
H_{max} - the maximum deepening size of the well profile from the day surface.
Then, deceleration at the third period equals:

$$\frac{d^2x}{dt_{III}^2} = -\frac{F_{res.II-III}}{(m_p + m_{ds} + m_{fc})}, \tag{9}$$

where v_{max} – the maximum speed of the cycle.
As in equation (8) $\frac{dx}{dt} = v_{\max}$, the breaking period length will be:

$$l_{III} = -\frac{v_{\max}^2}{2\frac{d^2x}{dt_{III}^2}}. \tag{10}$$

Thus, at the given maximum speed v_{max}, the length of the period that may be used for acceleration and further uniform motion is:

$$l_I + l_{II} = l_{dr} - l_{III}, \tag{11}$$

where l_I – the acceleration movement length;
l_{II} – the uniform motion length.

In order to determine the optimum period number in a cycle, let us analyze two different variants of the cycle progress. In the first variant the cycle will have two periods: the acceleration period with even acceleration a_I and length l_I and the breaking period with length l_{III}. The uniform movement period is absent in this variant. Then, the dependence of the pipeline movement velocity on the distance per two-period cycle will look like the one described in illustration (figure 3).

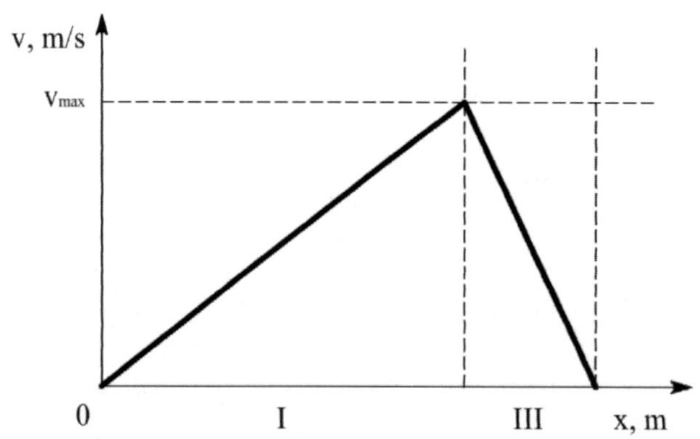

Figure 3. The dependence of the pipeline movement velocity on the distance per two-period cycle: I – acceleration period; III – breaking period; $\mathbf{v_{max}}$ – maximum cycle speed.

Therefore, acceleration at the acceleration period we will find from:

$$\frac{d^2x}{dt_I^2} = -\frac{F_{rig} + F_{res.I}}{(m_p + m_{ds} + m_{fc})}. \tag{12}$$

And the work to move the pipeline and drill string at the first period:

$$A_I = \frac{(m_p + m_{ds} + m_{fc}) \cdot v_{\max}^2}{2}. \quad (13)$$

The power needed to move the pipeline:

$$P_I = \frac{A_I}{t_I}. \quad (14)$$

Where the time to pass through the acceleration period will be found from:

$$t_I = \frac{v_{\max}}{a_I}, \quad (15)$$

where a_I – the acceleration at the first period.

In the second variant the cycle will have three periods: the acceleration period with length l_I, the uniform movement period with length l_{II} and the breaking period with length l_{III}. Therefore, the dependence of the pipeline movement velocity on the distance per cycle at its different periods in general will look like the one described in illustration (figure 2).

Introducing the uniform movement section, it is necessary to set its length l_{II}, then the acceleration at the acceleration period will be determined as:

$$\frac{d^2x}{dt_I^2} = \frac{v_{\max}^2}{2(l_{dr} - l_{II} - l_{III})}. \quad (16)$$

Then the work at the acceleration period will be found according to equation (13), and the work done at the uniform movement equals:

$$A_{II} = l_{II} \cdot \sum F_{resII-III}. \quad (17)$$

Let us assume that in both variants the pipeline movement before the beginning of the third period will go in equal periods, that is:

$$t_{I1} = t_{I2} + t_{II2}, \quad (18)$$

where t_{I1} - the time of passing the acceleration period in the first variant;
t_{I2} - the time of passing the acceleration period in the second variant;
t_{II2} - the time of passing the uniform movement period in the second variant.

Then, the saved amount of work will be found from equation (17). And the saved power will equal:

$$P_c = \frac{l_{II} \cdot \sum F_{resII-III}}{t_I + t_{II}}. \quad (19)$$

3. Results and Discussion

Now we have the opportunity to analyze the pipeline movement at different working speeds when performing pullback along the entire well profile length. Guided by the abovementioned, it is possible to determine the values of resistance forces that are necessary to overcome at different points of the well profile during pullback, as well as the values of the drilling rig power required for this at different parameters of the pullback process.

The dependences of the pullback resistance forces values on the well profile length in cycles at different maximum working speeds for steel pipelines are described in illustration (figure 4). The dependences are of the same nature and it shows that the pullback resistance forces will have the maximum value at the deepest point of the well profile.

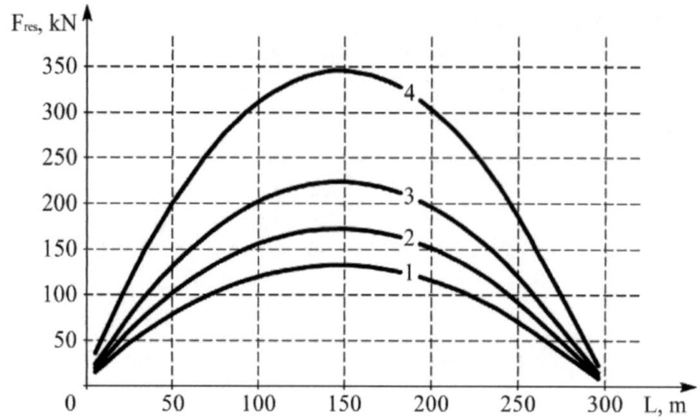

Figure 4. Values of the pullback resistance forces along the well profile length in cycles at maximum working speeds of 5 m/s for steel pipelines: 1 - ∅ 630 mm; 2 - ∅ 720 mm; 3 - ∅ 820 mm; 4 - ∅ 1020 mm.

The dependences of the mean power values required to accelerate the drill string in one cycle at different points of the well profile on the position of the pipeline in the well during pullback at different maximum working speeds for a ∅1020 mm steel pipeline are described at illustration (figure 5). The dependences show that the maximum working speed of the pipeline pullback is one of the most important parameters defining the values of energy consumption to perform the pipeline pullback [7].

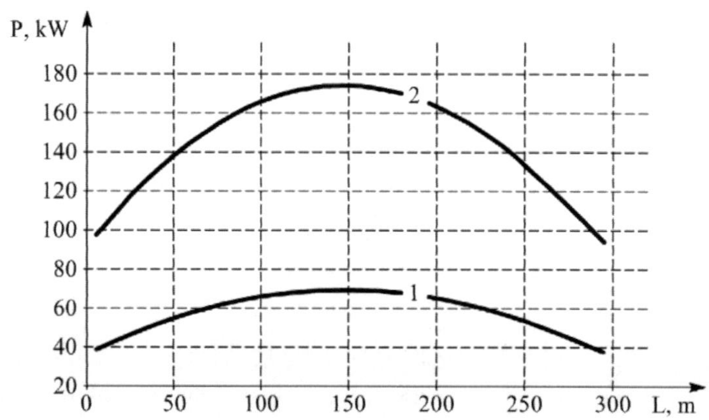

Figure 5. Mean power values required to accelerate the drill string in one cycle at different points of the well profile during the pullback process at different maximum working speeds for a ∅1020 mm steel pipeline: 1- 0.2 m./s, 2- 0.5 m/s.

As in this model the pipeline and the drill string are moving inertially to a standstill after acceleration in each cycle, the obtained power values are sufficient to move the pulled pipeline to a length equal to the length of one drill pipe.

4. Conclusion

The methods presented in this paper allow such dependences to be received and applied to any drilling equipment, various diameters of pipes made of steel and plastics and any working speed of the pipeline pullback process. This can be taken as a basis for pullback process management taking into account dynamic characteristics to minimize power consumption and the most accurate selection of equipment for horizontal directional drilling.

The division of the process model into cycles, each of which corresponds to the movement of the pipeline and the drill string to a length of one drill pipe, is a necessary condition of model's best fit to real construction conditions and allows developing practical recommendations on selecting the pipeline pullback model.

References

[1] Toropov E S, Toropov V S, Zemenkov Yu D and Seroshtanov I V 2015 Complications at crossings construction by horizontal directional drilling method *Territory "Oil and Gas"* **5** 32-7

[2] 2008 *Monitoring of hydrodynamic and technic characteristics of pipeline systems* ed Zemenkov Yu D (Tyumen; Vector Book) 432

[3] Kiryanov I E, Toropov V S and Toropov S Yu 2015 Determination of dependences of pipeline movement resistance forces in curve borehole from physical parameters and rheological model of using mud *Oil equipment* **8**

[4] Zemenkova M Yu, Shalay V V and Zemenkov Yu D 2015 Expert systems of multivariable predictive control of oil and gas facilities reliability *Procedia Engineering* **113** 312-315

[5] Toropov E S, Kiryanov I E and Toropov V S 2014 Methods of energy consumptions reduction in pipeline pullback process to the borehole at using horizontal directional drilling technology for industrial objects construction. *International conference "Oil and gas. ATR – 2014". Resources, transport, technologies*

[6] 2004 *Transporting and storage of oil and gas in examples and sums* ed Zemenkov Yu D (St. Petersburg; Nedra) 544

[7] 2008 *Technique and technologic processes at transport of energy resources* ed Yu.D Zemenkov (Tyumen; Vector Book) **1** 380

[8] Willoughby D 2005 *Horizontal Directional Drilling Utility and Pipeline Applications* (New York; The McGraw-Hill Companies) 264

IOP Conf. Series: Materials Science and Engineering **154** (2016) 012032 doi:10.1088/1757-899X/154/1/012032

Finding the Displacement of Elastically Bent Pipeline Disposed on Roller Supports Under the Action of the Axial Shear Force

E S Toropov, T I Vorobieva , S Yu Toropov and V S Toropov

Industrial University of Tyumen, 38 Volodarskogo St., Tyumen, 625000, Russia

E-mail: 264301@mail.ru

Abstract. The authors of this article have developed a spatial position determination method of a real steel pipeline free segment located above the ground surface during pipeline pullback using a pipeline pusher for the horizontal directional drilling technology. Values of elastically bent pipeline displacement were found in different points of the pipeline profile with the aim of defining the admissible values of the shear force applied to the pipeline by additional equipment during the pipeline pullback process. The developed methods allow finding the points of the curved pipeline profile in which it is necessary to place additional fixing connections above the pipeline, as this will expand the range of the shear force applied to the pipeline.

1. Introduction

When using the horizontal directional drilling technology, the values of the axial shear force applied to the pipeline as a subsidiary during pipeline pullback can have a wide range and depend on the potential of the equipment model used [2]. Nowadays, various types of approaches to applying the shear force to the pipeline exist. At the moment, there are several types of pipeline pushing machines [8] and obviously their number will continue to grow.

The behavior of elastically bent pipelines under the action of the axial shear force should be investigated in the following way. It is necessary to determine the pipeline stress-strain state considering the force application point and the values of roller supports height. This article solves the problem of finding different displacement points of elastically bent pipeline terminal section pushed by the pipeline shear force applied to the end of the pipeline [4].

2. Research object (model, process, structure, synthesis, experimental part, etc.)

We will use a system consisting of such interacted objects as pipeline, roller supports and pipeline pusher to solve our problem. Main physical parameters of a pipeline are taken into account during the solution process. They are elastic modulus of pipeline material, uniformly distributed load from its own weight, flexural stiffness, tensile stiffness, etc. A shear force application point and different values of the force are taken in account as well.

Being in a curved position, the pipeline researched, having in view the regulatory compliance of the bending values [5] and the most natural profile for real pipelines [1] of the terminal portion, is placed on roller supports. The problem is solved by dividing the terminal section into different parts and finding general solutions of equilibrium equation systems for each part located between the roller supports that separated these parts.

3. Methods.

A free pipeline segment located above the ground surface during pullback will have the profile shown in figure 1. It refers to a steel pipeline that has a much larger allowed elastic bending radius [6], in contradistinction to plastic pipelines [7]. We divided the elevated pipeline segment by *O, A, B, C* and *D* into individual parts with different values of its curvature.

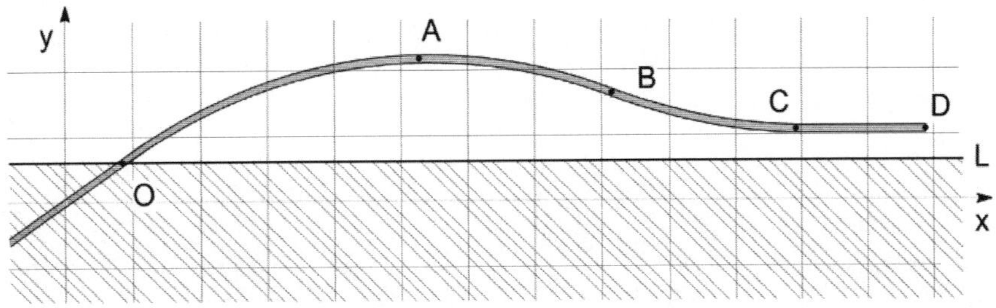

Figure 1. Free pipeline segment profile located above the ground surface.

At the same time, the pipeline is located on roller supports and we will consider that the supports are installed at equal intervals (figure 2). In this case, the pipeline in each section P_iP_{i+1} can represent a clamped rod loaded with its own weight, that is a uniformly distributed load q.

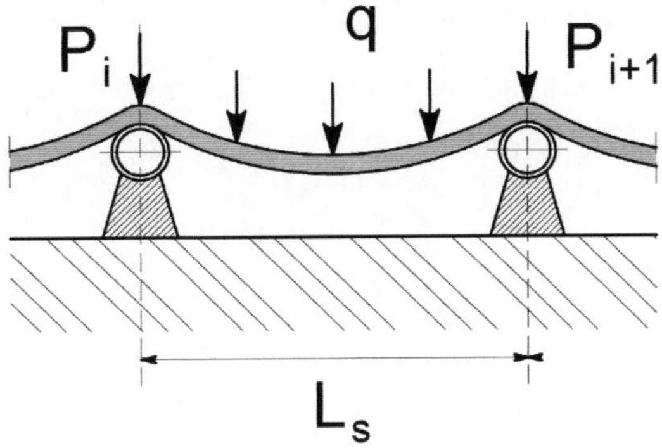

Figure 2. Design scheme of a pipeline segment on roller supports: L_s - interval between two supports.

Let us find the general solution of differential equilibrium equations for each pipeline section described above. For such a case, we will have six equations in a set [3]. The solution will depend on six specific constants. As we have divided the pipeline into n different parts, we have $6 \cdot n$ specific constants: $C_1 \dots C_{6n}$.

Now we have two boundary conditions at O, six matching conditions at intermediate points and two boundary conditions at D. In total, there are $6 \cdot n$ conditions for determination of constants $C_1 \dots C_{6n}$. Matching conditions for supports without connection and supports with connection will be different. Let K be the pipeline section number located to the left of the support, and $K + 1$ – to the right (figure 3).

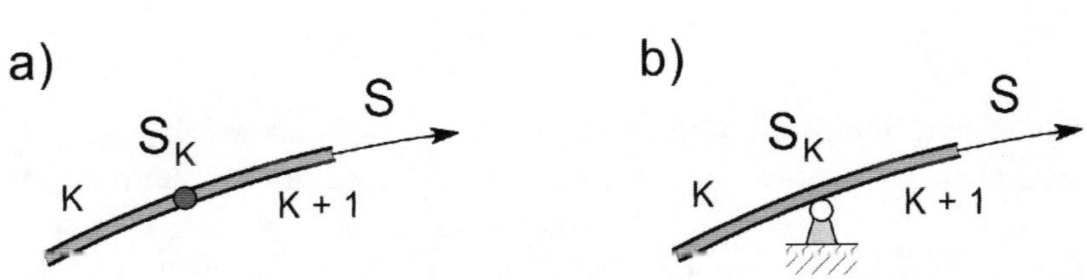

Figure 3. Supports without connection (a) and supports with connection (b) for a pipeline segment placed on roller supports.

Continuity conditions of all stress-strain state parameters must be met in those points where supports without connection are located:

$$\begin{aligned} U_K(S_K) &= U_{K+1}(S_K) \\ W_K(S_K) &= W_{K+1}(S_K) \\ \Theta_K(S_K) &= \Theta_{K+1}(S_K) \\ N_K(S_K) &= N_{K+1}(S_K) \\ Q_K(S_K) &= Q_{K+1}(S_K) \\ M_K(S_K) &= M_{K+1}(S_K) \end{aligned} \quad (1)$$

where S – the curvilinear coordinate;
S_K – the Kth support coordinate;
$U(S)$, $W(S)$ – the tangential and axial movements;
$\Theta(S)$ – the angle of rotation;
$N(S)$, $Q(S)$ – the axial and shear forces;
$M(S)$ – the bending moment.

In its turn, the following conditions should be met in supports with connections:

$$\begin{aligned} U_K(S_K) &= U_{K+1}(S_K) \\ W_K(S_K) &= 0 \\ W_{K+1}(S_K) &= 0 \\ \Theta_K(S_K) &= \Theta_{K+1}(S_K) \\ N_K(S_K) &= N_{K+1}(S_K) \\ M_K(S_K) &= M_{K+1}(S_K) \end{aligned} \quad (2)$$

The transverse force Q has a gap in the support, so the continuity conditions for it should not be met.

Now we will find the general solution of this problem. The equilibrium equations for pipeline segments curved convex up (figure 4) and curved convex down (figure 5) will be different.

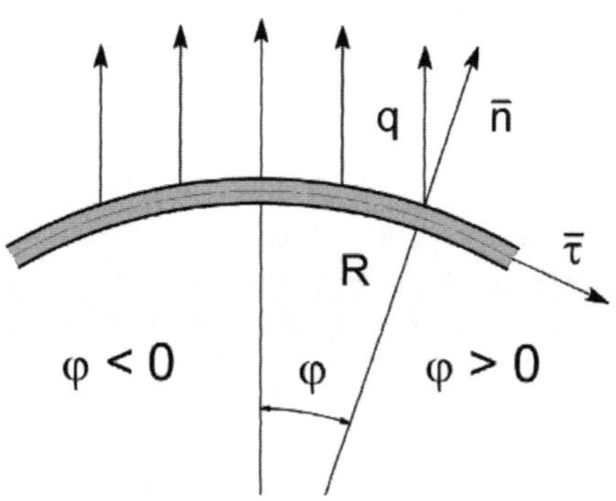

Figure 4. Design scheme of pipeline segments on roller supports curved convex up: q - distributed load; R – pipeline segment radius of curvature; φ - central arc angle; n - normal to the pipeline surface; τ - tangential force.

The equation for pipeline segments curved convex up will be as follows:

$$(N+1)\begin{cases}\frac{1}{R}\cdot\frac{dN}{d\varphi}-\frac{1}{R^2}\cdot\frac{dM}{d\varphi}=q\cdot Sin\varphi\\ \frac{N}{R}+\frac{1}{R^2}\cdot\frac{d^2M}{d\varphi^2}=q\cdot Cos\varphi\\ Q=-\frac{1}{R}\cdot\frac{dM}{d\varphi}\end{cases} \quad (3)$$

$$(N+2)\begin{cases}N=D\cdot\varepsilon\\ M=H\cdot\chi\end{cases} \quad (4)$$

$$(N+3)\begin{cases}\varepsilon=\frac{1}{R}\left(\frac{dN}{d\varphi}+W\right)\\ \Theta=\frac{1}{R}\left(\frac{dW}{d\varphi}-U\right)\\ \chi=\frac{1}{R^2}\left(\frac{d^2W}{d\varphi^2}-\frac{dU}{d\varphi}\right)\end{cases} \quad (5)$$

where q – the vertical load distributed uniformly;
ε - the axial deformation;
χ - the flexural deformation (curvature change);
$D = E \cdot A$ – the tensile stiffness;
E – the elastic modulus;
A – the cross sectional area of pipeline;
$H = E \cdot I$ – the flexural stiffness;
I – the pipeline cross-section moment of inertia.

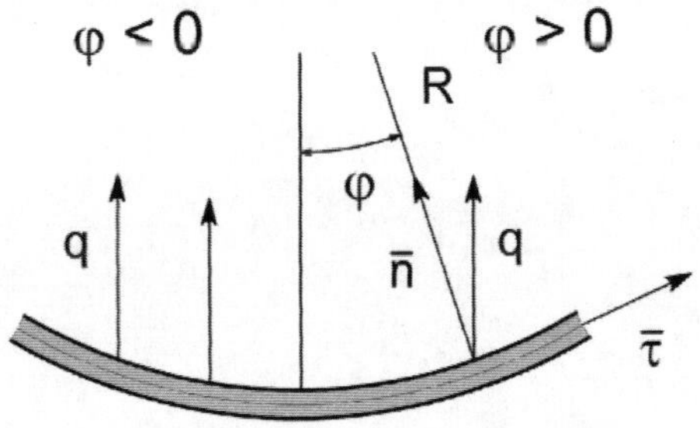

Figure 5. Design scheme of pipeline segments on roller supports curved convex down.

For pipeline segments curved convex down, the equation will be:

$$\begin{cases}\frac{1}{R}\cdot\frac{dN}{d\varphi}+\frac{1}{R^2}\cdot\frac{dM}{d\varphi}=-q\cdot Sin\varphi\\ \frac{N}{R}-\frac{1}{R^2}\cdot\frac{d^2M}{d\varphi^2}=q\cdot Cos\varphi\\ Q=-\frac{1}{R}\cdot\frac{dM}{d\varphi}\end{cases} \quad (6)$$

IOP Conf. Series: Materials Science and Engineering **154** (2016) 012032 doi:10.1088/1757-899X/154/1/012032

$$\begin{cases} \varepsilon = \dfrac{1}{R}\left(\dfrac{dU}{d\varphi} - W\right) \\ \Theta = \dfrac{1}{R}\left(\dfrac{dW}{d\varphi} + U\right) \\ \chi = \dfrac{1}{R^2}\left(\dfrac{d^2W}{d\varphi^2} + \dfrac{dU}{d\varphi}\right) \end{cases} \quad (7)$$

And for straight segments of a pipeline without tangential distributed loading the general solution is:

$$\begin{cases} \dfrac{dN}{dS} = 0 \\ \dfrac{d^2M}{dS^2} = q \\ Q = -\dfrac{dM}{dS} \end{cases} \quad (8)$$

$$\begin{cases} \varepsilon = \dfrac{dU}{dS} \\ \Theta = \dfrac{dW}{dS} \\ \chi = \dfrac{d^2W}{dS^2} \end{cases} \quad (9)$$

4. Results and discussion

The general solution of the system *(N+1)* ... *(N+3)* for pipeline segments curved convex up was derived.

The tangential movement for this case:

$$\begin{aligned} U = {} & C_1(A\cdot\varphi\cdot Sin\varphi + B\cdot Cos\varphi) + C_2(A\cdot\varphi\cdot Cos\varphi - B\cdot Sin\varphi) - \\ & - C_3\cdot\frac{R^3}{H}\cdot\varphi + C_4\cdot\frac{R^3}{H}\cdot Cos\varphi - C_5\cdot\frac{R^3}{H}\cdot Sin\varphi + C_6\cdot\frac{R^3}{H} - \\ & - q\cdot R\cdot\frac{R^3}{H}\cdot\frac{3\cdot(Sin\varphi - \varphi\cdot Cos\varphi)}{2} - \\ & - q\cdot R\cdot\frac{A}{4}\cdot(2\cdot\varphi\cdot Cos\varphi - 2\cdot\varphi^2\cdot Sin\varphi - Sin\varphi) \end{aligned} \quad (10)$$

The axial movement:

$$\begin{aligned} W = {} & -C_1\cdot A\cdot\varphi\cdot Cos\varphi + C_2\cdot A\cdot\varphi\cdot Sin\varphi + C_3\cdot\frac{R^3}{H} + C_4\cdot\frac{R^3}{H}\cdot Sin\varphi + \\ & + C_5\cdot\frac{R^3}{H}\cdot Cos\varphi + q\cdot R\cdot\frac{R^3}{H}\cdot\frac{\varphi\cdot Sin\varphi}{2} + \\ & + q\cdot R\cdot\frac{A}{4}\cdot(2\cdot\varphi\cdot Sin\varphi - 2\cdot\varphi^2\cdot Cos\varphi + Cos\varphi) \end{aligned} \quad (11)$$

The angle of rotation:

$$\Theta = -C_1 \cdot \frac{R^2}{H} \cdot Cos\varphi + C_2 \cdot \frac{R^2}{H} \cdot Sin\varphi + C_3 \cdot \frac{R^2}{H} \cdot \varphi - C_6 \cdot \frac{R^2}{H} + \\ + q \cdot R \cdot \frac{R^2}{H} \cdot (2 \cdot Sin\varphi - \varphi \cdot Cos\varphi) \tag{12}$$

The axial force:

$$N = C_1 \cdot Sin\varphi + C_2 \cdot Cos\varphi + q \cdot R \cdot \varphi \cdot Sin\varphi \tag{13}$$

The shear force:

$$Q = -C_1 \cdot Cos\varphi + C_2 \cdot Sin\varphi - q \cdot R \cdot \varphi \cdot Cos\varphi \tag{14}$$

The bending moment:

$$M = C_1 \cdot R \cdot Sin\varphi + C_2 \cdot R \cdot Cos\varphi + C_3 \cdot R + q \cdot R^2 (\varphi \cdot Sin\varphi + Cos\varphi) \tag{15}$$

In these equations:

$$\gamma = \frac{H}{R^2 D} << 1 \tag{16}$$

$$A = \frac{R^3}{2 \cdot H}(1+\gamma) \approx \frac{R^3}{2 \cdot H} \tag{17}$$

$$B = \frac{R^3}{2 \cdot H}(1-\gamma) \approx \frac{R^3}{2 \cdot H} \tag{18}$$

And the general solution for pipeline segments curved convex down was derived, too.
The tangential movement:

$$U = C_1(A \cdot \varphi \cdot Sin\varphi + B \cdot Cos\varphi) + C_2(A \cdot \varphi \cdot Cos\varphi - B \cdot Sin\varphi) + \\ + C_3 \cdot \frac{R^3}{H} \cdot \varphi - C_4 \cdot \frac{R^3}{H} \cdot Cos\varphi + C_5 \cdot \frac{R^3}{H} \cdot Sin\varphi + C_6 \cdot \frac{R^3}{H} + \\ - q \cdot R \cdot \frac{R^3}{H} \cdot \frac{3 \cdot (Sin\varphi - \varphi \cdot Cos\varphi)}{2} + \\ + q \cdot R \cdot \frac{A}{4} \cdot (2 \cdot \varphi \cdot Cos\varphi - 2 \cdot \varphi^2 \cdot Sin\varphi - Sin\varphi) \tag{19}$$

The axial movement:

$$W = C_1 \cdot A \cdot \varphi \cdot Cos\varphi - C_2 \cdot A \cdot \varphi \cdot Sin\varphi + C_3 \cdot \frac{R^3}{H} + C_4 \cdot \frac{R^3}{H} \cdot Sin\varphi + \\ + C_5 \cdot \frac{R^3}{H} \cdot Cos\varphi + q \cdot R \cdot \frac{R^3}{H} \cdot \frac{\varphi \cdot Sin\varphi}{2} + \\ + q \cdot R \cdot \frac{A}{4} \cdot (2 \cdot \varphi \cdot Sin\varphi - 2 \cdot \varphi^2 \cdot Cos\varphi + Cos\varphi) \tag{20}$$

The angle of rotation:

$$\Theta = C_1 \cdot \frac{R^2}{H} \cdot Cos\varphi - C_2 \cdot \frac{R^2}{H} \cdot Sin\varphi + C_3 \cdot \frac{R^2}{H} \cdot \varphi + C_6 \cdot \frac{R^2}{H} + \\ + q \cdot \frac{R^3}{H} \cdot (2 \cdot Sin\varphi - \varphi \cdot Cos\varphi) \tag{21}$$

The axial force:

$$N = C_1 \cdot Sin\varphi + C_2 \cdot Cos\varphi - q \cdot R \cdot \varphi \cdot Sin\varphi \tag{22}$$

The shear force:

$$Q = C_1 \cdot Cos\varphi - C_2 \cdot Sin\varphi - q \cdot R \cdot \varphi \cdot Cos\varphi \tag{23}$$

The bending moment:

$$M = -C_1 \cdot R \cdot Sin\varphi - C_2 \cdot R \cdot Cos\varphi + C_3 \cdot R + q \cdot R^2 (\varphi \cdot Sin\varphi + Cos\varphi) \tag{24}$$

And finally, for the case of straight pipeline segments the results are:

$$U = C_1 \cdot \frac{x}{D} + C_2 \cdot \frac{l}{D} \tag{25}$$

$$W = 0 + 0 + C_3 \cdot \frac{x^3}{6H} + C_4 \cdot \frac{lx^2}{2H} + C_5 \cdot \frac{l^2 x}{H} + C_6 \cdot \frac{l^3}{H} + \frac{qx^4}{24H} \tag{26}$$

$$\Theta = 0 + 0 + C_3 \cdot \frac{x^2}{2H} + C_4 \cdot \frac{lx}{H} + C_5 \cdot \frac{l^2}{H} + 0 + \frac{qx^3}{6H} \tag{27}$$

$$N = C_1 \tag{28}$$

$$M = 0 + 0 + C_3 \cdot x + C_4 \cdot l + 0 + 0 + \frac{qx^2}{2} \tag{29}$$

$$Q = 0 + 0 - C_3 + 0 + 0 + 0 - qx \tag{30}$$

5. Conclusion

The general solutions developed in this article are an important step in determining the pipeline stress-strain state when the force is exerted on the pipeline. The solutions can be applied to pipelines of different lengths and diameters for special cases of calculating the elastic bending and shear force combinations.

We get a possibility to find the points of the curved pipeline profile for placing additional fixing connections above the pipeline using the obtained solutions to find the pipeline displacement from the initial position. This will allow increasing the admissible shear force and will prevent the pipeline from falling off the roller supports in case of excessive increase in the axial shear force.

References

[1] Zemenkova M Yu, Shipovalov A N, Dudin S M and Zemenkov Yu D 2007 System analysis in oil and gas objects control processes *News of higher school. oil and gas* **5** 116-9

[2] Hammerhead Mole pneumoshock machine usage as auxiliary equipment in horizontal directional drilling [Network resource] / Ditch Witch Company Information Center, http://www.ditchwitch.ru/articles/ispolzovanie - pnevmoudarnoi - ustanovki - hammerhead-mole – v – kachestve – vspomogatelnogo – oborudovaniya – v - gorizontalnom-napravlennom-burenii-39.htm#, 2010.12.06.

[3] Pisarenko G S, Yakovlev A P and Matveev V V 1988 *Materials handbook* (Kiev; Naukova Dumka) 736

[4] Toropov E S, Kiryanov I E and Toropov V S 2014 Methods of energy consumptions reduction in pipeline pullback process to the borehole at using horizontal directional drilling technology for industrial objects construction. *International conference "Oil and gas. ATR – 2014". Resources, transport, technologies*

[5] 2008 *Technique and technologic processes at transport of energy resources* ed Yu.D Zemenkov (Tyumen; Vector Book) **1** 380

[6] Toropov V S, Osyka A V and Toropov E S. 2001 Pipeline pullback into 3d curved boreholes *Abstracts of inter-regional youth scientific conference "Severgeoekotech-2001"* 70-71

[7] Toropov S Yu and Toropov V S 1999 Features of polyethylene pipelines construction through water barriers *Collection of scientific papers "Maintenance and repair problems of field and main pipelines"* 26-27.
[8] Pipe Thruster. Mobile power reserves for pipeline construction. [Network resource] https://www.herrenknecht.com/en/products/additional-equipment/tunnelling-logistics/pipe-thruster.html.

IOP Conf. Series: Materials Science and Engineering **154** (2016) 012033 doi:10.1088/1757-899X/154/1/012033

Behavior of a Double-Layer Pipeline Under Static Load

S Yu Toropov, E S Toropov, A V Kozlov and V S Toropov

Industrial University of Tyumen, 38 Volodarskogo St., Tyumen, 625000

E-mail: 264301@mail.ru

Abstract. The authors investigated the features in the behavior of various components of "pipe in pipe" system when changing the internal pressure in the annular space. During the research, the dependence of plastic pipe creep modulus on tensile stress was received. Using a method of least squares, the authors calculated the empirical coefficients for dependences of plastic pipe creep modulus on pipeline service period and stresses in the pipe material caused by static load. They revealed an unequivocal relationship between the parameters of a double-layer pipeline with the maximum use of plastic or steel pipelines resistance.

1. Introduction

As a rule, the use of "pipe in pipe" system in construction practice means pulling a plastic pipe inside a steel pipeline [2]. Hereby we get a new double-layer pipeline consisting of pipes made from different materials. Modeling the behavior of such a pipeline under a load is complicated by non-linear deformation properties of a plastic pipe interacting with a steel pipeline [6], which in its turn has linear deformation properties.

Moreover, using this design, the repair technology allows us to get a pipeline, various layers of which seal tightly to each other [7], wherein clearance between them is virtually eliminated. Nevertheless, in practice there may be loose fit pipelines [4] and the presence of any size of the annular space between the internal and external pipelines.

The present article solves the problem of defining the rational parameters of a pipeline section, that is the ratio of diameters for given values of the outside diameter and wall thickness of a steel pipeline. We assume the existence of a gap between pipelines in "pipe in pipe" system.

2. Materials and methods

Strain properties of plastics are characterized by a non-linear relationship between strain and stress that causes it [1]. It is known that creep modulus E of the pipe material is a value that depends on stress σ. This dependence is manifested differently for various materials. Because of the insignificance it can be ignored for steel materials up to a certain limit - the elastic limit.

However, it is noticeable even at relatively low (compared to steel) stresses for plastic materials [9], and if we ignore this dependence, it could cause serious errors in mechanical calculations. As can be seen from the graph (figure 1) built according to table 4 of the Code SP 42-101-96, creep modulus E varies linearly with changes in stress. Tabulated function $E(\sigma)$ can be approximated by a linear dependence of the form:

$$E = E_0 - K\sigma, \quad (1)$$

where E_0, K – model parameters determined from the condition of the minimum sum of squared deviations.

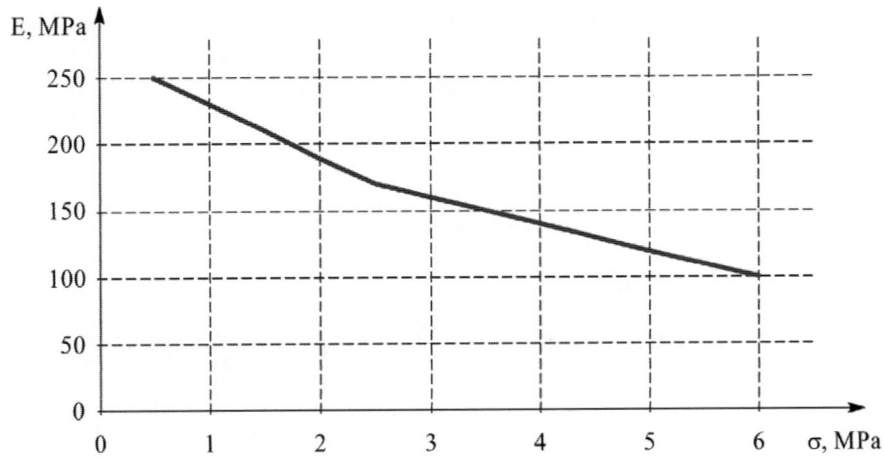

Figure 1. The dependence of plastic pipe creep modulus on tensile stress.

The sum of squared deviations:

$$\Delta = \sum_{i=1}^{n} \left(E_i - E_0 + K\sigma_i\right)^2. \quad (2)$$

Resolving equations of a method of least squares look like:

$$\left.\begin{aligned} nE_0 - K\sum_{i=1}^{n}\sigma_i &= \sum_{i=1}^{n} E_i; \\ E_0\sum_{i=1}^{n}\sigma_i - K\sum_{i=1}^{n}\sigma_i^2 &= \sum_{i=1}^{n} E_i\sigma_i. \end{aligned}\right\} \quad (3)$$

where n – the number of data points.

The system of equations (3) has the following solution:

$$E_0 = \frac{\sum_{i=1}^{n} E_i\sigma_i \sum_{i=1}^{n}\sigma_i - \sum_{i=1}^{n} E_i \sum_{i=1}^{n}\sigma_i^2}{\left(\sum_{i=1}^{n}\sigma_i\right)^2 - n\sum_{i=1}^{n}\sigma_i^2};$$

$$K = \frac{n\sum_{i=1}^{n} E_i\sigma_i - \sum_{i=1}^{n} E_i \sum_{i=1}^{n}\sigma_i}{\left(\sum_{i=1}^{n}\sigma_i\right)^2 - n\sum_{i=1}^{n}\sigma_i^2}. \quad (4)$$

Calculations were carried out for the data shown in table 4 of SP 42-101-96. And the following results were obtained for the dependence (1).

For the 10-year period of pipeline service:

$$E = 245.17 - 26.67\sigma. \quad (5)$$

For the 25-year period of pipeline service:

$$E = 224.60 - 24.36\sigma. \quad (6)$$

And for the 50-year period of pipeline service:

$$E = 217.67 - 25.93\sigma. \quad (7)$$

Parameter E_0 values are expressed in MegaPascals.

The nonlinearity of the strain properties of plastic pipes [5] may be accounted by successive approximations with a quasi-static formulation of the problem. In this case, creep modulus values should be calculated step by step according to (1) using the previous step stress value obtained by formulas for a linear dependence of stress and strain. Wherein the process of successive approximations terminates automatically when sufficient clarification of the required value is reached.

The stress-strain state of a plastic pipeline located inside a sealed steel pipeline can be schematically represented as follows (figure 2).

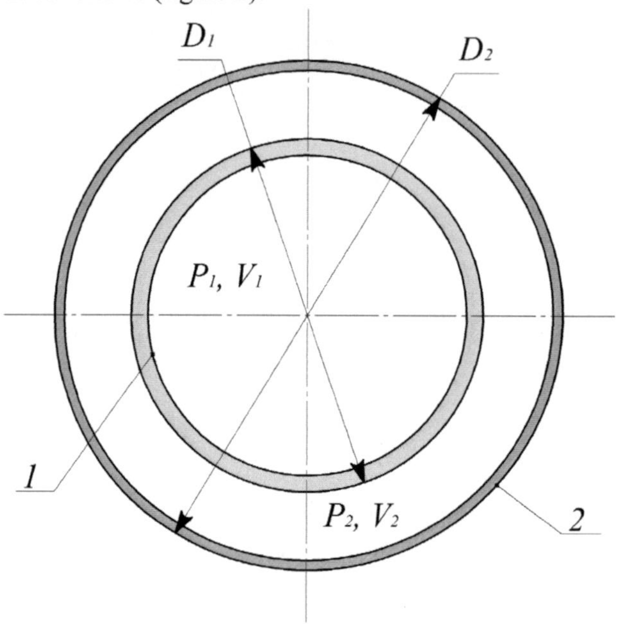

Figure 2. The cross section of "pipe in pipe" system:
1 - plastic pipe; 2 - steel pipe.

As shown in figure 2, there is a plastic pipeline inside a steel pipeline. The internal volume of the plastic pipe is designated as V_1 and the volume of the annulus is designated as V_2. The inner pressure P_1 causes the accordant radial stress σ_r and radial displacement in the wall of the plastic pipe, which in its turn causes a change in the volume of the annulus V_2 and pressure P_2 according to Clapeyron equation ($PV = const$) [9].

The total radial stress in the wall of the plastic pipeline may be expressed as:

$$\sigma_{r1} = \frac{(P_1 - P_2)D_{1in}}{2\delta_1}, \tag{8}$$

where D_{1in} – the inside diameter of the plastic pipe;
δ_1 – the wall thickness of the plastic pipe.

The corresponding radial displacement of the plastic pipe wall would be:

$$\Delta r_1 = \frac{\sigma_{r1} D_1}{2E_1} = \frac{(P_1 - P_2)D_{1in} D_1}{4\delta_1 E_1}, \tag{9}$$

where E_1 – the elastic modulus of plastics determined by (1);
D_1 – the median diameter of the plastic pipe cross section area.

The radial stress in the wall of the steel pipeline may then be calculated as:

$$\sigma_{r2} = \frac{P_2 D_{2in}}{2\delta_2}, \tag{10}$$

where D_{2in} – the inside diameter of the steel pipe;
δ_2 – the wall thickness of the steel pipe.

And the corresponding radial displacement of the steel pipe wall would be:

$$\Delta r_2 = \frac{\sigma_r D_2}{2E_2} = \frac{P_2 D_{2in} D_2}{4\delta_2 E_2}, \tag{11}$$

where D_2 – the median diameter of the steel pipe cross section area;
E_2 – the elastic modulus of steel, which is equal to 2,000,000 MPa.
After pipes are deformed, the volume of the annulus became equal to:

$$V_2' = \pi\left(\frac{D_{2in}}{2} + \Delta r_2 - \frac{D_{1out}}{2} - \Delta r_1\right)^2, \tag{12}$$

where D_{1out} – the outside diameter of the plastic pipe;
According to Clapeyron equation, the gas volume change will cause a corresponding change in gas pressure. The equation of gas state for the isothermal process looks like:

$$P_2 V_2 = P_2' V_2', \tag{13}$$

where P_2 and V_2 – the pressure and volume of gas in the annular space prior to pipe deformation;
P_2' and V_2' – the pressure and volume of gas in the annular space after pipe deformation.
In consideration of (12) and (13), we find a new gas pressure value:

$$P_2' = \frac{P_2 V_2}{V_2'}. \tag{14}$$

Then, we repeat the calculation of stresses and the corresponding radial displacements of the pipe wall using the new value of gas pressure in the annular space. Continuing the calculations (4) - (10), we obtain a new increment of gas pressure in the annular space, and each approaching will reduce these increments. We will continue the calculations as long as the increments are notable.

However, the actual conditions of pipeline construction demand the determination of rational parameters of the "pipe in pipe" system cross section.

Pipeline cross section parameters D_{1out} and δ_1 must satisfy the following conditions for given values of the outside diameter D_{2out} and wall thickness δ_2 of a steel pipe (figure 1):

At first, the radial stress in the plastic pipe σ_{r1} should not exceed the rated plastic resistance R_{1p}:

$$\sigma_{r1} \le R_{1p}. \tag{15}$$

Secondly, the radial stress in the steel pipe σ_{r2} should not exceed the rated steel resistance R_{1s}:

$$\sigma_{r2} \le R_{1s}. \tag{16}$$

Substituting (16) into (10) we obtain:

$$R_{1s} \ge \frac{P_2 D_{2in}}{2\delta_2}. \tag{17}$$

Consequently:

$$P_2 \le \frac{R_{1s} 2\delta_2}{D_{2in}}. \tag{18}$$

The pressure inside the plastic pipeline is limited by the strength of the plastic pipeline and may be calculated as:

$$P_1 \le P_2 + \frac{R_{1p} 2\delta_1}{D_{1out} - 2\delta_1}. \tag{19}$$

Taking the atmospheric pressure as the initial pressure in the annulus and observing Clapeyron equation, we find the initial volume of the annulus from the expression:

$$P_2' V_2' = V_2, \tag{20}$$

where V_2 and V_2' – the volume of the annular space before and after deformation, respectively;
P_2' – the pressure of gas in the annular space after pipe deformation.
In accordance with the condition of strength, the maximum permissible strain of the radial displacement of the plastic pipe wall will be:

$$\Delta r_1 = \frac{R_{1p} D_1}{2E_1}. \tag{21}$$

Having in mind that the value of the initial gap between the walls of the plastic and steel pipes cannot be less than the value determined by (21), we find:

$$D_{1out} = D_{2in} - 2\Delta r_1 \tag{22}$$

Substituting (21) into (22) we obtain:

$$D_{1out} = D_{2in} - \frac{R_{1p}}{E_1}\left(D_{1out} - \delta_1\right). \tag{23}$$

Therefore:

$$D_{1out} = \frac{E_1}{E_1 + R_{1p}}\left(D_{2in} + \frac{R_{1p}}{E_1}\delta_1\right). \tag{24}$$

On the other hand, a decrease in V_2 as a result of radial displacement of the plastic pipe wall should provide an increase in the annulus pressure from one atmosphere to a value determined by the expression (18). That is, according to Clapeyron equation, the volume of the annulus should be reduced P_2' times (see (20)).

$$V_2' = \pi\left(\frac{D_{2in}}{2} + \Delta r_2 - \frac{D_{1out}}{2} - \Delta r_1\right)^2 = \frac{V_2}{P_2'}. \tag{25}$$

From (25) we obtain:

$$\left(\frac{D_{2in}}{2} + \Delta r_2 - \frac{D_{1out}}{2} - \Delta r_1\right)^2 = \frac{1}{P_2'}\left(\frac{D_{2in}}{2} - \frac{D_{1out}}{2}\right)^2. \tag{26}$$

Consequently:

$$\frac{D_{2in}}{2} + \Delta r_2 - \frac{D_{1out}}{2} - \Delta r_1 - \sqrt{\frac{1}{P_2'}}\left(\frac{D_{2in}}{2} - \frac{D_{1out}}{2}\right) = 0. \tag{27}$$

That is:

$$\frac{D_{2in}}{2\sqrt{P_2'}}\left(\sqrt{P_2'} - 1\right) + \Delta r_2 - \Delta r_1 = \frac{D_{1out}}{2\sqrt{P_2'}}\left(\sqrt{P_2'} - 1\right). \tag{28}$$

Expressing D_{1out} from the equation (28), we get:

$$D_{1out} = \frac{2\sqrt{P_2'}}{\sqrt{P_2'} - 1}\left(\Delta r_2 - \Delta r_1\right) + D_{2in}. \tag{29}$$

Next, we combine (29) and (24):

$$\frac{E_1}{E_1 + R_{1p}}\left(D_{2in} + \frac{R_{1p}}{E_1}\delta_1\right) = \frac{2\sqrt{P_2'}}{\sqrt{P_2'} - 1}\left(\Delta r_2 - \Delta r_1\right) + D_{2in}. \tag{30}$$

3. Results and Conclusion

Thus, we obtain the plastic pipe wall thickness corresponding to the conditions (15) and (16):

$$\delta_1 = \frac{\left(E_1 + R_{1p}\right)}{R_{1p}}\left(\frac{2\sqrt{P_2'}}{\left(\sqrt{P_2'} - 1\right)}\left(\Delta r_2 - \Delta r_1\right) + D_{2in}\right) - \frac{E_1}{R_{1p}} D_{2in} \tag{31}$$

Equations (24) - (31) give us unambiguous double-layer pipeline parameters, taking into account the maximum resistance of both plastic and steel pipelines.

The results of this work can be used not only in the area of constructing urban infrastructure facilities, water supply and sewerage. Pipelines consisting of parts having different physical properties

IOP Conf. Series: Materials Science and Engineering **154** (2016) 012033 doi:10.1088/1757-899X/154/1/012033

may find application in the construction of underwater crossings by the "pipe in pipe" method, and it already concerns directly oil and gas pipelines, including pipes of average diameters.

The given work can also be used for the calculations of forecasting the service life of double-layer pipeline sections [3], selection of rational parameters for their design, and ensuring the maximum long service period of underwater crossings [8].

References

[1] Antipiev V N, Zemenkov Yu D and Shabarov A B 2002 *Technical and parametrical diagnostics in pipeline systems* (Tyumen; Vector Buk) 432

[2] Zemenkova M Yu, Shalay V V and Zemenkov Yu D 2015 Expert systems of multivariable predictive control of oil and gas facilities reliability *Procedia Engineering* **113** 312-315

[3] Ibragimov A A, Podorozhnikov S Yu, Shabarov A B, Medvedev M V and Zemenkova M Yu 2014 Calculation model and algorithm for determining residual resource of the pipeline under the periodic variations of stress and corrosion *Mining informational and analytical bulletin Scientific and technical journal* (Moscow; Gornaya kniga) **S4** 199-206

[4] 2008 *Monitoring of hydrodynamic and technical characteristics of pipeline systems* ed by Zemenkov Yu D (Tyumen; Vector Buk) 432

[5] Smirnov V V and Zemenkov Yu D 2013 Using the finite element method for calculation of intense deformed condition of above ground pipelines constructed on permafrost soil *Pipeline transport: theory and practice* **4** 18-23

[6] M Yu Zemenkova, A N Shipovalov and Yu D Zemenkov 2016 Mathematic Modeling of Complex Hydraulic Machinery Systems When Evaluating Reliability Using Graph Theory. IOP Conf. Series: Materials Science and Engineering 127 (2016) 012056 doi:10.1088/1757-899X/127/1/012056

[7] Toropov S Yu and Toropov E S 2006 Intratubal repair sleeves based on shape memory alloys *News of Higher Schools. Oil and gas* **6** 70-6

[8] Toropov S Yu and Toropov V S 1997 Methods of predicting the reliability of oil-gas transportation equipment *Abstracts for the conference " Tyumen oil yesterday and today"* 63-5

[9] 2004 *Transporting and storage of oil and gas in examples and sums* ed by Zemenkov Yu D (St. Petersburg; Nedra) 544

IOP Conf. Series: Materials Science and Engineering **154** (2016) 012034 doi:10.1088/1757-899X/154/1/012034

Interaction Between the Pipeline and Additional Equipment for Trenchless Technologies

V S Toropov, R M Temirbaev, E S Toropov and S Yu Toropov

Industrial University of Tyumen, 38 Volodarskogo St., Tyumen, 625000, Russia

E-mail: 264301@mail.ru

Abstract. In this article the authors defined the limits of applicability of the pipeline pusher as additional equipment for pipeline construction when using trenchless methods. In this case, the pushing force is applied to the free end of the pipeline section located on the day surface. The authors obtained analytical dependences for determining the stress-strain state of the pushed pipeline when using the pusher in a particular case. In addition, they identified values of axial pushing force that are considered dangerous, because they can cause pipeline dropping down from the roller supports during pipeline pullback.

1. Introduction

Such pipeline construction scheme that provides pipe displacement into the borehole or horizontal well by means of a drilling rig or other power unit is relevant for trenchless technologies in general. It involves the application of forces to the pipeline for its movement within the wellbore with the aim of placing it in its operating position. Note that according to the classical scheme the force applied to the pipeline is pulling [2].

Nevertheless, as additional equipment, pipeline pushers may be used to set pipeline in motion by a wide range of forces [5]. Moreover, the force of the pusher can be applied to different points of the pipe. This dramatically affects the general picture of stress distribution in the pipe material [1]. The article attempts at studying interaction between the pushed pipeline and the pusher in some special cases.

Research on the interacting pipeline and pushing equipment implies finding correlation between the force of a pusher and the stress appearing in a spatially curved pipeline. It is also necessary to take into account the primary stress in the wall of the pipeline from elastic bending, so the problem has a specific solution only in a particular case that implies a certain disposition of roller supports, a range of pipe diameters and radiuses of elastic bending, as well as a ratio of different pipeline sections lengths.

Due to the fact that the pipeline section already situated in the well at the moment is limited in its displacements by the well walls, we are most interested in analyzing the behavior of the pipeline section that is laid on roller supports at the entrance to the well. It is this interaction between the pipeline section and the pusher that will limit the possible force that may be applied to the pipe to intensify the pullback process.

The principal possibility to push the pipe by the drilling rig using a reverse thrust can base on specially developed instructions or technological charts. These instructions must be based on calculations of stability of drilling rods of different types when pushing them into wells of different diameters using a variety of drilling rigs.

For steel pipelines of large diameters, it is necessary also to keep in mind limits of load on launch way supports when pushing the pipeline out of the borehole in the opposite direction [6]. These limits must be based on appropriate calculations, too.

2. Materials and methods

At the entrance of the well, the pipeline is subject to elastic deformation with the bend radius $R = 1000 \cdot D$, where D – the outside diameter of the pipeline (figure 1). The left part of the pipeline section profile on the day surface is illustrated below. The pipeline profile is shown as it should look in real life.

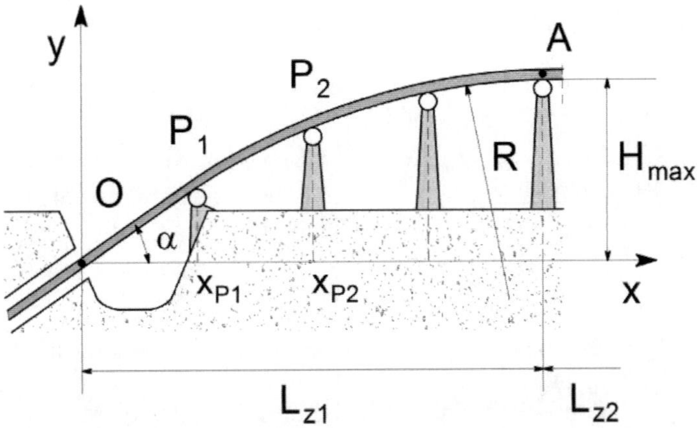

Figure 1. The design scheme of the aboveground section of the pipeline during pullback into the well using a trenchless method (left part): O – the entrance to the well; x_{p1}, x_{p2}, x_{pn} – coordinates of the supports, R – the radius of curvature of bent sections; L_{z1}, L_{z2}, L_{z3}, L_{z4} – lengths of different sections horizontally.

This scheme of elastic bending is most suitable for real steel pipelines. The right part of the design scheme is shown in figure 2. The AD section will be straight only for plastic pipelines [8].

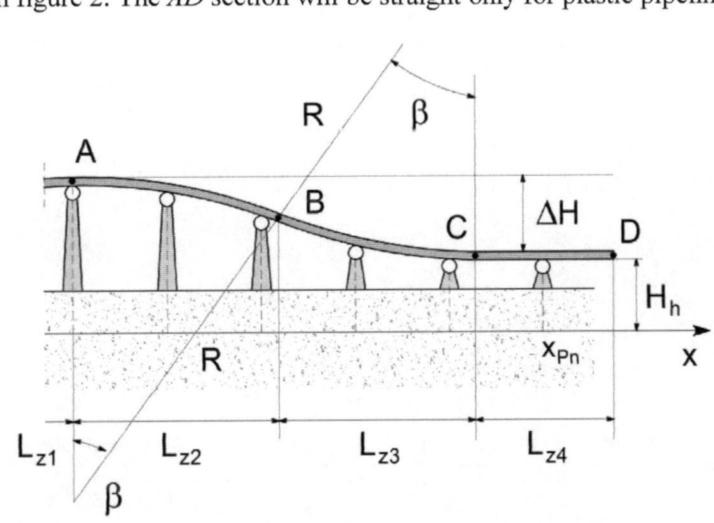

Figure 2. The design scheme of the aboveground section of the pipeline during pullback into the well using a trenchless method (right part): ΔH – the level difference between the maximum height and the height of the horizontal section; β - the central angle of the arch of the free section.

The pipeline profile is supported by roller supports P_1, P_2 ... P_n ensuring free movement of the pipeline at a tangent to the pipe axis. We consider that distances between supports ΔP are equal. It is desirable that H_{max} was as small as possible for practical reasons. For this purpose, the bending radius

R should be minimally permissible, that is it should be constant throughout the curved OC section (figures 1, 2).

The height will be measured from the zero level. We will assume that the zero level is the point of the well entrance that begins in the foundation pit. It is recommended to choose the entry angle into the well within 5÷20° and it should be kept in mind that an increase in α causes a significant increase in the value of H_{max}. The pipeline is curved convex up in the OB section, it is curved convex down in the BC section and the CD section is straight. In addition, we designate H_h as the height of the horizontal straight section.

Let us determinate the height of supports above the zero level that provides the circular arc shape in sections OB and BC. The maximum height would be:

$$H_{\max} = R \cdot (1 - Cos\alpha) \tag{1}$$

As the maximum height should be $H_{max} \geq H_h$, the following equation must be fulfilled:

$$\alpha \geq \arccos\left(1 - \frac{H_h}{R}\right) \tag{2}$$

In case $H_{max} = H_h$, the AD section of the pipeline would be straight and horizontal. Probably, it is the best variant but it is possible only for polyethylene pipelines of small diameters and it does not make sense to consider this variant.

The scheme of the pipeline elastic bending that is most suitable for real steel pipelines must be accepted for practical use. This scheme is described in figure 1. Les us find the angle β available on the scheme. In the picture, we can see that:

$$\Delta H = 2R \cdot (1 - Cos\beta) = 4R \cdot Sin^2\frac{\beta}{2} \tag{3}$$

This implies:

$$Sin^2\frac{\beta}{2} = \frac{\Delta H}{4R} \tag{4}$$

$$Cos^2\frac{\beta}{2} = 1 - Sin^2\frac{\beta}{2} = 1 - \frac{\Delta H}{4R} \tag{5}$$

Therefore:

$$Sin\beta = 2 \cdot Sin\frac{\beta}{2} \cdot Cos\frac{\beta}{2} = 2\sqrt{\frac{\Delta H}{4R}} \cdot \sqrt{1 - \frac{\Delta H}{4R}} \tag{6}$$

$$Cos\beta = Cos^2\frac{\beta}{2} - Sin^2\frac{\beta}{2} = 1 - \frac{\Delta H}{4R} - \frac{\Delta H}{4R} = 1 - \frac{\Delta H}{2R} \tag{7}$$

Now let us define the length of the pipeline sections:

$$L_{Z1} = R \cdot Sin\alpha \tag{8}$$

$$L_{Z2} = L_{Z3} = R \cdot Sin\beta \tag{9}$$

The equation of the pipeline axis in the OB section in the xOy coordinate system would look like:

$$Y_{OB} = \sqrt{R^2 - (x - L_{Z1})^2} - (R - H_{\max}) \tag{10}$$

For the BC section:

$$Y_{BC} = -\sqrt{R^2 - (x - L_Z)^2} + (R + H_h) \tag{11}$$

where L_Z –the length of the aboveground section of the pipeline without a straight horizontal section. It may be calculated as:

$$L_Z = L_{Z1} + L_{Z2} + L_{Z3} \tag{12}$$

Then the height of supports from the zero level will have the following values.

For the OB section:

IOP Conf. Series: Materials Science and Engineering **154** (2016) 012034 doi:10.1088/1757-899X/154/1/012034

$$H_{Pi} = Y_{OB}(x_i) \tag{13}$$

For the *BC* section:

$$H_{Pi} = Y_{BC}(x_i) \tag{14}$$

For the *CD* section:

$$H_{Pi} = H_h \tag{15}$$

If the shape of the curved axis exactly coincides with the described one, the maximum stress from bending with the bend radius *R* would be:

$$\sigma_R = E \cdot \frac{D}{2R} \tag{16}$$

where *E* – Young's modulus.

If we have conditions that $E = 2 \cdot 10^{11}$ Pa and $R = 1000 \cdot D$, we will have:

$$\sigma_R = \frac{2 \cdot 10^{11} \cdot D}{2 \cdot 1000 \cdot D} = 100 \text{ MPa} \tag{17}$$

However, pipeline sections sag between supports under the gravitation force. Hence, it appears that stresses can fundamentally differ from the theoretical ones. Obviously, the bigger the distance between supports, the greater the difference. [4].

Let us define the relationship of additional stresses to the distance between supports. If we place roller supports at regular intervals, then we can consider the P_iP_{i+1} section of the pipeline as a clamped rod loaded with an evenly distributed load *q* – its own weight (figure 3).

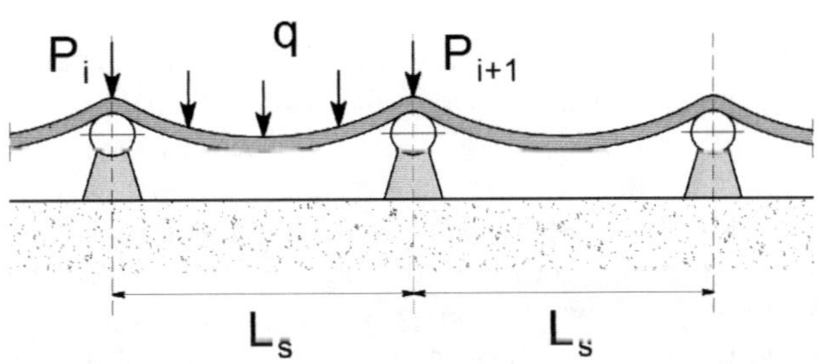

Figure 3. The design scheme of the section of the pipeline laying on supports under the action of its own weight: L_s – the distance between supports.

Then maximum bending moment in the pipeline section laid on supports will appear on its ends and will equal [3]:

$$M_q = \frac{q \cdot L_S^2}{12} \tag{18}$$

The maximal stress from this moment caused by loading action *q*, will be:

$$\sigma_q = \frac{M_q D}{2 \cdot I} = \frac{q \cdot L_S^2 D}{24 \cdot I} \tag{19}$$

where *I* – the moment of inertia of the pipeline cross-section.

At the *OB* section, the direction of the moment M_{max} coincides with the direction of the bending moment of the pipeline, therefore stress from bending over the radius *R* and stress from the load *q* will be summarized:

$$\sigma_{cur} = \sigma_R + \sigma_q \tag{20}$$

It means that the maximum stress will exceed the stress σ_R determined according to (16). Now let us find the distance L_s between the supports from:

$$\sigma_q = \sigma_R \cdot J,\ \% \tag{21}$$

For this purpose, we express L_s from (19):

$$L_S = \sqrt{\frac{24 \cdot I \cdot \sigma_q}{q \cdot D}} = \sqrt{\frac{24 \cdot I}{q \cdot D} \cdot \sigma_R \cdot J} \tag{22}$$

As a result, we will obtain the value of bending stresses for steel 09G2S (A 516-55, A 516-55, A 516-65, A 561 Gr 70 - USA; SM41B, SB49 - Japan) that equals 400 MPa. Therefore, we can set the distance between the supports ~ 40 m taking into account the safety factor ~ 2. At the same time, boundary conditions at O is not exactly a fixed-end condition and even more so freely supporting, $M_O = 0$, because the diameter of the well is much bigger than the diameter of the pipeline. Consequently, stresses at O can grow stronger than in the middle part of the pipeline ground section. Namely, the distance between the supports must be ~ 20 m.

Next, let us consider the possibility of using equipment for pushing a pipe by means of the axial force applied at D (figure 2) to intensify the pipeline pullback process. Application of the axial force T at D causes the appearance of compressive stresses:

$$\sigma_T = \frac{T}{A} \tag{23}$$

where A – is the cross-sectional area.

This stress is not very big. For example, application of the axial force $T = 10^6$ H to a pipe with the diameter $D = 530$ mm and the wall thickness $\delta = 10$ mm increases the stress by about 61 MPa. On the other hand, the axial force will cause the appearance of the excess bending moment in the curved OC section and the stress from this bending moment may be significant. [7].

We will consider a pipeline in a deformed condition to determine the stress-strain state under the action of the axial force. This pipeline is described in figures 1 and 2.

The supports under the pipeline are unilateral constraints, since they do not prevent the pipe from tearing off them. Accounting of the unilateral constraints considerably complicates solving the problem. In this work, the following method is applied. At first, the problem is solved without the constraints. The constraints are restored in the points with a negative bent W and the problem is solved again. Further, we check the deflections. And if at all points the deflections are positive, then the obtained results are accepted for the solution of the problem. Note that the direction of the deflection W and hence the connections placement depends on the value of the force T for a given value of the pipeline weight. Therefore, we have to repeat all the calculations for each value of the force T.

Thereby we found the general solution of the differential equation of equilibrium for every section between the supports.

3. Results and Discussion

The solutions of these equations were obtained. Analyzing the obtained solutions, we can make a conclusion. Let us take as an example a special case where the axial force of the pusher is applied to the elastically bent pipeline that has a diameter $D = 0.53$ m, a wall thickness $\delta = 0.01$ m and a radius of curvature $R = 1000 \cdot D$ (figure 4).

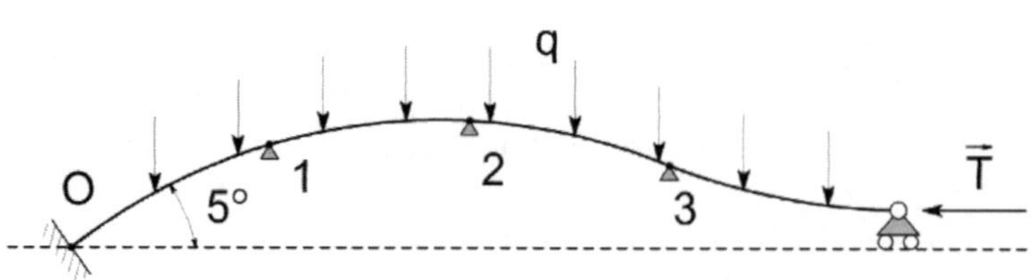

Figure 4. A special case of determining the stress-strain state of a pipeline when using a pusher.

A complete absence of connections at intermediate supports is shown in a scheme in figure 4.

IOP Conf. Series: Materials Science and Engineering **154** (2016) 012034 doi:10.1088/1757-899X/154/1/012034

The calculation results show that deflections at all points are less than zero, under the action of the axial force $T = 6 \cdot 10^5$ N or less. This means that for this value of the force T the pipeline should remain lying on the supports.

When increasing the axial pushing force to $T = 7 \cdot 10^5$ N, the pipeline loses contact with the support at B, but at other intermediate points the deflections are still less than zero. That is, while $T = 7 \cdot 10^5$ N, the pipeline will lay on all supports except point B.

In case of increasing the axial force to $T = 8 \cdot 10^5$ N, the pipeline loses contact with all supports at intermediate points. Wherein the lifting will be up to 0.4 m at B. This pipeline lifting may be dangerous and it could cause pipeline falling during the pullback process. If at the same time we put a connection at B, the pipeline will rise above the support at *1* (figure 4), but it will be a very small value, about 1.5 mm.

If the axial force reaches $T = 10^6$ N, it will cause pipeline rising above all supports up to 0.3 meters. Herewith at *1* the pipeline will rise above the support up to 0.13 meters, and the stress in the pipeline will reach a value of up to 340 MPa. These stresses will be caused mainly by bending, whose value can be reduced noticeably by adding retaining connections in intermediate points of the bent pipeline.

4. Conclusion

Thereby, we can apply to the pipeline the axial force of a pusher up to $6 \cdot 10^5$ N without using the additional upper support devices. These dependences were obtained for the pipeline entrance angle into the borehole $\alpha = 5°$. Larger values of angles can be investigated as well, but it is obvious that an increase in the entrance angle will cause a decrease in the permissible value of the axial pushing force applied to the pipeline.

Applying the force of a pusher to the free end of the pipeline located on the day surface is the most easily implemented way in the practice of pipeline construction. Significant results can be achieved using this most economic and universal method of pullback.

References

[1] Antipiev V N, Zemenkov Yu D and Shabarov A B 2002 *Technical and parametrical diagnostics in pipeline systems* (Tyumen; Vector Buk) 432

[2] Zemenkova M Yu, Shipovalov A N and Zemenkov Yu D 2016 Mathematic Modeling of Complex Hydraulic Machinery Systems When Evaluating Reliability Using Graph Theory *IOP Conf. Series: Materials Science and Engineering* **127** (2016) 012056 doi:10.1088/1757-899X/127/1/012056

[3] Pisarenko G S, Yakovlev A P and Matveev V V 1988 *Resistance of materials* reference book (Kiev; Naukova dumka) 736

[4] Smirnov V V and Zemenkov Yu D 2013 Using the finite element method for calculation of intense deformed condition of above ground pipelines constructed on permafrost soil *Pipeline transport: theory and practice* **4** 18-23

[5] 2008 *Technique and technologic processes at transport of energy resources*: two volume instructional aid. **Vol 1** ed by Zemenkov Yu D (Tyumen; Vector Book) 380

[6] Toropov E S, Toropov V S, Zemenkov Yu D and Seroshtanov I V 2015 Complications at crossings construction by horizontal directional drilling method *Territory "Oil and Gas"* **5** 32-7

[7] Toropov S Yu and Dorofeev S M 2005 Derivation of rod stability equation *Collection of scientific papers: Mathematical and information modeling* (Tyumen; TSOGU) 8-14

[8] Toropov S Yu and Toropov V S 1999 Distinctive features of pipe laying under water obstacle. *Collection of scientific papers: Problems of exploitation and repair of field and trunk pipelines* (Tyumen: TSOGU) 26-7.

[9] Zemenkova M Yu, Zemenkov Yu D, Pimnev A L and Kurushina E V 2016 System of Controlling the Reliability of Hydraulic Machinery in Oil and Gas Facilities *IOP Conf. Series: Materials Science and Engineering* **127** (2016) 012055 doi:10.1088/1757-899X/127/1/012055

IOP Conf. Series: Materials Science and Engineering **154** (2016) 012035 doi:10.1088/1757-899X/154/1/012035

The Development of the Software to Optimize Geophysical Field Oil and Gas Exploration

R K Akhmadulin and S K Turenko

Industrial University of Tyumen, 38 Volodarskogo Str., Tyumen, 625000, Russia

E-mail: ark@tyuiu.ru

Abstract: The quality of planning and the results of geophysical oil and gas exploration largely affect the efficiency of exploratory research, the quality of static models and, ultimately, the efficiency of oil-and-gas fields operation. One of the ways to improve the quality of geophysical exploration is the use of specialized software at its planning stage. In this work the analysis of the existing software in the scope of geophysical exploration planning is made; the urgency of development of the given class software on the basis of Russian package is noted. The ways of development of the software to optimize geophysical field oil and gas exploration are suggested.

1. Introduction

The quality of planning and the results of geophysical oil and gas exploration largely affect the efficiency of exploratory research, the quality of static models and, ultimately, the efficiency of oil and gas fields operation. The methods, which enable to enhance geophysical exploration, comprise such ones as the use of modern information technologies, including specialized software, at the stage of exploration planning.

In this article, based on the analysis of the existing software products to optimize geophysical field exploration, development of Russian package of application software PLEKS used in geophysical oil and gas exploration planning, is suggested.

2. Research

Today the information technology market comprises the products in the scope of geophysical exploration planning, which contain both similar and partly different functions implemented at different technological levels. For example, the software products of such companies as ION (MESA), GeoSeisControl (Pikeza) and others are well known. Let us regard them in detail.

MESA is a software product designed for planning seismic acquisition. It was developed by GreenMountainGeophysics Company, which in 1997 was in turn acquired by ION.

MESA is used in onshore and offshore projects to optimize the exploration scheme and plan. Its advantage is clear, user-friendly interface. The main functions of the product are [1]:

- exploration planning and analysis of the subsurface area coverage for 3D/3C seismic survey;
- static modeling, ray tracing and illumination analysis;
- chronological tracking of the project and statistics of seismic crews' work.

The MESA software packages complex includes:

- MESA Field is ideal for planning three-dimensional seismic survey on the basis of advanced technologies with the use of the basic analytical package;
- MESA Professional provides for a full set of tools for planning and analysis of three-dimensional full-wave survey;
- MESA Expert is a tool for planning three-dimensional surveys with maximum opportunities including static modeling and enhanced ray tracing.

IOP Conf. Series: Materials Science and Engineering **154** (2016) 012035 doi:10.1088/1757-899X/154/1/012035

This product is one of the most popular in its segment and is used in many large geophysical companies.

The Pikeza software package has been developed by CJSC NPTS "GeoSeisControl" since 1999 and is positioned as the software for planning and methodical-technological control of field seismic surveys [1].

The Picasa-4 package, as well as MESA, is widespread in Russia and is used in many projects. An example is the work [2], which describes the technologies which enabled to obtain high-quality results despite the occurrence of the survey target under complex natural, landscape and administrative conditions.

The Pikeza-4 version comprises a number of software, including those intended for [1]:

- survey plan visualization, color charts, seismic SEGY files viewing; control of design parameters;
- conversion of maps, association of maps;
- common points coordinate transformation;
- coordinates and heights generation, which performs the functions of data interpolation and data separation with a specified interval.

There are other software products in this area, for example Gedco Omni and Tessaral, enabling model calculations for planning areal shooting schemes.

At the same time, there are less known products in this field, for example the PLEKS application software package (ASP).

The PLEKS ASP is a system of scientific research type, designed to optimize seismic exploration, related to calculation and analysis of interference systems and networks. The package is designed for use at the stage of seismic operations planning, and enables task setting, solving and evaluating the efficiency of solving the task.

The algorithmic content of the PLEKS package can be regarded as an advantage. The following functions have been implemented here:

- evaluation of the areal shooting scheme noise immunity in three-dimensional seismology methods.
- rendering of optimal shooting schemes of the three-dimensional seismology method.
- calculation and rendering of KND diagrams characterizing the potential noise immunity of the grouping complex in the three-dimensional seismology method;
- calculation of spectral and integral characteristics of both individual groups of free structures and their complexes;
- summation of the seismic field, both synthetic and experimental according to the grouping law with selected parameters;
- calculation of optimized parameters of the groupings, etc.

Special attention was drawn to adaptation of the package to the South of the Tyumen region. A distinctive feature of these areas, in contrast to the sparsely developed Northern territories, is the occurrence of a large number of engineering facilities (roads, settlements etc.). This circumstance restricts the operation of regular survey network, as well as the application of certain research methods. The activity of the engineering facilities is an additional source of hindrances that must also be considered in planning and implementation of geophysical exploration. Under such conditions use of piecewise-regular and irregular survey networks is the most appropriate variant.

The main disadvantage of the package is its level of technological development, inferior to other modern packages of this class. To date, the package is the software, implemented for the Windows operating system that enables to enter data as numbers and the text, and present the results in the form of the text and numbers, graphs (Figure 1), diagrams and maps.

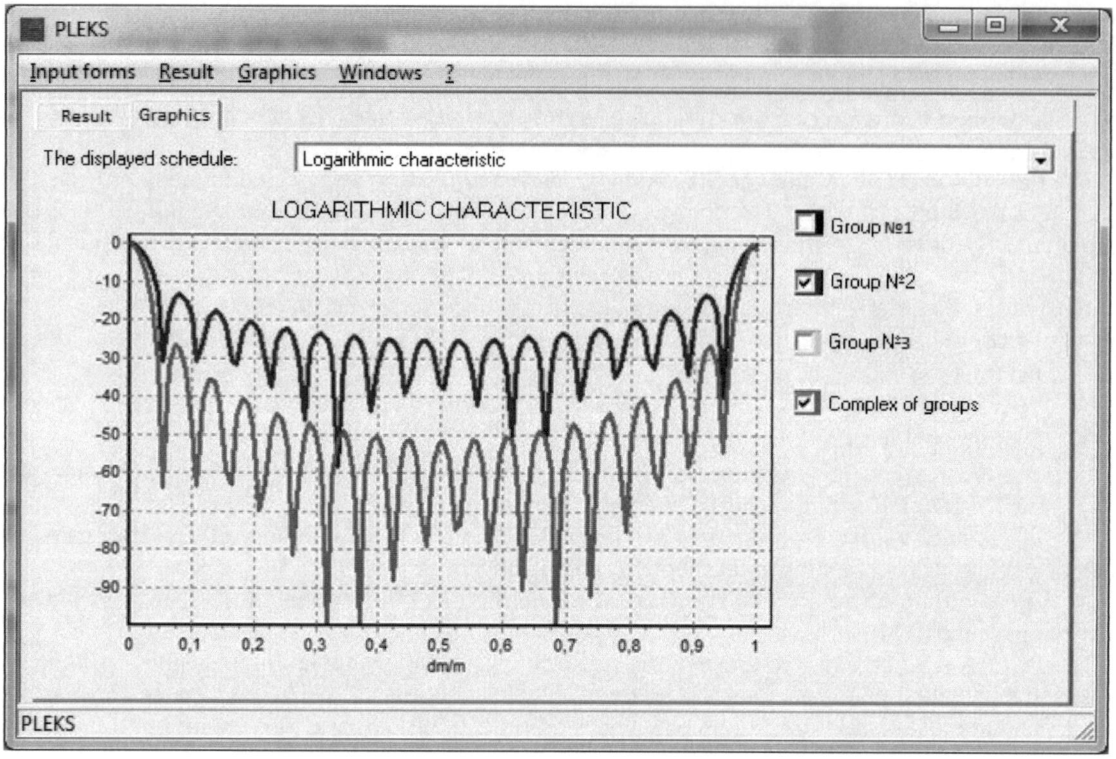

Figure 1. Example of calculation result as a graph.

When adapting the package to the modern technological level the following basic principles have been formulated:

1) the whole calculation part is implemented in Fortran and C and is used in the new version of the software "as it is", i.e. without changes. This will enable to speed up the process of adapting the software and avoid the occurrence of errors in previously debugged code. It is implemented as **dll**;

2) all interfaces are developed using modern tools and principles of programming;

3) the control module, which is a software shell, is responsible for the interaction between calculation parts, data and interfaces. It is the software shell which displays different interfaces, launches the calculation module and provides for data exchange between them.

As a result, the architecture implemented in the latest version of the PLEKS package (Figure 2), enables further development of the package and includes the following elements:

1) the program-monitor for launching the required calculation module;

2) the database and knowledge base, necessary for normal operation of the software and storage of the source and result data;

3) the decision maker, providing for methodological support of work with the package;

4) calculation modules designed to solve specific application tasks.

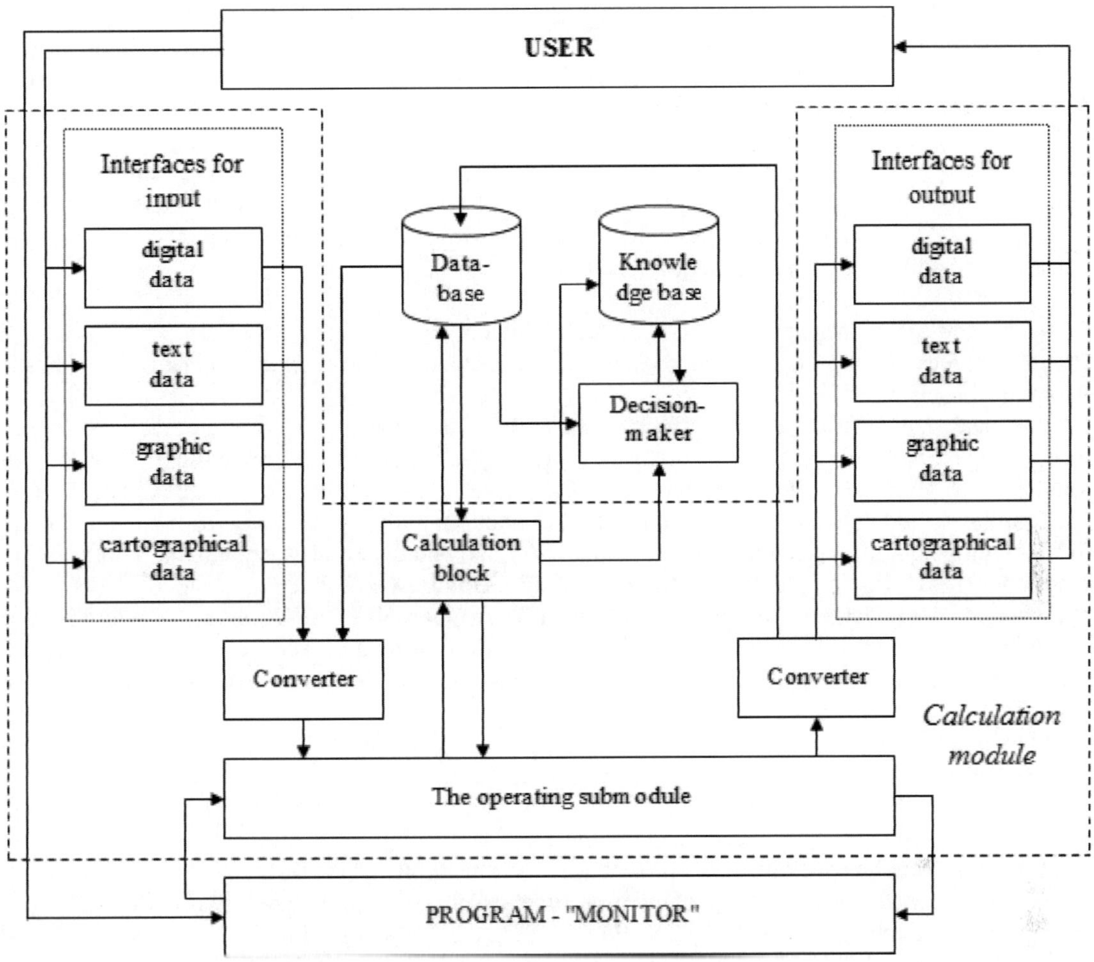

Figure 2. Generalized architecture of the PLEKS software complex.

Such architecture enables to present the package as a kind of shell in which calculation modules are built. This approach enabled to create a package using both the existing software libraries that were part of the earlier versions of the product, and newly developed software modules created using new tools.

The advantage of this architecture is the flexibility of the developed software product that will enable its further development in the future.

The knowledge base and the decision-making block, along with a functional content, is a distinctive feature of the package and enables to accumulate experience and knowledge and effectively use calculation modules in the solution of practical problems. The decision-making block can be used both when solving weakly-formalized problems and building the solution graph by using the existing software modules. When developing this module the experience of creation and use of expert systems for solving problems in various fields of activities is taken into account. For example, in one such work [3] the issue of development of neural expert systems, in which, apart from other things, neural networks are used, was considered.

The development of the new version of the package was carried out using the principles of object-oriented approach. Given that it was based on the existing mathematical apparatus, implemented in earlier programming languages, the paradigm of object-oriented programming was adopted to implement the possibility to present data in different forms, depending on the user's needs at a specific

IOP Conf. Series: Materials Science and Engineering **154** (2016) 012035 doi:10.1088/1757-899X/154/1/012035

point of time. This approach is relevant due to the fact that the basis of geological and geophysical software is often made up of complex algorithms, often developed based on authors' methods, but they may not have a user-friendly graphical interface satisfying the user. This issue, in terms of the software under study and other tasks in Geology and Geophysics, is presented in the work in more detail [4].

Thus, taking into consideration the urgency of further development of the software to optimize geophysical field oil and gas exploration, as well as the importance of development of domestic technologies in Russia, the development of the PLEKS package for planning and quality assessment of geophysical oil and gas exploration appears expedient.

The tasks to be solved are:

a) adaptation of the package to the modern technological level;

b) development of the algorithmic content of the package based on the analysis of the existing software of this class.

In this case a top priority task, according to the authors, is precisely the technological development of the package. So, within this task, apart from other things, we are to determine the most appropriate technologies and tools for the development of the package, as well as the architecture, which will enable painlessly add modules with new algorithmic functionality.

The development of specialized software can be implemented as both a separate software package and a module (modules) supplementary to the existing specialized software.

In case of the development as a separate software complex the resulting product will be independent and will not require any additional specialized software. The implementation can take place using one of the existing integrated software development environments such as Microsoft Visual Studio or Embarcadero RAD Studio.

Implementation as a supplementary module can be performed, for example, for Petrel of Schlumberger.

Petrel is an integrated platform that provides users with the ability to cope with various challenges from seismic interpretation to dynamic modeling [1]. To develop additional subroutines and launch them directly in Petrel a specialized development environment Ocean is used.

Ocean enables to create a variety of plugins for the Petrel complex. The plugins themselves can then spread among the users of the system. Plugins created by other developers may also be purchased and used. There are a number of works devoted to the subject, where, for example [5], we consider the development of a plugin for automatic correlation of well sections. Some plugins developed by both the very company and third-party developers, are presented on the site of Ocean Store (Figure 3).

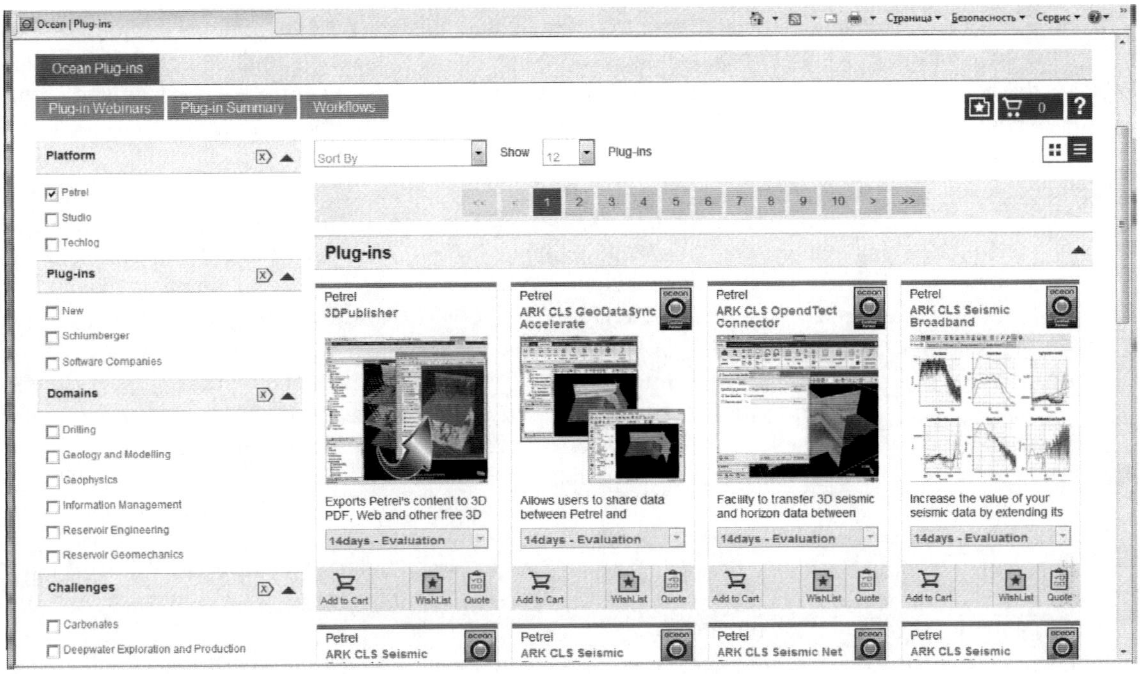

Figure 3. Site of Ocean Store.

Apart from Petrel there are other similar products. One of them is Russian software complex TimeZYX, developed by the TRAST Information Technology Laboratory. This complex is designed for static and dynamic modeling and monitoring of oil-and-gas field development.

TimeZYX is represented by three blocks of application modules on the single open platform of the same name: static modeling, dynamic modeling, monitoring and optimization of field development. Great attention is paid to the design and development of software intended for automation of the processes of formation, storage, verification and transmission of the output geophysical data in various formats to third parties [1]. Among the articles dedicated to this software product, we can emphasize the work [6], which considers the current domestic software system of the DV series that are intended for 3D seismic and static modeling, entering the line of software modules of the open platform TimeZYX. Their specification is given in comparison with major foreign analogues. It is shown that their application by specialists in various fields at all stages of geological-geophysical exploration and development of oil and gas fields, as well as in building of regional static models enables to efficiently perform a comprehensive analysis of multidisciplinary data and obtain three-dimensional static and dynamic models.

However the TimeZYX system does not provide for a tool for the development of functionality through independent development and introduction into the routines complex.

3. Conclusion

Thus, in this paper the expediency of the development of the PLEKS software for planning and assessment of the geophysical field oil and gas exploration quality has been proved. Other existing software products of this class have been reviewed. Possible ways of development of the application software have been observed.

Based on the analysis of the material the following conclusions have been made:

- to plan geophysical field exploration mainly imported software is used;
- at the same time, there are domestic products which are of certain interest;

- further development of the software to optimize geophysical field oil and gas exploration on the basis of the PLEKS package is suggested;
- the development should go the two following ways: as a separate software product, and as an adaptation of individual modules to the existing systems of static model quality assessment.

References

[1] Turenko S K and Akhmadulin R K 2015 Development of the software to optimize geophysical field oil and gas exploration *Geology, geophysics and development of oil-and-gas fields* **8** 30-3
[2] Leontiev I V and Krutov A L 2016 Conquest of Siberia, or seismic survey in the delta of the Yenisei river *Oil. Gas. Innovations* **1** 30-4
[3] Bakanovskaya L N 2015 Revisiting the development of neural network expert systems *Materials of the XI International scientific and practical conference "Conduct of modern science" Economic science* **3** 11-14 DOI:org/10.17686/sced_rusnauka_2016-1785.
[4] Ahmadulin R K 2016 Experience of applying the principles of object-oriented programming when developing applications in Geology and Geophysics *Automation, Teleautomation and Communication in Oil Industry* **1** 38-42
[5] Lapkovsky V V, Istomin A V, Kontorovich V A and Berdov V A 2015 Correlation of well logs as a multidimensional optimization problem *Russian Geology and Geophysics* **56(3)** 487-92 DOI: 10.1016/j.rgg.2015.02.009.
[6] Larin G V 2010 Effective computer tools of the Geologist and Geophysicist in the Exploration of Oil and Gas Fields *Geophysics* **3** 3-15

IOP Conf. Series: Materials Science and Engineering **154** (2016) 012036 doi:10.1088/1757-899X/154/1/012036

Influence of Heat Treatment on Magnetic and Mechanical Properties of Steel 10HSND

V F Novikov, Yu V Paharukov, A M Chehunova, F K Shabiev[1] and K R Muratov[2]

Industrial University of Tyumen, 38 Volodarskogo st, Tyumen, 625000, Russia
[1]E-mail: faridshab@mail.ru
[2]E-mail: muratows@mail.ru

Abstract. The effect of heat treatment on the mechanical and magnetic properties of steel 10HSND and morphology of carbides transforming under heat treatment, on mechanical and magnetic properties was studied. Method of determining the structural changes during the rapid testing on changing of the magnetoelastic part of the magnetic field of scattering was proposed.

1. Introduction
During steel operation, there happens their natural aging. Aging of metals causes changes in their mechanical, physical and chemical properties [1]. Carbon steels loading process is accompanied by changes of mechanical stresses over a large range, and hence, under the circumstances the elastic stresses and thermal heating may change the structure of material due to carbon reallocation and defects, which leads to artificial aging [2]. In a study [3], a concept of structural carbide construction of steels on the micro level, which allows the formation of a number of visco-strong steel properties. Traditional methods of determining the mechanical properties do not provide reliable results on the material state. Therefore, it is necessary to develop methods that allow determining the state of carbon steels. In this paper the artificial aging process was achieved by steel heat treatment.

The objective of the study: research of the heat treatment effect on the magnetic and mechanical properties of steel 10HSND and development of practical recommendations for non-destructive testing of these properties on changing the magnetic parameters of the material.

2. The methodology of the study
For studies, samples in shape of plates of 10 mm thickness and 75 × 75 in size mm were prepared. All samples were quenched from 950 ° C and tempered at 600, 640, 680 and 720 ° C with following cooling in water. After the heat treatment, magneto-mechanical tests were conducted: coercive force, a tangential component of the magnetic field intensity before and after the shock loading, hardness specific resistance and magnetostriction were measured. Dependence of ultimate tensile strength and yield point on tempering temperature was taken from the steels grade guide.

To monitor the change in carbide-ferrite phase, photomicrographs at the optical microscope with an increase in 1000 were obtained.

3. Results and discussion
Figure 1 shows the austenitic structure of steel, quenched from 950 ° C.

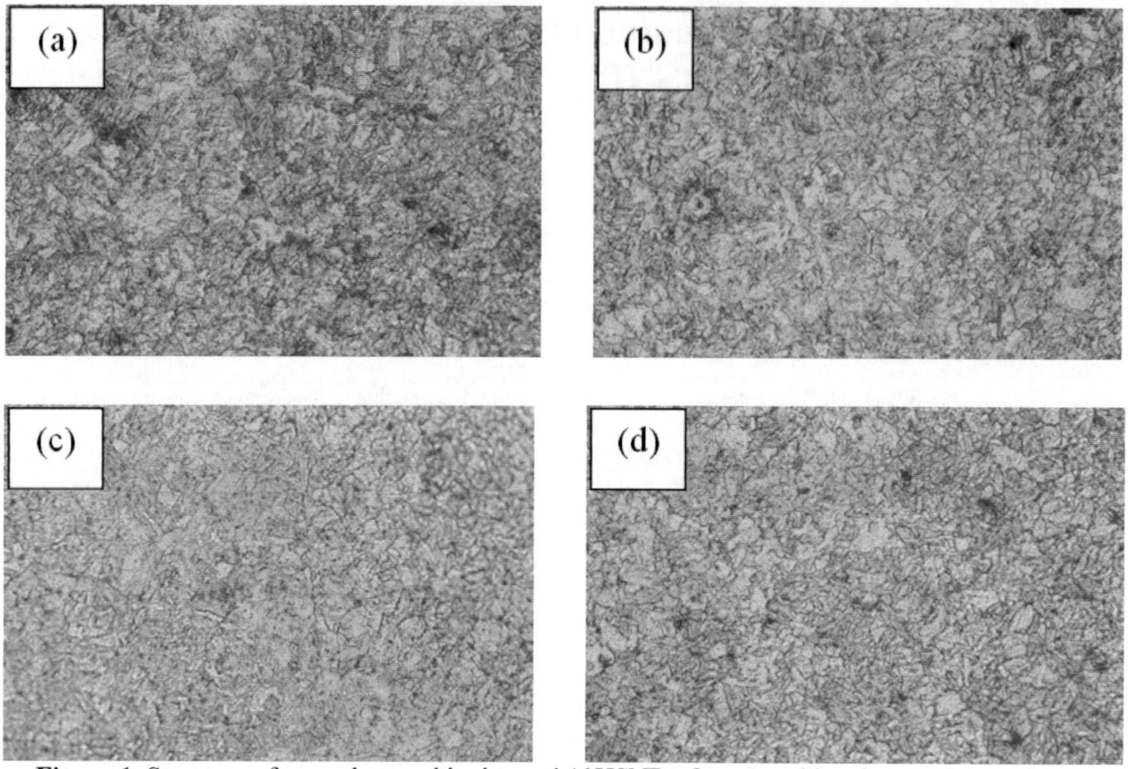

Figure 1. Structure of secondary sorbite in steel 10HSND after quenching from 950 ° C and tempering to a) 600°C; b) 640°C; c) 680°C; d) 720°C.

The data in Figure 1 shows that the structure of steel after high temperature tempering changes forming secondary sorbite. At higher tempering temperatures, structure - martensite breaks up into ferrite-carbide homogeneous mixture containing carbide inclusions. The evolution of the size of carbide inclusions is clearly seen in Figure 1, and in Figures 1a - 1d improving of structural reallocation of carbides can be seen, i.e. carbides areas become more distributed and decrease in size. Secondary sorbite structure is the most homogenous in Fig. 1d, with further increase of tempering temperature, coalescence process is observed - minor carbide monocrystals "disappear" being absorbed by larger ones, and structural restructuring of carbides become more considerable in size again.

Figure 2 shows the dependence of the coercive force on tempering temperature. It can be seen that up to temperatures 600 ° C takes place almost monotonic decrease of a parameter up to tempering temperature 720 ° C, and with further increase in temperature, there is an abrupt jump.

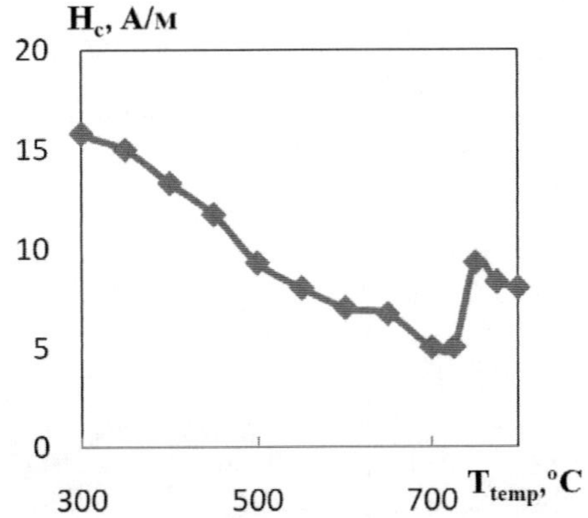

Figure 2. The dependence of the coercive force Hc on tempering temperature of quenched at t = 950°C and tempered in a temperature range from 300 ÷ 800 °C steel 10HSND [4].

Such a parameter as specific resistivity behaves in a similar way (Fig. 3).

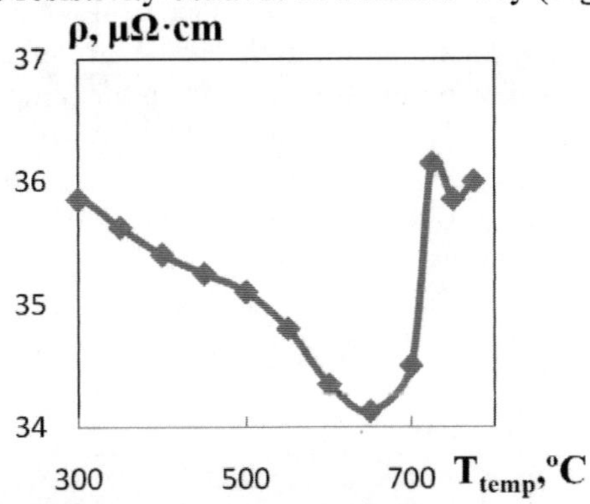

Figure 3. Dependence of specific resistivity on tempering temperature of steel 10HSND quenched at t – 950 °C and tempered at temperatures from 300 ÷ 800 °C.

Resistivity contrast jump at high tempering temperatures indicates the role of thermal defects. One can assume that the carbon atoms are involved in this process. This conclusion is in agreement with a decrease in the magnetostriction at increase in tempering temperature (Figure 4). The mechanism of these phenomena in steels underexplored.

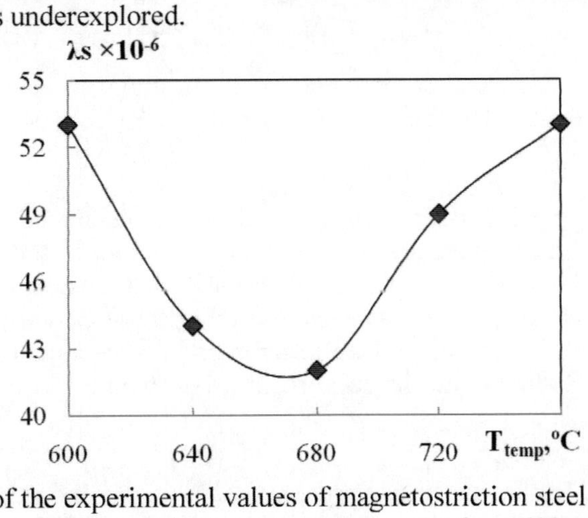

Figure 4. The curve of the experimental values of magnetostriction steel 10HSND quenched at t = 950°C and tempered in the temperature range from 600 ÷ 760°C.

Normally, at heat treatment, the control parameter is the hardness, however, as shown by the data in Figure 5 a, hardness, with an increase in tempering temperature, decreases almost monotonously and does not "feel" the reallocation carbides and their sizes, as well as ultimate tensile strength and yield strength, changes of which with the tempering temperature are shown in Figure 5 B.

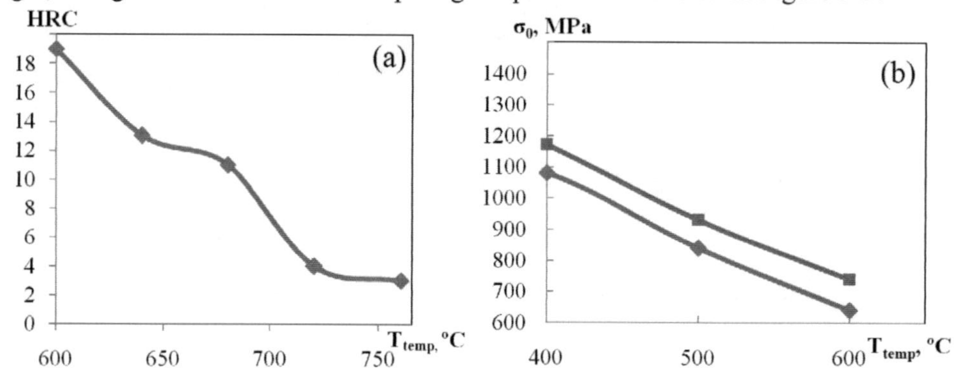

Figure 5. Dependence of hardness (fig. a); ultimate tensile strength σ_β and yield strength $\sigma_{0,2}$ of steel 10HSND quenched from 900 ℃ (fig. b) from the tempering temperature.

In the graph (Figure 6), the dependence of the changes of the tangential component of the magnetic field intensity on the tempering temperature is shown. On the curve, a maximum at tempering temperature of 640 ° C is clearly visible, then a sharp decline in a parameter with an increase in tempering temperature can be seen, therefore, one can judge on the restructuring the restructuring involving carbon and thermal defects by changing the tangential component of the stray magnetic field.

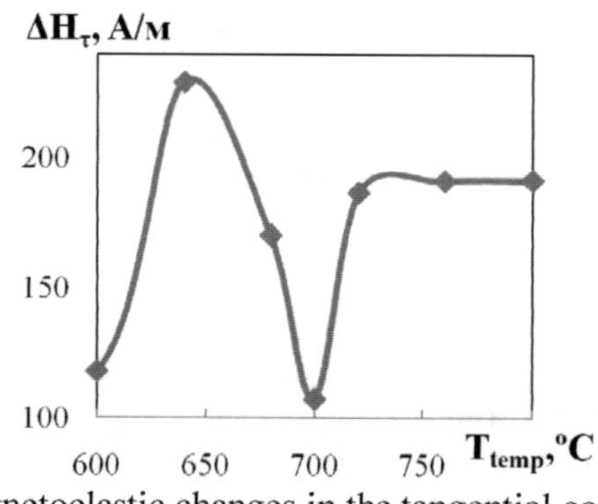

Figure 6. Dependence of magnetoelastic changes in the tangential component of the stray magnetic field tension on the tempering temperature for steel 10HSND.

4. Conclusions

The paper shows that, along with the known mechanisms of the material structure changes with increase in tempering temperature, there are other mechanisms in which the role of thermal defects is significant. Identifying the role of the latter in structural transformations requires additional studies. Such magnetic parameter as a change in the tangential component of the stray magnetic field monitors the change in the structure due to the carbon reallocation and can be used in the rapid analysis as an indicator of non-destructive testing of mechanical properties of steel.

References

[1] Nechaev Yu S 2008 Metallic materials for the hydrogen energy industry and main gas pipelines: complex physical problems of aging, embrittlement, and failure *Phys. Usp.* **51** 681–97

[2] Gorynin V I, Kondratyev S Yu, Olenin M I and Rogozhkin V V 2014 The concept of carbide constructing of steels with increased cold resistance *MiTOM* **10** 32-8
[3] Gorynin V I, Kondratyev S Yu, Olenin M I and Rogozhkin V V 2013 Improvement of resistance to brittle fracture of pearlite and martensitic steels under thermal impact on the morphology of carbide phase *MiTOM* **10** 22-9
[4] Novikov V F and Bacharev M S 2001Magnetic diagnostics of mechanical stresses in ferromagnetic materials (Tyumen; VectorBook) 219
[5] Pacharukov Yu V, Oksengendler B L, Yunusov M S and others 1986 Basic atomic processes and electronic structure of defects in semiconductors (Tashkent; Fan) 174

IOP Conf. Series: Materials Science and Engineering **154** (2016) 012037 doi:10.1088/1757-899X/154/1/012037

Repair and Insulating Works Using the Grouting Material "FORT" by "SNK" LLC

A V Kozlov, S Yu Toropov, V S Toropov and S V Volkova

Industrial University of Tyumen, 38 Volodarskogo st., Tyumen, 625000, Russia

E-mail: nashdoc@yandex.ru

Abstract: The paper deals with the technological schemes of conducting repair and insulating works using the grouting material "FORT" by "SNK" LLC. The sequence of preparing the grouting material and the main stages of work organization are presented. Using "FORT" material allows providing the insulation of water influx successfully.

1. Introduction

The water encroachment of oil layers and wellbores evolving in the conditions of water displacement is a regular and unavoidable process. The operation of separate wells and the entire deposit is complicated by the presence of bottom water that is conically drawn into the bottom-hole zone and come into the well. In this case the well water cut progresses.

Under conditions of a layered structure of producing layers their most permeable, separate and narrow intervals are exposed to prime development. Naturally, they are sources of the well water cut. Besides, water emergence in oil wells is caused by defects in design, first of all, a low-quality cement ring.

Irrespective of the reasons water emergence in the production of oil wells leads to the decrease in the oil output. There are cases when separate wells are filled up with water completely while the most part of a producing layer still remains oil-saturated. Under these conditions to ensure the fullest production of layers and the reduction in the volumes of the extracted water it is necessary to isolate the already developed and watered intervals or to improve the low-quality cement ring [1, 2].

Despite a variety of conditions of the well water cut, distinctions in a geological structure of layers and various service conditions, almost the only material currently applied when carrying out repair and insulating works is grouting cement. Low filtration characteristics and high hygroscopic behavior can refer to the main shortcomings of the grouting material. Besides, the cement stone formed by Portland cement in hostile environment and mineralized water is a subject to high corrosion reducing its durability. Alternative grouting mixtures on the binding basis don't have the majority of these shortcomings. The grouting material "FORT" has been used as such a mixture.

2. Research

In February 2016 trials of the "FORT" mixture were carried out in the wells of Western Siberia fields.

The grouting material "FORT" includes the following components: FORT-B is a mixture of inorganic binding substances, FORT-C is a regulator of setting time and FORT-P is a grouting fluid.

FORT-B represents a mixture of mineral binding substances.

FORT-C represents a mixture of inorganic and organic acid salts.

FORT-P represents a water solution of inorganic and organic acids salts and surface active substances.

When mixing all "FORT" components a stable suspension is formed that possesses the ability to harden forming a strong stone. The hardening time of the grouting material "FORT" depends on the temperature conditions and is regulated with the quantity of the entered setting time retarder. The optimal time for the loss of mixture mobility is from 2.0 to 4.0 hours. Full hardening is from 1 to 2 days depending on the layer temperature.

The formulation of the working mixture and the amount of chemical reagents for preparation of 1 m^3 of "FORT" is specified every time relative to specific geological field conditions of the treated well.

Preparation of the grouting material "FORT" is carried out in the following sequence:

- the design quantity of FORT-P is collected in the cement surge tank;
- FORT-C and FORT-B are entered in the mixing mode;
- the received mixture is blended for 10-15 minutes until a uniform mixture is obtained and then the mixture is pumped in.

Thus prepared composition has the density of 1500-1800 kg/m^3. The strength development time of "FORT" after pumping is from 1 to 2 days depending on the layer temperature. As a result of reaction the "stone" is formed serving as an impenetrable barrier.

During operations and implementation of the technological process on preparation and pumping "FORT" the following equipment is used:

- an equipment set is necessary for connection with the wellhead assembly;
- a cement surge tank is necessary to prepare the mixture;
- a pump unit is necessary to pump the mixture;
- an oil tank car and a loader crane are necessary to deliver chemical reagents.

To prevent the influence of the increased pressure on the production string when repair and insulating works are carried out, the technology provides for the use of additional service equipment:

- a mechanical packer;
- a packer – retainer.

The technology of repair and insulating works using the grouting material "FORT" includes preparing the mixture on the surface, pumping it in the isolation zone, waiting on cement hardening and putting the well into operation. The general organization of works includes three main stages:

1. Hydrodynamic studies are carried out in the well; their aim is to receive initial data for planning the insulating works [3].

2. A pumping design is developed and a grouting material formulation is also selected, the necessary volume of the mixture and technological parameters of pumping are calculated [4].

3. The third stage includes all types of works connected with the preparation and pumping of the grouting mixture [5].

Before repairing and insulating works the following preparatory operations are carried out:

- raising the downhole equipment;
- identifying the source of water cut or the interval of production casing leakage;
- preparing the wellbore for insulating works;
- determining the injection capacity of a well at various modes.

After arranging the process equipment and briefing the personnel on safety measures by a person in charge the following types of works are carried out:

-connecting the equipment with the wellhead assembly;

pressure testing the delivery line for 1.5 multiple operating pressure;

-preparing and pumping the necessary volume of the "FORT" mixture according to the technical plan for carrying out the repair and insulating works.

Upon termination of waiting on mixture time the following works are performed:

- drilling and normalization of the bottomhole (if necessary);
- downhole logging;
- secondary opening of the producing horizon;
- development of a well and lowering the downhole equipment.

Technological schemes of carrying out insulating works in wells using the grouting material "FORT" are constructed by analogy to schemes of insulating works with the application of grouting Portland cement.

The grouting material "FORT" is a non-selective material. Hardening happens both in water-saturated, and in oil-saturated parts of a layer. The hardened material is insoluble both in water, and in oil. Improving the integrity of the cement ring is carried out with the help of pumping the mixture behind casing in the places of violations. Shutting separate layer is carried out by pumping the mixture in the bottom-hole zone of a layer. Irrespective of the purpose of the carried out works the wellbore and, first of all, the intervals of isolation have to be carefully washed from various deposits, products of corrosion and products of acid reactions.

The well must be filled with the well-killing fluid. At the same time, the difference in the well-killing fluid density in tubing and the annual space isn't allowed, as it can lead to the emission of the pumped mixture if tubing breaks off and is raised and it can mix with the liquid filling the well.

If the well injection capacity is greater than 20 m^3/(day·MPa), at first, the absorption intensity is reduced. The formulation of the grouting material "FORT" is specified for each new batch for the conditions of expected temperatures and pressure for a specific layer and technology. The time of the beginning of mixture hardening has to be not less than the required time for pumping it in the isolation zone. The preparation of working solutions is carried out only after the implementation of all the preparatory work performed by the well workover crew. After preparation, to control the time of hardening a sample of the mixture is selected. The well-killing fluid is used as a displacement fluid. All equipment used for preparing and pumping working solutions is thoroughly washed out with water after the use. Improving the integrity of the cement ring is carried out by pushing the mixture through the existing perforation interval. The scheme of processing is given in figure 1.

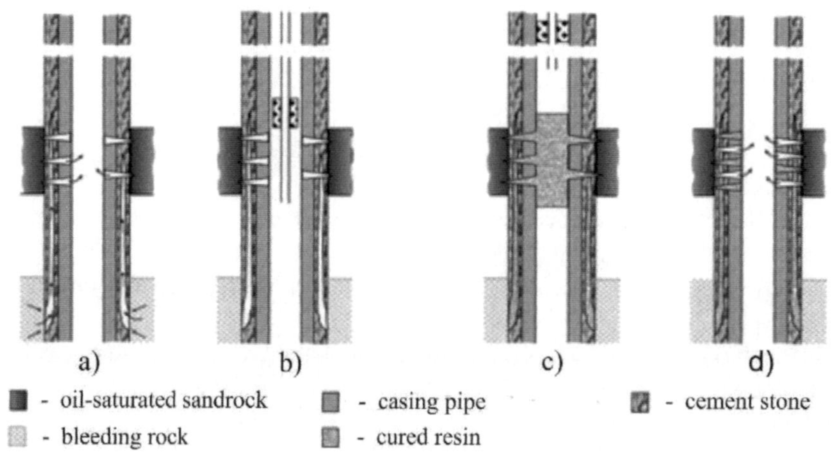

Figure 1. The process scheme of improving the integrity of the cement ring by squeezing the mixture through the existing perforation interval.

Improving the integrity of the cement ring to eliminate behind-the-casing overflows of formation fluids is carried out by a plugging under pressure method. Raising the tubing from the well and the subsequent wellbore drifting in the isolation interval is obligatory. Determining the well injection capacity is carried out in three pumping modes. If necessary, measures to increase the well injection capacity are taken (acid treatment, drainage, etc.). If the well injection capacity is greater than 20 m^3 (day·MPa), at first, the absorption intensity is reduce. When improving the integrity of the cement ring located above the producing formation, additional preparatory operations are conducted.

Depending on the size of the expected squeezing pressure [figure 1a] in the oil well tubing is lowered with or without a packer [figure 1b]. The mixture in a given volume is brought to the isolation zone and squeezed into the well with an unfitted packer and an open annular space. Then the tubing is lifted above the possible level of the mixture in the well, a control mixture section is cut, a packer is set (if available) and the mixture is squeezed behind the casing in a set volume, the well is sealed under the pressure of 40-60% of the production casing test pressure for the required hardening and strength development time [figure 1c]. Figure 1d shows drilling of the hardened mixture "FORT".

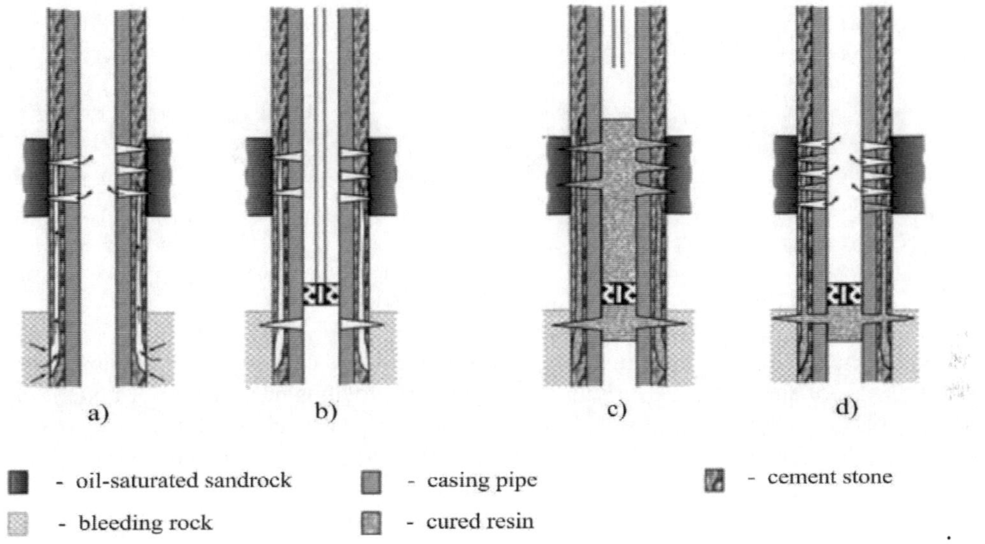

Figure 2. The process scheme of improving the integrity of the cement ring by squeezing the mixture through special holes.

To improve the cement ring [figure 2a] the string is additionally opened in a narrow interval of 0.5-0.8 meters [figure 2b]. The interval of additional opening is chosen so that between it and the interval of opening of a producing layer there was a possibility to reliably fit a packer-retainer. The packer-retainer is lowered and installed between the interval of opening of a producing layer and special holes in the string [figure 2b]. Fluid circulation is caused by forcing the pressure in the subpacker zone through breaks in a cement ring behind the string.

The design volume of the mixture is pumped via tubing through special holes behind the production string at an opened annual space. Detaching from the packer-retainer is performed and the volume of the mixture required for installing the bridge and overlappings of the perforation interval is pumped in the production string. Lifting of tubing is performed on a safe depth, backflush of tubing and sealing the wellhead is carried out at the waiting on mixture time [figure 2c]. Drilling of the hardened mixture, opening the layer and development of the wall are carried out. It is preferable to leave the interval of special holes closed with the grouting mixture column [figure 2d]. In some cases, perforation is made opposite an interbedded layer watering the well [figure 2b]. In these cases, it is advisable to create special holes in a roofing part of the watering interbedded layer, and besides filling the breaks in the cement ring, use them to create an impenetrable interbedded layer. To create an impenetrable interbedded layer the mixture is squeezed through special holes with a closed annual space and a set packer.

Figure 3 shows two general process schemes of isolating the bottom interbedded layer [figure 3, I, a] and the top interbedded layer [figure 3, II, a].

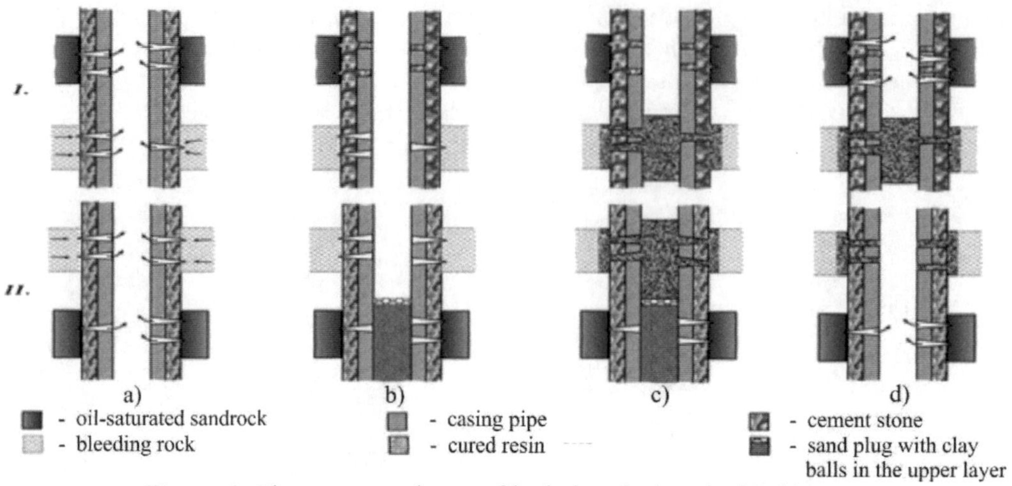

Figure 3. The process scheme of isolating the interbedded layers.

Injection of the mixture "FORT" in the perforation interval is carried out both using a technological packer in the case of shutting the overlying interval (with previous back filling of underlying interval) and using a packer-retainer when isolating the underlying interval. In the second case, a packer-retainer is set above the bottom perforation interval. The isolation zone must be carefully washed from different deposit, products of corrosion and products of acid reactions [6].

Building up the cement ring behind the uncemented casing string is made by plugging under pressure in the following cases:

- to protect the casing strings against corrosion from aggressive formation fluids;
- to eliminate or prevent the overflow of formation fluids in the uncemented annular space;
- to fill the annular space with the grouting material in the area of a casing defect or producing formations.

Pumping the grouting material into the annulus is carried out through special holes. If the well injection capacity of the absorption zone is greater than 20 $m^3/(day \cdot MPa)$, at first, the absorption intensity is reduced. The main method of eliminating the leakage of the casing string and annular space is grouting under pressure with the use of a technological packer of a packer-retainer. If there is a perforation interval below the unsealed casing, works are made to isolate it (a cement bridge, a bridge plug, an explosive packer). If the casing damage zone is located more than 100 m above the perforation interval, it is recommended to set up an additional cement bridge. Its capacity should be not less than 5 m at the 20 to 50 meter distance below the defect. If there are a few defects in the string successive grouting of every defect is performed from top down. Before grouting below the next defect at the 20 to 30 meter distance a dividing bridge is set up. Its capacity is not less 5 meters. If the string defect injection capacity is greater than 20 $m^3/(day \cdot MPa)$, a mixture with fillers is used or the absorption intensity is reduced. For the required hardening and strength development time a well is pressurized under the pressure of 40-60% of the production casing test pressure.

3. Conclusion

Thus, the grouting material "FORT" possesses sufficient hardening and setting times to get the work don;, the cement stone formed with this solution is resistant to corrosion. The use of the specified grouting material for repair and insulating works allows providing the isolation of water inflows.

References

[1] Strizhnev V A 2009 New technological approaches to the decision of a problem of repair and insulating works *Oil economy* **11** 54-7

[2] Presnyakov A U, Strizhnev V A, Umetaev V C and Sakhan A V 2012 New technologies, repair and insulating works in complicated conditions *Petroleum engineering* **7** 47-51
[3] Kozyar N V 2008 Assessment of quality of cementation of columns and cuts of wells by results of acoustic of wells by results of acoustic researches *Oil economy* **9** 24-7
[4] Fyodorov K M 2009 Design of repair and insulating works *Oil economy* **7** 108-11
[5] Meling K V 2006 Restoration of hermeticity of producing columns with profile shutters *Oil economy* **3** 72-5
[6] Strizhnev V A 2009 Choice of technology of repair and insulating works on shutdown of top and intermediate layers *Petroleum engineering* **7** 42-5.

IOP Conf. Series: Materials Science and Engineering **154** (2016) 012038 doi:10.1088/1757-899X/154/1/012038

Specifics of Using Two-Electrode Method of Intensive Measurements for Determining the State of Corrosion-Resistant Coating of Main Pipelines of the Western Siberia

D V Pyalchenkov, R N Tryascin, V V Gostev

Industrial University of Tyumen, 38 Volodarskogo st., Tyumen, 625000, Russia

E-mail: pjalchenkovdv@tyuiu.ru

Abstract. Integrity of anti-corrosion coating of the main pipeline is a prerequisite for system survival of the entire oil and gas transmission system. Currently, most of the main pipelines of Western Siberia are being operated for 25-35 and more years, and therefore relevance of the problem of controlling their state of isolation is not doubted. Enterprises operating gas and oil pipeline network of great extent, no doubt are interested in qualitative monitoring and adequate interpretation of the results.

This paper represents some diagnostic testing results of insulation coating state of main pipelines by a two-electrode intensive measurements method, on the basis of which, recommendations on optimization of this type of work in the conditions of Western Siberia are presented.

1. Introduction

Only the direct losses from corrosion in the world are so huge that the society has to spend dozens and even hundreds of billions dollars annually to fight it. Total direct corrosion losses in the United States (in Russia, the statistics, unfortunately, is not conducted) is about 70 billion dollars per year, i.e. more than 4% of the gross domestic product. And that's not taking into account the problems arising from the accident which occurred because of corrosive destruction in different constructions [1].

Since it is not possible to stop the corrosion process completely, one of the most important environmental and economic issues, the most relevant for the fuel and energy complex of the country in general and Western Siberia in particular, is slowing down the metal corrosive destruction rate to an acceptable level.

Practically for cessation on outer surface of a pipelines of corrosion process either it is necessary to create a flow on a pipeline of electric current or maximally increase resistance of external circuit of current of corrosion couple. First method is implemented by imposing on pipeline of external current, second – by applying on outer surface of the pipeline of insulation coating. Use of first method is called active protection, second one – passive. Therefore, insulation coatings of pipelines fulfill the role of passive protection, and cathodic polarization – active.

Cathodic polarization of pipeline is carried out by an external source of constant current – special units (in other words stations) of cathodic protection (CPS) or by connecting sacrificial anodes (protectors), and for providing sufficient efficiency must be conducted on the full length (protection in length) and incessantly (protection in time). As for active, as well as passive corrosion protection of underground pipelines must meet certain requirements, given in appropriate State Standards and regulatory and technical documents.

Protection from corrosion of underground pipelines is implemented as the complex, as specified: use of insulation coating (basic type) and cathodic polarization (subsidiary type) for protection primarily locations of bare pipe (surface of pipe in locations of pores, crack and other defects in insulation coatings) by supporting necessary value of protection potential.

Process of corrosion protection of steel underground pipelines presents complex multi-factor process, to provide the sufficient efficiency of which should be regularly controlled, taken in account and regulated different parameters. Control of condition (quality) of insulation coatings and efficiency of cathodic protection on underground pipelines is reached by executing appropriate electrometric surveys (measurements).

Currently, there are two basic ways to protect underground metal constructions from corrosion - the so-called active (electrochemical) and passive (using insulation coatings) protection that are used on most of the pipelines in the complex.

Operating principle of the passive protection is based on reducing corrosive environment access to the metal surface, which implies that the less insulation coating is damaged, the better its operating efficiency is.

The essence of the method reduces to measurements in pitches of 5-10 meters enable and disable potentials and potentials of their gradients (Uen; Udis; dUen; dUdis) for fixed operating modes of cathode protection systems (CPS) that affect the security the surveyed pipeline section, and further algebraic manipulation with the results of these measurements [2].

Proper application field of this method is the pipelines with insulation, built meeting the requirements [3], which allows localizing individual through faults amid the high-quality insulation coating and evaluating the pipeline safety in these defects.

This scheme is recommended for use the majority of manufacturers of diagnostic equipment, as well as regulatory documents of organizations, operating pipelines [4-5].

However, this scheme does not fully consider the peculiarities of pipeline transportation system work under conditions of Russia. In particular - the presence of 1 to 20 parallel pipelines in one pipeline right-of-way with the studied pipeline, as well as challenging geographical and climatic conditions in the places of diagnostic operations.

2. Methods of research

As objects of study, were chosen: main pipeline UBKUA (Ust-Balik-Kurgan-Ufa-Almetyevsk) in the area 641-889 km, main pipeline NKK (Nizhnevartovsk-Kurgan-Kuibyshev) in the area 879-1129 km, and main pipeline T-Yu (Tyumen Yurgamysh) in the area 0-251 km located in one pipeline right-of-way. The diagnostic study was carried out during the period of 2013 to 2014 using the measuring complex «Modata» of the German company «WeilekesElektronik», and universal diagnostic measuring instrument "Diakor" of CJSC "KhimServis". During measurements, external reference electrode was placed according to the scheme in Figure 1, i.e. perpendicular to the examinee pipeline at a distance of 5-10 m.

3. Results

Figure 1 shows one of the schedules of intensive measurement by two-electrode method in the area 1 km, obtained by «Modata» measuring complex.

In locations of probable damages of anticorrosion insulation coating of pipeline (through damages or complete loss of insulation) are indicated values of switch-on potential (UON) and switch-off potential (UOFF) as well as value of “funnels of voltage” in moments of switching off cathodic protection and difference of “funnels of voltage” in moments of switching on and off of cathodic protection.

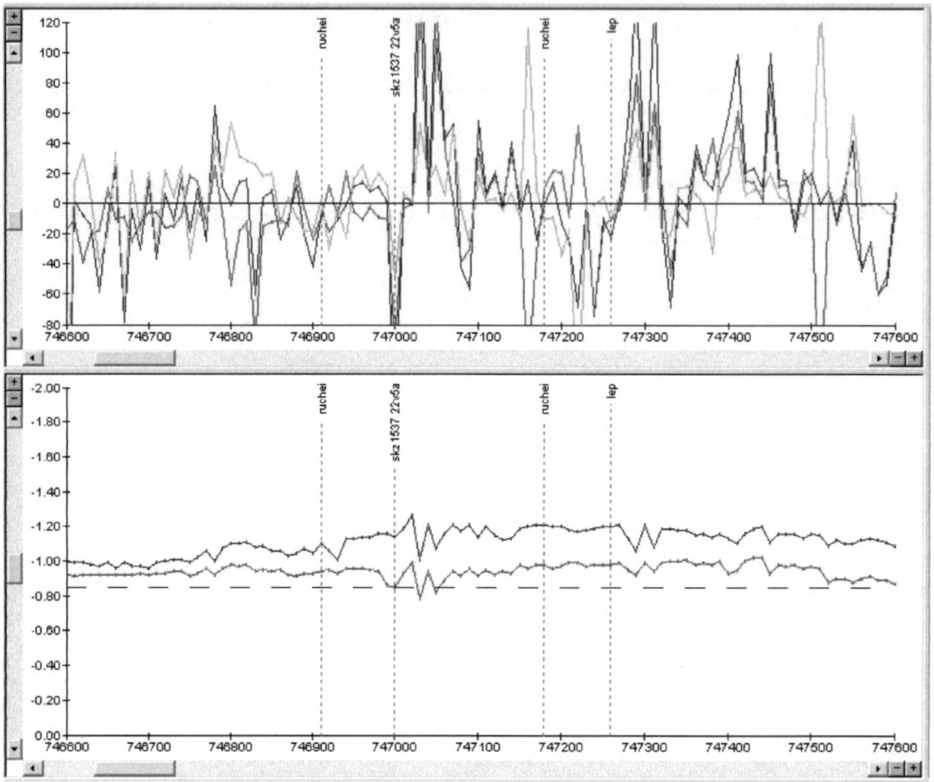

Figure 1. Schedule intensive two-electrode measurement method with 3 pipelines in one pipeline right-of-way with the perpendicular installation of the external reference electrode.

As can be seen, with placement of eternal reference electrode perpendicular to the pipeline, measuring ON and OFF "funnel" gives no clear results, because while spacing the electrodes at a distance of 5-10 m, neighbour pipelines, also having defects in the insulation coating and joint protection, fall in the measuring area. As a result, it seems impossible to interpret measurement results clearly and to indicate defects in corrosion-resistant coating exactly in the inspected pipeline. To clarify the state of the insulation in the given sector, an additional diagnostic examination was carried out using alternating current (Pearson's method). As a result, a defect of insulation was found for 1 meter long and with total area of 4m2, confirmed by check pit sampling. Photo of check pit is shown in Figure 2.

Figure 2. Check pit on the pipeline UBKUA made based on the results of the diagnostic test.

Later on, during examination, the modified scheme of measurements was applied in some areas to clarify the data. External electrode (EE) with 5-10 meter step was placed not perpendicular but parallel to the axis of the pipeline. The results of comparison of the data obtained are shown in Table 1.

Table 1: Comparison of results of measurements perpendicular and parallel placement of MEA.

Quantity of defects	UBKUA			T-Yu		
	Area 641-646 km	Area 740-750 km	Area 882-889 km	Area 7-14 km	Area 150-160 km	Area 224-227 km
MEA. perpendicular	14	22	15	3	22	14
MEA. parallel	14	23	17	5	23	14

Chart of intensive measurement with a parallel installation of the MEA is shown in Figure 3. The graph clearly shows corrosion-resistant coating defects detected during the diagnostic testing.

IOP Conf. Series: Materials Science and Engineering **154** (2016) 012038 doi:10.1088/1757-899X/154/1/012038

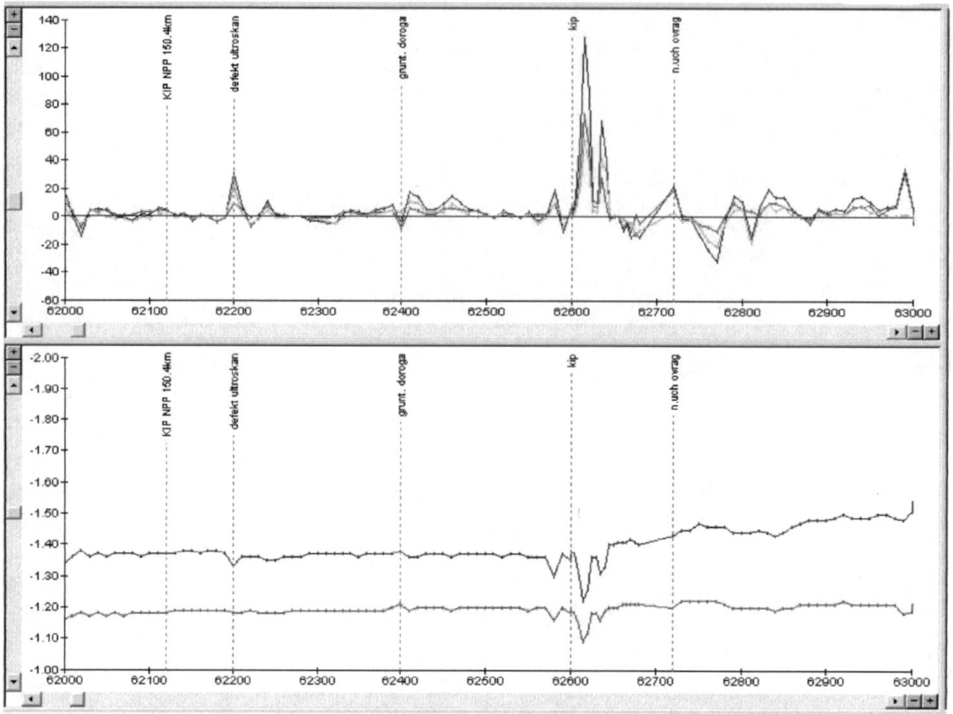

Figure 3. Chart of intensive two-electrode method measurements by three pipelines in one pipeline right-of-way with parallel installation of MEA.

As can be seen from Table 1, at a parallel placement of reference electrodes in areas of the total length of 42 km, it became possible to identify 6 insulation defects additionally. Some of them were confirmed by check pit sampling, such as defect on the pipeline T-Yu with coordinates 153km + 140m. Photo of a check pit is shown in Figure 6.

The only disadvantage of this method, identified empirically, was the worst (as compared with a perpendicular arrangement MEA) accuracy in determining the location of the defect. The mistake was in the range of ± 5 m. However, if a defect is detected, it is sufficient to carry out additional measurements with the removal of the electrode perpendicular, which will increase accuracy up to ± 1 m.

Figure 4. Check pit on the pipeline T-Yu made based on the results of the diagnostic test.

4. Conclusions

1. On the basis of the submitted data, it was revealed that a parallel installation of reference electrodes (MEA) during the intensive two-electrode method measurement on pipelines in multiline version, provides a more plausible information about safety and state of researched pipeline insulation coating, and if placing the 2nd electrode in place of the 1st, it gives a 100% inspection of length.
2. An additional and important advantage of using this method is the easier passing of areas with overgrown shrubs (trees), areas with a complex profile, and also to the marshland by a diagnostic team.
3. Based on the results obtained, it is recommended to conduct extensive measuring in combination. This will enhance the quality and significantly improve the speed of the diagnostic works.

References

[1] Tkachenko V.N., Electrochemical protection of pipeline networks. Tutorial. 2nd ed., Rev. and ext. M.: Stroyizdat, 2004.-320 pp.
[2] Ostapenko V.N., Yakubovskaya L.N., Lukovich V.V., etc. Electrochemical protection of pipelines against corrosion /. Kiev: Sciences, 1988.-245 pp.
[3] Unified corrosion and aging protectionsystem. Underground constructions. General requirements for corrosion protection. All Union State standard ISO 9.602-2005.
[4] Organization of corrosion surveys of Gazprom objects. Basic requirements of STO Gazprom 2-2.3-310-2009.
[5] Trunk pipeline transport of oil and oil products. Examination of the corrosive condition of the main oil and product pipelines RD-29.200.00-KTN-047-14.

IOP Conf. Series: Materials Science and Engineering **154** (2016) 012039 doi:10.1088/1757-899X/154/1/012039

On Choosing Multicomponent Multiphase System Separation Progress Optimization Criteria

A M Kulikov

Industrial University of Tyumen, 38 Volodarskogo st., Tyumen, 625000, Russia

E-mail: lis-prepod@mail.ru

Abstract. The article describes the existing criteria of separation processes optimization. The possibility of application of the general theory of separation for calculation of technological schemes is being justified. The technological, thermodynamical, statistical and economic criteria of separation processes, methods of their analysis and comparison are briefly described. The method of selection of optimization process criteria with technology research, installation engineering schemes and under automatic control of separation processes, as well as statistical planning of extreme experiments was offered.

1. Introduction

Recently, an issue of efficiency criteria and parameters technological processes optimization has been widely discussed. Evaluation of the technological process results is usually carried by several criteria. The interest to general criterion, clearly expressing efficiency of the process, at the first stage was associated with the natural human desire to evaluate the results of their activities.

With the formulation of the problem of optimal automatic control of technological processes and development of cybernetic methods of research, the overall efficiency criterion becomes technically necessary, since it formally defines the direction and point of optimization, the essence of the study.

The approach to the object of study enables us to apply statistical methods of planning of extreme experiments for the prompt identification of optimal process conditions and obtaining its mathematical model.

The validity of this approach is proved by complexity and probabilistic nature of most separation processes.

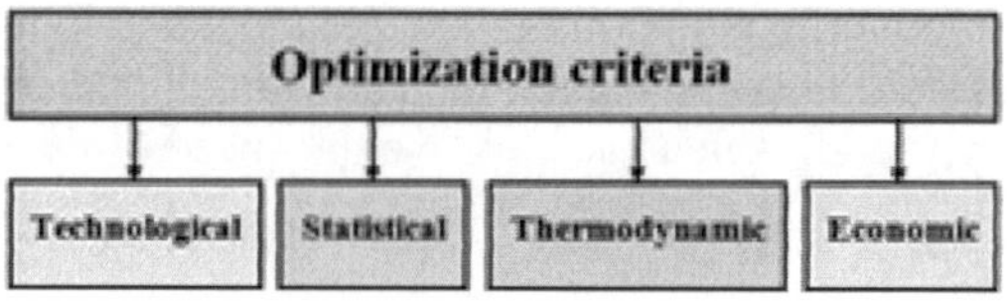

Figure 1. Criteria for technological processes optimization.

In some ways, the choice of optimization criterion determines the results of the study. Because these results, at the beginning of the study, are the question of the future and it is impossible to imagine clearly enough all the consequences of optimization, the choice of criteria is very difficult. It is connected not only with erudition, level of knowledge on the subject under study, but also with the psychology of researcher. Naturally, the criterion may vary during the study. In the final part of the article the authors attempted to provide a general approach to the selection criteria consistent in solving individual problems of the separation process.

Technological, thermodynamical, statistical and economic optimization criteria can be referred to as the most frequently used. At the thermodynamic approach to the separation process using the theory of separation it is possible to develop the application of the principle of minimality of dilution to a wide range of separation processes. This approach makes it possible to select the optimization criteria of separation schemes in terms of distribution and circulation of intermediate products calculation the number of stages and their efficiency in the separation cascade of operations.

Criteria for differences of physical and chemical properties of the separated materials or environments (wettability, density, magnetic permeability, melting point, boiling point, and others), associated with the concept of separability, should be considered together with a description of the relevant processes. Criterion of separation perfection on a particular device and kinetic criteria are also very important by themselves or in combination with other criteria.

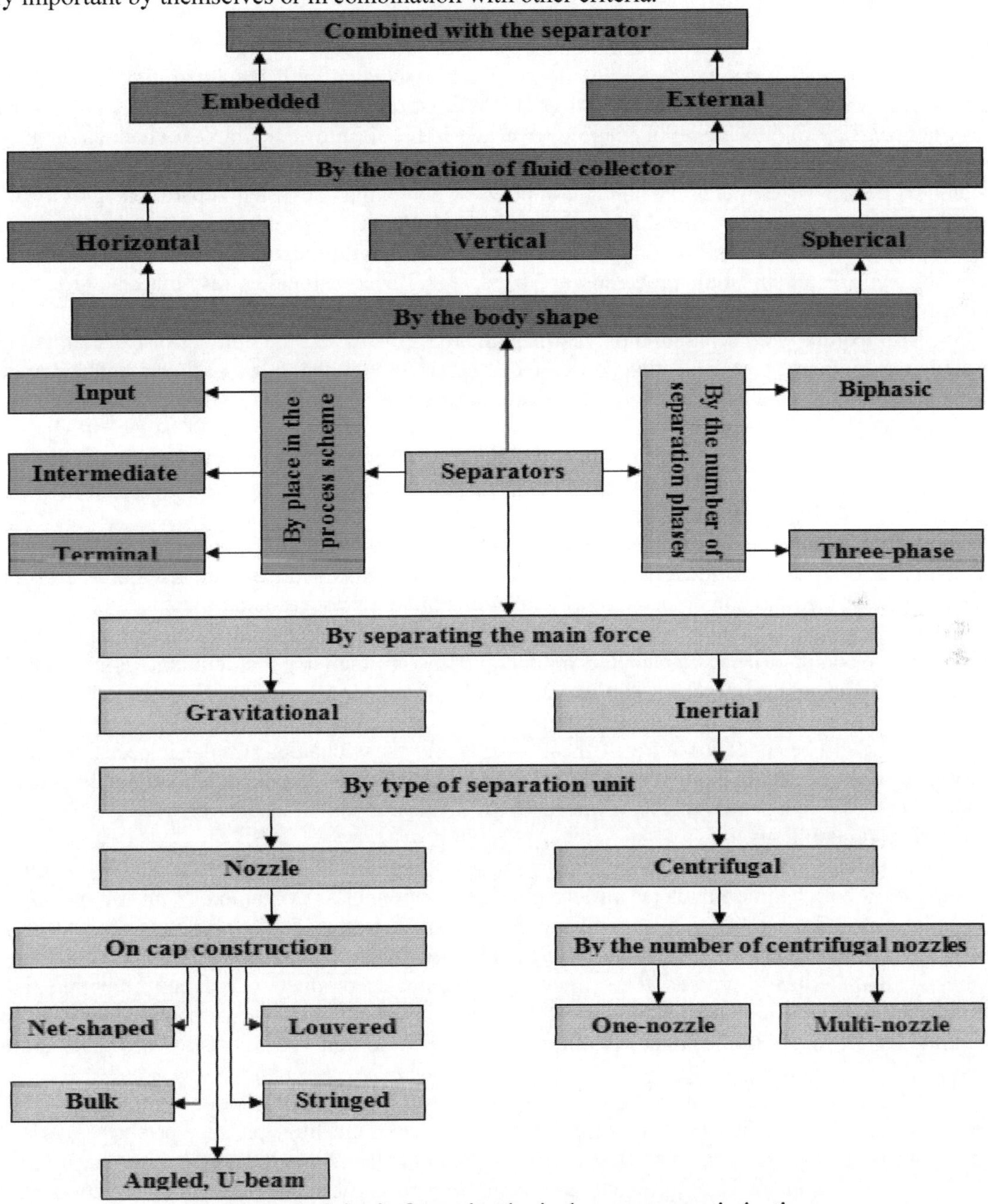

Figure 2. Criteria for technological processes optimization.

Industrial exploitation of natural resources is somehow associated with the separation process. The vast majority of materials used by mankind does not exist in nature in finished form, and to obtain them, it is necessary to allocate all of their components in the form of the simplest chemical elements or compounds. There is a variety of separation processes based on physical, chemical and physico-chemical properties of elements and their compounds; screening - by size separation, gravity - by density, flotation - by water wettability, magnetic and electric separation, hydrometallurgy - by solubility, liquid extraction, ion exchange, distillation, pyrometallurgical processes, diffusion isotope separation, etc... The separation processes are used in a variety of industries: mineral processing, metallurgy, oil, food, chemical, nuclear and others.

2. Research

Industrial exploitation of natural resources is somehow associated with the separation process. The vast majority of materials used by mankind does not exist in nature in finished form, and to obtain them, it is necessary to allocate all of their components in the form of the simplest chemical elements or compounds. There is a variety of separation processes based on physical, chemical and physico-chemical properties of elements and their compounds; screening - by size separation, gravity - by density, flotation - by water wettability, magnetic and electric separation, hydrometallurgy - by solubility, liquid extraction, ion exchange, distillation, pyrometallurgical processes, diffusion isotope separation, etc... The separation processes are used in a variety of industries: mineral processing, metallurgy, oil, food, chemical, nuclear and others.

A common feature of all separation processes is the mass character: the number of basic units to be separated (e.g., mineral particles, molecules of chemical compounds, atoms of elements or grains, etc.), are presented in countless numbers. In separation processes, particularly thin and complex in its technology, there is always some probability that even with a clear distinction, some of separated particles properties, some of them do not get in their product and hence there will be some contamination of cleaned substance with alien particles on the one hand, and under-recovery particles of substance into the product - on the other.

At these probabilistic characteristics of the separation process, assessing the quality and extraction, more errors caused by imperfections in separation units are imposed, as well as the existence of particles with intermediate properties - isomorphic atoms minerals splices and others.

Thus, as a result of an industrial separation process, from the initial mixture, not pure substances are obtained, but only enriched or depleted products. Although the term "enrichment" is used only to refer to the separation of minerals, it can be used equally well for all industrial separation processes, since in all cases, at least two of the above three reasons for the impossibility of clear separation play an important role. The random nature of these causes and mass character make it necessary to use probabilistic characteristics to evaluate the separation. All the technological characteristics of separation processes are therefore not more than integrated or averaged characteristics of the corresponding probabilities.

As already mentioned, in the industrial separation processes, it is impossible to achieve complete separation, only enriched and depleted in any product component are obtained. With the increase in the completeness of separation, cost of the process greatly increases and performance of the device significantly reduces. Therefore, at some point, it becomes more profitable to recycle an additional amount of raw materials to produce the same output of finished products which can be obtained at a lower total extraction of the initial amount. Completeness in the extraction of valuable component into concentrate (or its losses in the waste) is one of the most important characteristics of the separation process.

Since the extraction efficiency is associated with separating installation, and ultimately it is only required to obtain a specific output of finished products, then the final product performance is the second most important criterion of separation. It is economically connected to the first one by prime cost of raw material and finished products and technologically, these two criteria are in antagonistic contradiction to each other: with productivity growth, extraction falls and vice versa.

The third characteristic of the separation process is the quality of concentrate (content of valuable component) becomes especially important in those processes (mostly intermediate), which allow wide variations in this parameter. The quality of the concentrate can be the higher, the lower the performance (or prime cost of processing) and extraction is. Quality and recovery are related to each other through the function extracting content from initial raw material in processes of adjustment or further processing of concentrate.

Thus, all three main criteria of separation process are interconnected so that increase in process efficiency in terms of any of them reduces, to some extent, the efficiency the other two. This greatly complicates the evaluation of the efficiency of the separation process. Additional difficulties arise when evaluating separation of multicomponent systems with obtaining several products, the value of which is different. In comparing experiments or optimization of the separation process of complex products, it is necessary to consider several parameters, which have a difficult economic and technological dependency. Automatic adjustment of a process or statistical planning of the experiments in such conditions is impossible, if the functional connections between of optimization criteria are unknown.

In this regard, the problem of choosing the optimization criterion or efficiency of the separation process arises. A significant number of papers, relating not only to the separation process, and to a variety of fields - from economics to systems engineering, are dedicated to the choice of efficiency criteria. Initially, these studies were of a purely informative value and corresponded to the natural human need to somehow evaluate the results of their work.

With the development of mathematics, and then automating, the choice of optimality criterion has become a technical necessity, determining not only the accuracy of setting a specific task, but also the speed of its solution. Mathematical optimization unit is developed quite fully: mathematical analysis, calculus of variations, functional analysis and dynamic programming allow finding extreme values in any tasks, if you know the criterion of efficiency.

The efficiency criterion, designed for technology research or process control, in particular the separation process, should meet several requirements:

1. It should characterize the process efficiency in terms of the ultimate goal of this production. It is clear that the recovery cannot be a criterion for the separation excluding quality of concentrate, as on such criteria it is the most profitable not to enrich the ore, extraction in this "process" is equal to 100%. But at constant quality of concentrate, extraction can be considered as a criterion of separation, with reservations.

2. The criterion should be quantitative and one-figure, i.e., expressed in by one number. The numeric values of the criterion are not necessarily have to be in a proportional or other dependence from each other: in the simplest case numbers expressing the efficiency may indicate just ordinal number of separation process mode when the ranking of technological parameters on several grounds. However, this criterion is not very convenient, since the statistical operations with such numbers are quite limited. Ideally, the numerical values of the criterion should be proportional to the efficiency and uniformly distributed over the entire range of changes of a criterion. As it will be shown below, this property is possessed only by economic criteria, and separation entropy.

3. The fullness or universality of the efficiency criterion. For example, economic criteria (without corresponding corrections) do not have this property, since they do not allow or restrict processes which are unprofitable but necessary in terms of technology, strategic considerations and technical policy. On the contrary, technological criteria often do not take into account the important economic reasons.

4. Statistical efficiency of a criterion is to some extent antagonistic to demand of sensitivity to changes in process parameters. The physical meaning of statistical efficiency - insensitivity to small random fluctuations, of mathematical evaluation - dispersion, standard deviation of criterion values for the same tests characterizing the scatter about the average result.

5. The criterion should have a physical meaning, not just be an abstract number. It can express the cost, amount of valuable components (absolute or relative) in the products of separation and others. When considering the efficiency criteria, it is necessary to specify their physical meaning. Convenience of criteria that have physical meaning is in less probability of serious errors in their

application, as well as the possibility of rationing. For such a criterion one can always find its value for an ideal division, compare it to the actual and thus find out what are the areas of improvement of the separation unit work. Rationing, which consists of calculating the ratio of the real and the ideal separation, allows to identify the interval of criterion changes from zero, corresponding to the absence of separation to unity (or 100%), corresponding to the ideal division, which gives greater visibility to criterion.

6. Simplicity of calculation criterion is required only in the case when it does not contradict with fullness and uniqueness.

Obviously, the greatest number of requirements satisfies the economic criteria, although in many cases there are no sufficient data for calculation, since it is impossible to establish the price of all the intermediate products produced in one or another manufacturing operation. Moreover, economic efficiency is subject to strong changes associated with technological progress. Nevertheless, the economic evaluation is the most complete, universal and accurate, and in some cases - the only possible one.

In most particular cases, quite satisfactory criteria of separation efficiency can be recommended. They can be used for various purposes related to technology, economics, automation, process research theory, reproducibility of experiments and so on. It is not necessary that the same criteria are applied to the entire system in general and each of its parts.

According to justification of the withdrawal and purpose, all of the proposed separation criteria for our purposes are convenient to classify into the following groups: technological, thermodynamical, kinetic, statistical, economic, of automatic control, of natural separability, instrumental.

Introduce the following notations:

α - content of valuable component in the raw material;

β - the same in enriched product (concentrate, matte, distillate, etc...);

υ - the same in the depleted product (tails, wastes, slag, etc...);

β_T - theoretically possible maximum content of valuable component;

γ - output of the enriched product;

$(1-\gamma)$ - output of depleted product;

e - extraction of valuable components,

$$\varepsilon = \frac{\gamma\beta}{\alpha} \tag{1}$$

R - extraction of non-useful part into the concentrate,

$$R = \frac{\gamma(1-\beta)}{1-\alpha} \tag{2}$$

$\varepsilon_н$ — extraction of nonuseful part into the ends,

All values are expressed in decimals, not in percentage.

The following relations are derived from the conditions of the balance of valuable component in raw materials and separation products:

$$\varepsilon\alpha = \gamma\beta \tag{4}$$

$$\alpha = \gamma\beta + (1-\gamma)\vartheta \tag{5}$$

For the case of allocation of n of separation products and wastes

$$\alpha = \sum_{i=1}^{n} \gamma_i \beta_i + \left(1 - \sum_{i=1}^{n} \gamma_i\right)\vartheta \tag{6}$$

The degree of reduction

$$i = \frac{1}{\gamma} = \frac{\beta}{\varepsilon\alpha} \tag{7}$$

The degree of enrichment

$$i = \frac{\beta}{\alpha} = \frac{\varepsilon}{\gamma} \tag{8}$$

When carrying out a large number of parallel experiments we can calculate mean square deviation (dispersion) and coefficients of variation for the different criteria. Obviously, the criterion is statistically the more effective, the smaller the coefficient of variation is. Using a variety of experiments, it is possible to build a distribution and calculate the dispersion and coefficient of variation for a number of criteria, as well as the main parameters of separation. The coefficient of variation in this case indicates the sensitivity of the criterion: the greater is the sensitivity, the greater the coefficient of variation is, if the reproducibility error on this criterion is not too large. It is necessary to take into account the sensitivity and the statistical efficiency of separation criteria during the following choosing them for research process.

The task of establishing a minimum number of criteria, completely characterizing the process, arises from the fact that some of the basic criteria can be correlated. In this case, it can turn out that with a change in separation conditions, any two indicators, such as isolation and kinetics of the process or the quality of concentrate and economic efficiency, are measured in the studied ranges in the same direction with symbatic curves. To establish this fact, it is necessary to calculate all the known criteria for a large number of experiments (at least 20-30) and to conduct the correlation analysis, by calculating pair correlation criteria coefficients for each pair. One should select one from each pair of correlated criteria and discard the second.

3. Conclusion

Experimental studies of the separation process, as well as the study of the work of industrial separation unit provide material for the expression of the test parameters of the process in the form of various separation criteria. According to the same parameters, e.g. α, β, γ, υ different criteria are calculated - technological and thermodynamic; kinetic and economic criteria are calculated by duration of the process or cost of materials, respectively.

The calculation of the various criteria is associated with a specific purpose or stage of a research. Since there is no universal common separation criterion, the experiments should be analyzed in terms of several, although contradicting, parameters. This approach involves the use of the experience of researcher and his subjective evaluations. Considering the number of indicators, a researcher in many cases can quite accurately give preference to the results of one experiment to the results of another. In some cases this can be done more confidently and objectively, otherwise - intuitively. As a result of the pairwise comparison, of all the experiments conducted they can be arranged in a series of consecutive increasing or decreasing of the separation process quality. Experiment number in order of size is thus a kind of general separation criterion.

It should be noted that the use of such a criterion is associated with certain inconveniences. Statistical processing of results in this case is impossible, and the accuracy of the criterion is quite small, which is predetermined by the subjective approach to the ranking of experiments. Therefore, the degree of approximation to the optimal mode is not known and can not be assessed. Analysis of statistical efficiency (sensitivity) and the choice of a minimum number of criteria are produced by methods of mathematical statistics.

References

[1] Bakhmat G V 1981 Investigation degassing prcess of multicomponent gas saturated liquids. *NTS. Development and exploitation of gas and offshore oil fields* (Moscow; VNIIEGazprom) 34-7

[2] Zemenkov Yu D, Markov L M, Prokhorov AD and Dudin S M 2009 *Collection and preparation of oil and gas* (Moscow; Academy)

[3] Sinai E G, Lapiga E Ya, Zaitsev YuV 2002 *Separation of multiphase multicomponent systems* (Moscow; Nedra) 620

[4] Trivus N A and Seidali-zade B M 1974 Optimum modes of condensate degassing stepped or efficiency *Development and exploitation of gas and gas condensate fields VNIIEGazprom* **6** 17-22

IOP Conf. Series: Materials Science and Engineering **154** (2016) 012040 doi:10.1088/1757-899X/154/1/012040

Laboratory Reactor for Processing Carbon-Containing Sludge

I O Korovin and A V Medvedev

Industrial University of Tyumen, 38 Volodarskogo st., Tyumen, 625000, Russia

E-mail: kio79@list.ru

Abstract. The paper describes a reactor for high-temperature pyrolysis of carbon-containing sludge with the possibility of further development of environmentally safe technology of hydrocarbon waste disposal to produce secondary products. A solution of the urgent problem has been found: prevention of environmental pollution resulting from oil pollution of soils using the pyrolysis process as a method of disposal of hydrocarbon waste to produce secondary products.

1. Introduction

Oil extraction process results in contamination of the soil, not only within the fields themselves, but also in the surrounding areas. The problem of oil and oil products getting into the soil exists at all stages of field development and production: drilling, recovery, gathering and transportation of produced fluids.

Oil and oil products cause almost complete depression of the functional activity of flora and fauna. Emergency spills on oil pipelines lead to high concentrations of oil in confined areas, resulting in the formation of depression-oil sectors of terrain characterized by strong oil contamination, disturbance of the lower levels of the geosystem.

The high toxicity and flammability of oil and oil products considerably exacerbate the implications of oil pollution.

The problem of protection and restoration of land contaminated with oil is very relevant, especially for the northern regions, since it is here that the main oil production complexes are concentrated and the climate of this region does not allow restoring the land after the oil pollution in the shortest time.

2. Methods, results and discussion

One of the possible solutions to the problem of disposing of sludge containing organic compounds is proposed in Industrial University of Tyumen (oil sludge, contaminated soils, etc.) [3,5]. The essence of this technology lies in the high temperature pyrolysis of the organic waste component in an environment similar to that created in furnaces with a "fluidized" bed. In this method, to sustain the pyrolysis process an electric arc of direct or alternating current is proposed to be used as a source of heat in the reactor.

The laboratory unit includes a reactor designed for the disposal of sludge containing organic compounds and is an electric arc furnace of indirect action. In this case, the arc burns between the horizontally arranged electrodes, and the salvage material acquires heat from the arc by radiation, convection and conduction [4]. The reactor *8* is fabricated as a sealed cylindrical metal tank in the top part of which is a waste loading hatch *7*, and at the bottom there is a hatch for discharging ash residue *15*. Hatch covers are removable; their tight closure is done by bolting. The reactor has a cooling water jacket 1 designed to control the desired temperature for the reaction process. Figure 1 shows the internal arrangement of the pyrolysis reactor laboratory unit for studying the process of organic sludge disposal.

Inside the reactor has two electrodes *3, 10* connected to the tap of the welding transformer which initiate the pyrolysis process. Electrodes are supplied manually with the help of a screw feeder *9*.

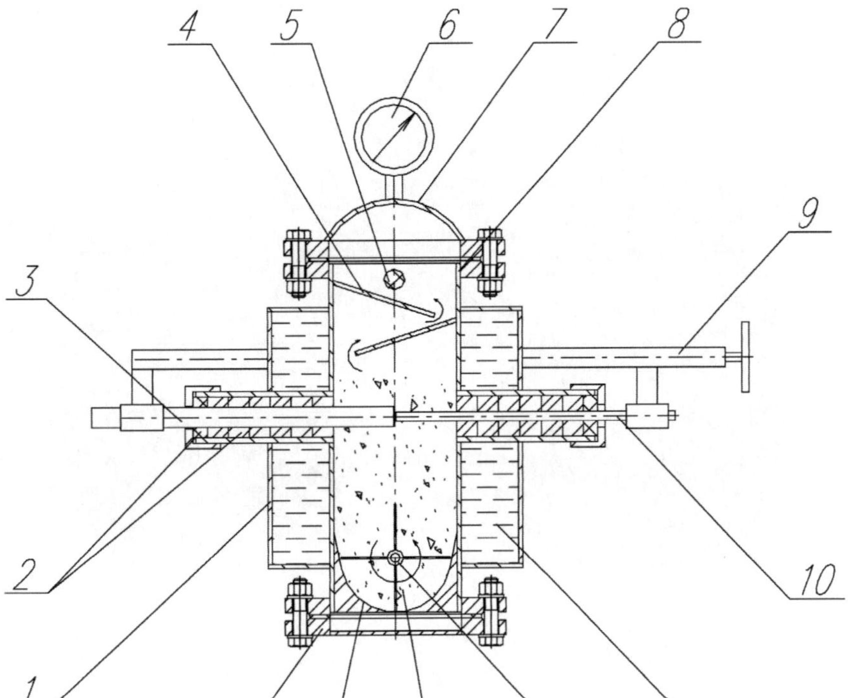

Figure 1. The internal arrangement of the reactor.

1 – cooling jacket; 2 – dielectric insulating rings; 3 – thick electrode; 4 – dash plate; 5 – mechanical coarse filter; 6 – pressure gauge; 7 – loading hatch; 8 – reactor vessel; 9 – thin electrode feeder; 10 – thin electrode; 11 – coolant; 12 – activator wheel; 13 – salvage material; 14 – bunker bowl; 15 – bottom (discharge) hatch.

Along the height of the reactor wall sockets for temperature sensors are located. The first thermocouple located in the middle of the unit at a height of 155 mm from the bottom of the reactor is designed to determine the temperatures in the reaction zone; the second one at a height of 300 mm is used to control the temperature of the pyrolysis gas. Temperature measurement is possible at various points of the unit by moving the thermocouple socket relative to the heat source.

This unit allows us to carry out research on the physical and chemical processes of waste disposal.

Parameters of the electrical current supplied to the electrodes contribute to the high temperatures of about 3000 K. This prevents the formation of highly toxic compounds in the exhaust pyrolysis gas and, therefore, reduces the impact on the atmosphere [4].

Figure 2 shows the isotherms of the temperature field built according to the experimental data obtained during sludge disposal. In this case, the electrodes 14 mm in diameter were used; the power of the supplied electric current was ≈ 5400 W.

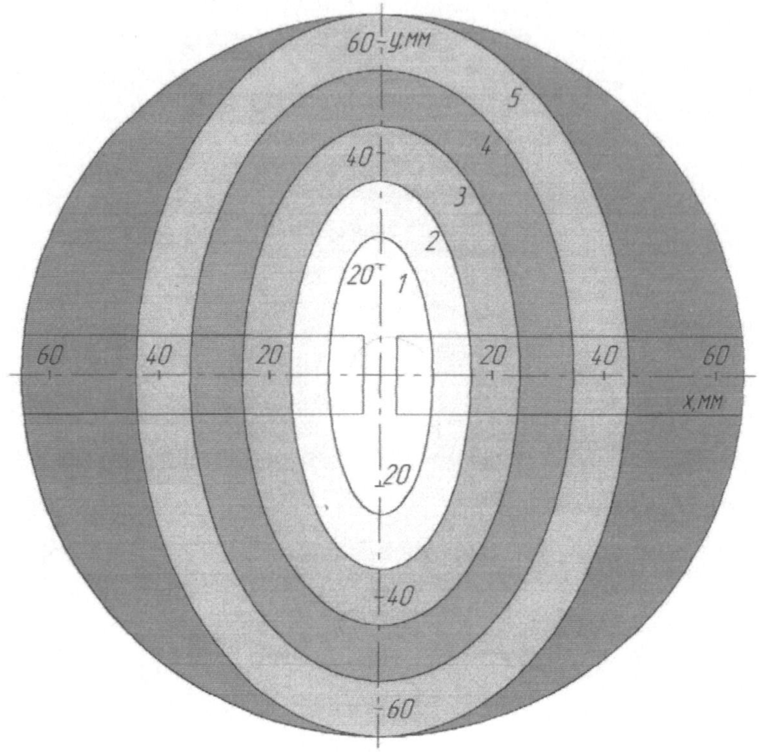

Figure 2. The approximate view of the isotherms of the temperature field in the pyrolysis reactor. 1 – 5 –isotherms at 1200, 1125, 880, 810 and 725 °C, correspondingly.

The reactor is a batch operation unit in which processing of sludge is carried out in batches. Loading of the reactor is carried out through the top (loading) hatch. The volume of salvage material is determined based on the degree of filling of the reactor - the loading level should be below the electrode placement. Thus the possibility of mineral waste component sintering at the initial stage of the electric arc operation is eliminated. When disposing of organic sludge the mass of the salvage material loaded simultaneously is an average of 200 - 400 g. An important condition is that loading of the unit must be done with the short-circuited electrodes in the central part of the reactor. This eliminates the likelihood of occurrence of the dielectric layer from the salvage sludge between the electrodes. Pretreatment of sludge (grinding, fractionating, drying) is not required. Then, tight closing of the loading hatch is performed.

The electric motor drive of the activator wheel is launched. When the electric current passes from the transformer through the electrodes and the electric arc discharge occurs in the interelectrode space, the sludge recycling process begins. The temperature of the arc is determined by the power characteristics of the electric arc.

The activator rotation causes uplifting of material portions from the bottom of the bunker bowl and tossing them up. The activator wheel speed is determined experimentally for each type of sludge, based on the maximum residence time of the salvage material in the high-temperature "near-electrode" zone. Thus, an effect is achieved similar to that on which the work of the furnace with a fluidized (boiling) bed is based. However, unlike the latter, where this effect is achieved by supplying air or inert gas into the unit, there is no dilution of the produced, as a final product, fuel gas with foreign impurities and polluting toxic compounds resulting from partial oxidation of the salvage material by air oxygen.

Application of the technology of the so-called "fluidized" bed in the laboratory unit is due to the fact that as the test salvage material was selected oil-contaminated soil with a high content (over 90%) of the mineral part which consists primarily of silicates. As a consequence, when it enters the area of high temperature (more than 1800 K) the mineral component passes into the melt that forms a solid glassy mass when cooled. In a stationary mode of sludge pyrolysis this leads to the formation of the so-called "gin poles" that "glue" electrodes together, envelop them with a dielectric film and form "sinters". Thus, if in the sinter itself the degree of material processing (in organic part and humidity) approaches one hundred percent, outside the sintered area due to its low thermal conductivity the level of processing does not exceed 10-15%. Application of the "fluidized" method of disposal leads to the vitrification of only individual sand grains and prevents them from sticking together to form "sinters". Small pieces of the sintered material adhering to the electrodes are easily degraded. The degree of sludge processing (in organic part and humidity) reaches 97-99%.

Combustible gas formed during pyrolysis of the organic part of the sludge rises to the top of the reactor. It passes through the labyrinth channel of inclined dash plates so that it is cleaned from large solids. The gas temperature downstream of the reactor in its upper part is 380 - 400 K. Then, the gas through the fine mesh filter, where small dust-like inclusions are separated, enters the refrigerator, an air-cooled helical tube. In the refrigerator gas desiccation takes place by separating water vapor and a resinous component which, when cooling down, condense and together with the gas enter the condensate tank where they remain. And the gas passing through the fiber filter and finally clean from residual moisture and solids enters the gas line. Through it the gas enters the gas meter to account for its amount and then is fed to the torch where it is burnt.

The main product of the sludge processing in the reactor of this type is a combustible gas, whose calorie content is 12 - 13 MJ. The share of gas output is 0.5 l/g, due to high levels of mineral components content.

A by-product of processing is a solid mineral residue containing less than 3% of the organic part, which means that it can be used as an aggregate for asphalt mixtures and for other construction purposes.

3. Summary

In Industrial University of Tyumen scientists have developed technologies that reduce the impact of the oil industry on the environment, with obtaining the secondary products of processing (fuel gas). They used an integrated system approach to the problem of oil-contaminated hydrocarbon waste disposal by a high-temperature pyrolysis process.

References

[1] Medvedev A V, Korovin I O and Shantarin V D 2002 A study of optimum parameters of the pyrolysis process of organic wastes in the installation with arc heating *Proceedings of the scientific and technical conference "Oil and Gas: Challenges of subsoil use, production and transportation" (Tyumen)* 156

[2] Voychenko A A, Medvedev A V and Oparin V V 22.07.2002 Installation for the thermal processing of sludge. Patent for invention RUS 2229060

[3] Medvedev A V, Korovin I O and Bagabiev R R 2003 The reactor for processing organic sludge *Proceedings of the higher educational institutions. Oil and gas (Tyumen: Publishing House TSOGU)* **3** 111-14

[4] Medvedev A V, Bagabiev R R and Korovin I O 2003 Installation for recycling organic sludge with a high content of mineral components *Proceedings of the higher educational institutions. Oil and gas (Tyumen: Publishing House TSOGU)* **3** 109

[5] Medvedev A V 2014 Study of the possibility of using the pyrolysis method for the disposal of waste oil *Online magazine Naukovedenie* **5(24)** 10

IOP Conf. Series: Materials Science and Engineering **154** (2016) 012041 doi:10.1088/1757-899X/154/1/012041

Technology of Ultrasonic Treatment of High-Viscosity Oil from Yarega Oilfield to Improve the Rheological Properties of Oil

Y D Zemenkov, M Y Zemenkova, V I Berg and E F Gordievskaya

Industrial University of Tyumen, 38 Volodarskogo st., Tyumen, 625000, Russia

E-mail: allbert-@mail.ru

Abstract. The article investigates the possibility of applying ultrasonic treatment oil from Yarega oilfield to improve of rheological properties, reduce oil viscosity in Russian pipeline transportation system, and increase its efficiency and performance. Created laboratory test bed of ultrasonic waves.

1. Introduction

World petroleum industry is characterized by the decrease of low viscosity oil recovery and by the increase of hard-to-recover viscous oil and bitumen. The world reserves analysis of viscous oil with a viscosity more than 25 mPa·s has shown, that the main reserves are located in the countries such as Canada, Venezuela and Russia. According to the various sources, global reserves of viscous oil are estimated at least 1 trillion tons [1].

The necessity to reduce production and transportation costs for viscous oil is a topical problem for the oil companies. Currently, such methods as oil heating, addition, magnetic and ultrasonic field effects are applied to improve the rheological properties Ultrasonic processing of viscous oil is a promising and economically profitable method of oil processing. However, the issue connected with the complex influence of ultrasonic treatment and depressor additives has not been covered properly. The integrated impact can have high treatment efficiency and a prolonged recovery time of oil properties.

The aim of this work is to conduct experimental research and to point out the correlations between ultrasonic treatment and viscous oil depressor additives integrated application.

2. Materials and methods

In the process of the research, Yarega field viscous oil treatment of 3000 ml and viscosity of 500 mPa·s with depressor additives DPN and DPN-1R, also ultrasonic transducer Ultrasonic with a frequency up to 25 kHz and 20 μm amplitude has been carried out. The processed oil was placed in the special thermostats, and was processed by ultrasound and depressor additives both separately and independently [2].

Ultrasonic treatment has been carried out using a rod waveguide that is placed in oil at depth of 50 mm. Depressor additives have been heated up to 60 °C before processing. The sample thermostation by special thermostats has been carried out before and after oil processing. The viscosity has been measured by viscometer VUB-1R by measuring 50 ml outflow time [3].

Figure 1 illustrates the model of the test bed.

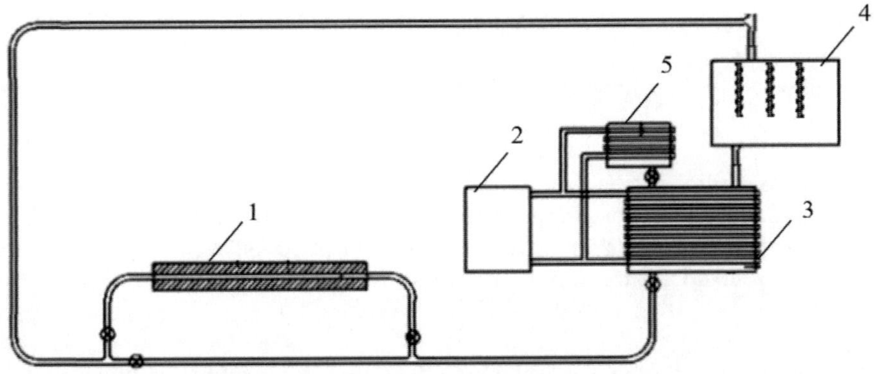

Figure 1. Model of the test bed.
1 – thermostat; 2 – oil tanks; 3 – pump; 4 – ultrasonic rods; 5 – additives input unit.

To determine the effective area and the values of ultrasonic treatment intensity, both experimental analysis and Minnaert's linear theory of ultrasonic treatment effectiveness evaluation in inviscid cavitating liquids have been used.

For the latter we describe the dynamics of the cavitation bubble occurrence and the cavitation bubble radius dependence on time $R(t)$ using Nolting-Naires equation. We write the wave resistance of the processed liquid[4]:

$$\rho_1 c_1 = \rho_0 c_0 \left(\frac{(1-\bar{K}) + \frac{\rho_2}{\rho_0 \bar{K}}}{(1-\bar{K}) + \frac{\beta_2}{\beta_0 \bar{K}}} \right)^{0.5}, \tag{1}$$

where ρ_0, ρ_1, ρ_2 - density of a perfect fluid, density of a cavity and density of raw liquid;
$\bar{K}$ - average index of cavitation;
$\beta_0, \beta_1, \beta_2$ - compressibility factor of a perfect fluid, a cavity and raw liquid;

The average index of cavitation is determined according to the following formula:

$$\bar{K} = \frac{\frac{1}{t}\int_0^t R^3(t)\partial t}{R_{MAX}}, \tag{2}$$

where t - period of ultrasonic wave oscillation;
$R(t)$ - cavity radius time-dependent;
R_{MAX} - maximum cavity radius;

And to determine the R_{MAX} linear theory generalization is used:

$$f = \frac{1}{2\pi R_{MAX}} \sqrt{\frac{1}{\rho}\left(3\gamma\left(\frac{2\sigma}{R_{MAX}} + p_0\right) - \frac{2\sigma}{R_{MAX}} - \frac{2\mu}{\rho R_{MAX}{}^2}\right)}, \tag{3}$$

where γ - adiabatic index of air;
σ - surface tension;

μ - dynamic viscosity;

Numerical solution of the resulting equation for R_{MAX} defines the maximum average radius of the cavity.

3. Results and Discussion

The effective area and ultrasonic processing intensity value of viscous oil which is equal to 12 W/cm^2 have been determined in the research process The processed oil viscosity measurements and rheological properties relaxation time were carried out every 12 hours at the following time periods: 12 hours, 24 hours, 36 hours, 48 hours, 60 and 72 hours (figure 2-4).

Figure 2 illustrates the result of experiments with the ultrasonic treatment application.

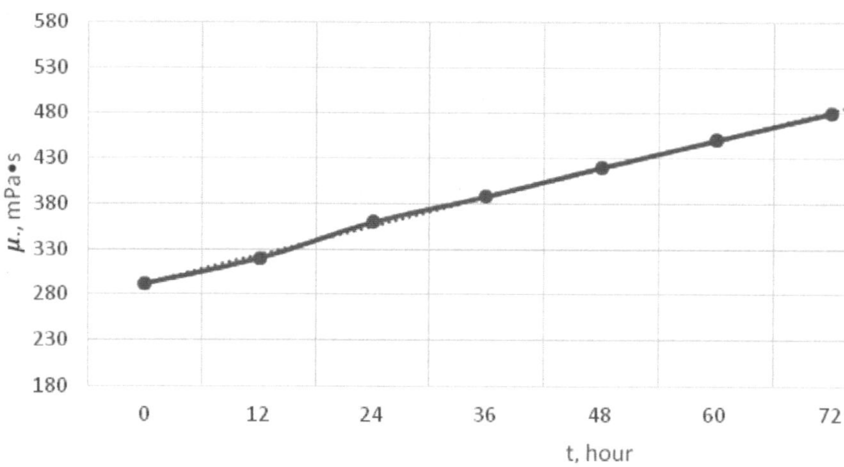

Figure 2. The dependence of oil dynamic viscosity on the sample storage time after ultrasonic treatment.

Figure 3 shows the result of experiments with the use of depressor additives.

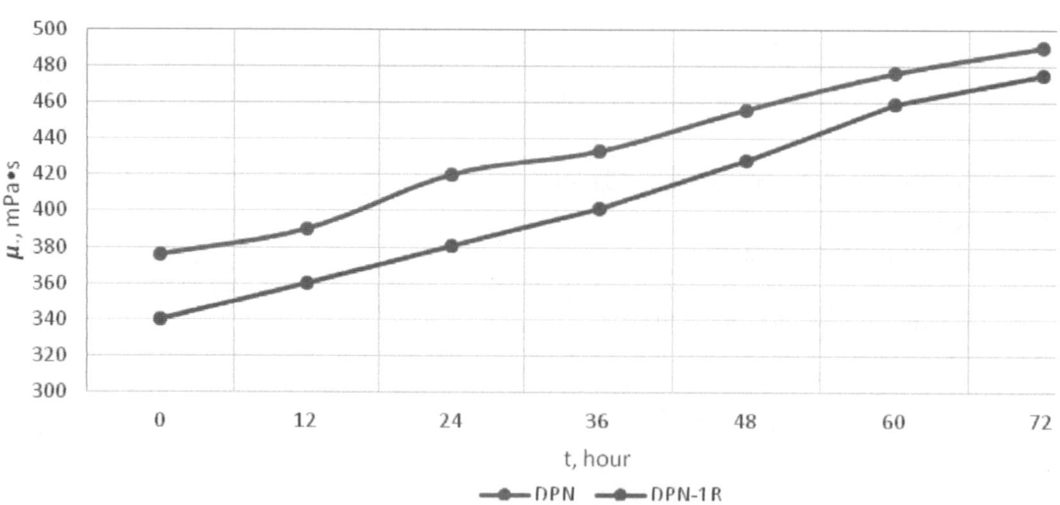

Figure 3. Oil dynamic viscosity dependence on the sample storage time after applying depressor additives.

Figure 4 shows the result of viscous oil complex processing experiments

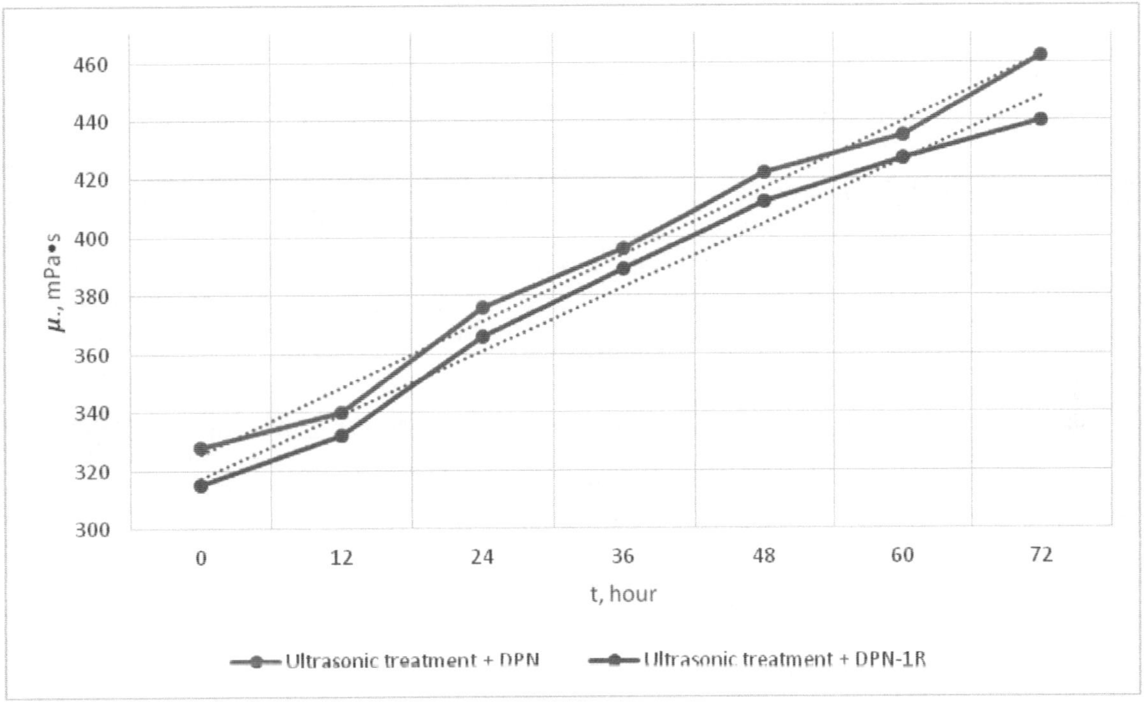

Figure 4. Depicting oil dynamic viscosity dependence on the sample storage time after combined processing method.

4. Conclusion

Oil dynamic viscosity has increased in comparison with the baseline values after treatment in the study process. This phenomenon is explained by the fact that primary and secondary molecular bonds have been destroyed by ultrasonic treatment, resulting in increased number of free molecules that has made their coagulation possible. This phenomenon is more noticeable after 72 hours of sedimentation. Research data show that exponential growth of viscosity is a characteristic of original oil.

The processing (treatment) results have showed that application of combined method of viscous oil processing is effective and can be used for short-term viscous oil processing at the tank farms or railway tanks at low temperatures.

References

[1] Brand A E, Vershinina S V, Vengerov A A and Mostovaya N A 2015 Applying the technology of hydrodynamic cavitation treatment of high-viscosity oils to increase the efficiency of transportation *IOP Conf. Ser.: Mater. Sci. Eng.* **93 012005**

[2] Ershov M A., Mullakaev M S and Baranov D A 2012 Mathematical processing of results of experiments on ultrasonic influence on the viscosity of the oil *Bulletin of Saratov state technical University* **2 (64)** 250-253

[3] Vengerov A A and Brand A E 2014 Oil cavitation treatment to prevent formation of paraffin deposits *J. Natur. and intellect. resource. of Siberia.* **15** 36-37

[4] Ilyin E G and Ivantsova S G 2014 *J. Crude oil pipeline* **6** 67-70

Institute of Physics
Dirac House, Temple Back
Bristol BS1 6BE UK

ISSN: 1757-8981
ISBN 978-1-5108-3177-3

9 781510 831773

CURRAN ASSOCIATES INC.
proceedings.com